Birkhäuser

Modeling and Simulation in Science, Engineering and Technology

Tayfun E. Tezduyar • Kenji Takizawa

Space–Time Computational Flow Analysis

A Chronological Catalog of Unconventional Methods and First-of-Its-Kind Solutions

Birkhäuser

Tayfun E. Tezduyar
Mechanical Engineering
Rice University
Houston, Texas, USA

Kenji Takizawa
Modern Mechanical Engineering
Waseda University
Tokyo, Japan

ISSN 2164-3679 ISSN 2164-3725 (electronic)
Modeling and Simulation in Science, Engineering and Technology
ISBN 978-3-031-88726-0 ISBN 978-3-031-88727-7 (eBook)
https://doi.org/10.1007/978-3-031-88727-7

This book is published under the imprint Birkhäuser, www.birkhauser-science.com by the registered company Springer Nature Switzerland AG
The registered company address is: Gewerbestrasse 11, 6330 Cham, Switzerland

Preface

The Space–Time Computational Flow Analysis (STCFA) started in 1990 with the inception of the Deforming-Spatial-Domain/Stabilized Space–Time (DSD/SST) method. The DSD/SST was introduced as a moving-mesh method for flows with moving boundaries and interfaces, which is a wide class of problems that includes fluid–particle interactions, fluid–structure interactions (FSI), and free-surface and multi-fluid flows. The first 3D computations were reported in 1992. The original DSD/SST method is now called "ST-SUPS," reflecting its stabilization components. As the STCFA evolved, advanced mesh moving methods, FSI coupling methods, and problem-class-specific methods were introduced to increase its scope, and the ST Variational Multiscale (ST-VMS) was introduced to upgrade its stabilization components to the VMS. The ST-VMS, introduced in 2010, serves as the core computational method. Complementary general-purpose methods developed in the evolution of the STCFA include the ST Isogeometric Analysis (ST-IGA) and ST Topology Change (ST-TC) and ST Slip Interface (ST-SI) methods. The ST-IGA, introduced in 2010, delivers superior accuracy through IGA basis functions not only in space but also in time. The ST-TC, introduced in 2013, enables high-fidelity moving-mesh computations even in the presence of topology changes in the fluid mechanics domain, such as an actual contact between moving solid surfaces. The ST-SI, introduced in 2015, enables high-fidelity moving-mesh computations even over meshes made of patches with nonmatching meshes at the interfaces between those patches.

With all these and other unconventional methods, the STCFA brought first-of-its-kind solutions in many classes of problems, ranging from fluid–particle interactions in particle-laden flows to FSI in parachute aerodynamics, flapping-wing aerodynamics of an actual locust to ventricle-valve-aorta flow analysis to car and tire aerodynamics with near-actual geometries, road contact, and tire deformation. With the success we see in so many classes of problems, we can conclude that the STCFA has reached a level of remarkable sophistication, scope, and practical value. We present a chronological catalog of the methods and solutions in the STCFA. We do that in two parts. In *Part I*, we focus on the methods and solutions in finite element analysis, and in *Part II*, the methods and solutions in isogeometric analysis.

We are fortunate to have friends whose work motivated some of our work in the STCFA. We are also fortunate that we had many students and associates who contributed to the methods and solutions in the STCFA. The work on the DSD/SST was motivated by Thomas Hughes's work on space–time finite element methods for elastodynamics. The work on the ST-VMS and ST-IGA was motivated by the VMS and IGA work of Thomas Hughes and Yuri Bazilevs. The students and associates who contributed to both the methods and solutions in the STCFA in the last 25 years include Yasuo Osawa, Sunil Sathe, Yuto Otoguro, Takashi Kuraishi, and Takuya Terahara. Some of the recent work included in the book was supported in part by International Technology Center Indo-Pacific (ITC IPAC).

Houston, Texas, USA
Tokyo, Japan
June 2024

Tayfun E. Tezduyar
Kenji Takizawa

Contents

Chapter 1
Introduction

The Space–Time Computational Flow Analysis (STCFA) started in 1990 with the inception of the Deforming-Spatial-Domain/Stabilized Space–Time (DSD/SST) method [1–3]. The DSD/SST was introduced as a moving-mesh method for flows with moving boundaries and interfaces (MBI). This class of problems includes fluid–particle interactions and fluid–structure interactions (FSI), and in a more general sense, moving solid surfaces. It also includes free-surface and multi-fluid flows. Our work on the DSD/SST was motivated by Thomas Hughes's 1987 work [4] on space–time finite element methods for elastodynamics. The original DSD/SST method is now called "ST-SUPS." The acronym reflects the stabilization components of the method. Those components are the Streamline-Upwind/Petrov-Galerkin (SUPG) method [5, 6] and the Pressure-Stabilizing/Petrov-Galerkin (PSPG) method [1, 7]. In a broader interpretation of the terminology, as pointed out in [8], the ST-SUPS includes the stabilization component coming from Least-Squares on the Incompressibility Constraint (LSIC), and that is how it was for the DSD/SST in its form in [9]. Altogether, the ST-SUPS has three stabilization terms: SUPG, PSPG, and LSIC.

As the STCFA evolved, advanced mesh moving methods [10, 11], robust FSI coupling methods [12, 13], and problem-class-specific methods were introduced to increase its scope, and the ST Variational Multiscale (ST-VMS) [14–16] was introduced to upgrade its stabilization components to the VMS. The VMS components are from the residual-based VMS (RBVMS) method [17–20]. The ST-VMS, introduced in 2010, serves as the core computational method. It subsumes its precursor ST-SUPS and has two more stabilization terms. Both the ST-SUPS and ST-VMS are residual-based. Because the ST-SUPS and ST-VMS serve as moving-mesh methods in MBI computations, they move the fluid mechanics mesh to follow the interfaces. That enables mesh-resolution control near the interfaces and, consequently, high-resolution boundary-layer representation near fluid–solid interfaces. They would be favored methods even in flow computations without MBI, because the ST framework brings higher-order accuracy (see [14, 15]).

T. E. Tezduyar, K. Takizawa, *Space–Time Computational Flow Analysis*,
Modeling and Simulation in Science, Engineering and Technology,
https://doi.org/10.1007/978-3-031-88727-7_1

As moving-mesh methods, the ST-SUPS and ST-VMS are comparable to the Arbitrary Lagrangian–Eulerian (ALE) method, which is older and more commonly used. Its finite element version goes at least as far back as 1981 [21]. Its use in 3D finite element flow computations with residual-based stabilized methods like the SUPG is newer compared to the ST-SUPS. The earliest 3D computation with the ALE-SUPS was the ram-air parachute FSI analysis in 1999 [22]. Among the moving-mesh methods with the VMS stabilization, the ALE-VMS [23, 24] precedes the ST-VMS. Clearly, the ST-SUPS, ALE-SUPS, ALE-VMS, and ST-VMS have much in common (see [25]). Therefore, the classes of problems computed with these methods since their inception also have much in common. The classes of problems computed with the ALE-SUPS, RBVMS, and ALE-VMS include wind turbines [26–29], turbomachinery [30–32], stratified flows [33, 34], bridges [35], marine applications [36, 37], free-surface flows [38, 39], two-phase flows [40–42], additive manufacturing [43], aircraft applications [44], hypersonic flows [45], parachutes [22], cardiovascular medicine [46], mixed ALE-VMS Immersogeometric Analysis (IMGA) computations [47] in the framework of the Fluid–Solid Interface-Tracking/Interface-Capturing Technique [48], and IMGA FSI and flow analysis [49].

Complementary general-purpose methods developed in the evolution of the STCFA include the ST Isogeometric Analysis (ST-IGA) [14, 50, 51] and the ST Topology Change (ST-TC) [52, 53] and ST Slip Interface (ST-SI) [54, 55] methods. The ST-IGA, introduced in 2010, delivers superior accuracy through IGA basis functions not only in space but also in time. The ST-TC, introduced in 2013, enables high-fidelity moving-mesh computations even in the presence of topology changes in the fluid mechanics domain, such as an actual contact between moving solid surfaces. That in turn enables flow analysis with actual-contact and high-resolution boundary-layer representations. The ST-SI, introduced in 2015, enables high-fidelity moving-mesh computations even over meshes made of patches with nonmatching meshes at the interfaces between those patches. Methods that are syntheses of the complementary general-purpose methods play a key role in the STCFA. The most comprehensive one is the ST-SI-TC-IGA [56], which combines the ST-IGA, ST-SI, and ST-TC, and their desirable features, in a well-integrated fashion.

With all these and other unconventional methods, the STCFA brought first-of-its-kind solutions in many classes of problems, ranging from fluid–particle interactions in particle-laden flows [57] to FSI in parachute aerodynamics [58–60], flapping-wing aerodynamics of an actual locust [50] to ventricle-valve-aorta flow analysis [61] to car and tire aerodynamics with near-actual geometries, road contact, and tire deformation [62]. With the success we see in so many classes of problems, we can conclude that the STCFA has reached a level of remarkable sophistication, scope, and practical value. This book will serve as a chronological catalog of the methods and solutions in the STCFA. The material is an expanded version of the chronological catalog in two recent journal articles [63, 64]. We present the material in two parts. In *Part I*, we focus on the methods and solutions in finite element analysis, and in *Part II*, the methods and solutions in isogeometric analysis. Many of the methods covered in *Part I* would also work in isogeometric analysis. Similarly, some of the methods covered in *Part II* would also work in finite element analysis.

Part I
Methods and Solutions in Finite Element Analysis

Chapter 2
Early Stabilized Methods

2.1 Compressible-Flow SUPG Method, 1982

The compressible-flow SUPG method was introduced, in the context of conservation variables, in a 1982 NASA report [65], with a concise version published as a 1983 AIAA paper [66] and a more thorough version with additional examples as a 1984 journal paper [67]. That 1982 method is now called "$(\mathrm{SUPG})_{82}$." Several SUPG-like compressible-flow methods were developed after that. The compressible-flow SUPG formulation [65–67] is written as

$$
\begin{aligned}
&\int_{\Omega} \mathbf{W}^h \cdot \left(\frac{\partial \mathbf{U}^h}{\partial t} + \mathbf{A}_i^h \cdot \frac{\partial \mathbf{U}^h}{\partial x_i} - \mathbf{S}^h \right) \mathrm{d}\Omega \\
&+ \int_{\Omega} \frac{\partial \mathbf{W}^h}{\partial x_i} \cdot \mathbf{K}_{ij}^h \cdot \frac{\partial \mathbf{U}^h}{\partial x_j} \mathrm{d}\Omega - \int_{\Gamma_{\mathrm{H}}} \mathbf{W}^h \cdot \mathbf{H}^h \mathrm{d}\Gamma \\
&+ \sum_{e=1}^{n_{\mathrm{el}}} \int_{\Omega^e} \left(\boldsymbol{\tau}_{\mathrm{SUPG}} \cdot \left(\frac{\partial \mathbf{W}^h}{\partial t} + \frac{\partial \mathbf{W}^h}{\partial x_k} \cdot \mathbf{A}_k^h \right) \right) \cdot \mathbf{R}^h \mathrm{d}\Omega \\
&+ \sum_{e=1}^{n_{\mathrm{el}}} \int_{\Omega^e} \nu_{\mathrm{SHOC}} \frac{\partial \mathbf{W}^h}{\partial x_i} \cdot \frac{\partial \mathbf{U}^h}{\partial x_i} \mathrm{d}\Omega = 0.
\end{aligned}
\tag{2.1}
$$

Here Ω is the spatial domain, n_{el} is the number of elements, Ω^e is the domain for element e, $\mathbf{U}$ is the vector of conservation variables, $\mathbf{W}$ is the test function vector associated with that, $\mathbf{S}$ is the term that represents all other components that might enter the equations, including the external forces, and $\mathbf{H}$ represents the natural

T. E. Tezduyar, K. Takizawa, *Space–Time Computational Flow Analysis*,
Modeling and Simulation in Science, Engineering and Technology,
https://doi.org/10.1007/978-3-031-88727-7_2

boundary conditions enforced on the Γ_H part of the boundary Γ. The matrices $\mathbf{A}_i$ and $\mathbf{K}_{ij}$ are defined as

$$\mathbf{A}_i = \frac{\partial \mathbf{F}_i}{\partial \mathbf{U}}, \qquad \mathbf{K}_{ij} \cdot \frac{\partial \mathbf{U}}{\partial x_j} = \mathbf{E}_i, \tag{2.2}$$

where $\mathbf{F}_i$ and $\mathbf{E}_i$ are the Euler and viscous flux vectors, and

$$\mathbf{R}^h = \frac{\partial \mathbf{U}^h}{\partial t} + \mathbf{A}_i^h \cdot \frac{\partial \mathbf{U}^h}{\partial x_i} - \frac{\partial}{\partial x_i}\left(\mathbf{K}_{ij}^h \cdot \frac{\partial \mathbf{U}^h}{\partial x_j}\right) - \mathbf{S}^h \tag{2.3}$$

is the residual of the system of equations representing the conservation of mass, momentum, and total energy. The expressions for $\mathbf{U}$, $\mathbf{F}_i$, $\mathbf{E}_i$, $\mathbf{A}_i$, $\mathbf{K}_{ij}$, and $\mathbf{S}$ are given in Appendix A. The last term in Eq. (2.1) is the shock-capturing term, which is also called "discontinuity-capturing (DC) term." The SUPG stabilization matrix $\boldsymbol{\tau}_{\mathrm{SUPG}}$ and the shock-capturing parameter ν_{SHOC} can be found in [68, 69].

Remark 2.1 The computations reported in [65] were for inviscid flows. In $(\mathrm{SUPG})_{82}$, the stabilization parameter had a scalar form and is now called "τ_{82}."

Remark 2.2 At its inception, $(\mathrm{SUPG})_{82}$ was not used in combination with a shock-capturing term. That was added in 1991 [70, 71], with a shock-capturing parameter that is now called "δ_{91}." This shock-capturing parameter was derived from the one given in [72] for the entropy variables.

Remark 2.3 The computations reported in [70, 71] were for steady-state inviscid flows. The shock-capturing parameter δ_{91} was based on the steady-state inviscid-flow residual.

Remark 2.4 The τ used in [70] with $(\mathrm{SUPG})_{82}$ was a slightly modified version of τ_{82}. Subsequent minor modifications of τ_{82} took into account the interaction between the DC and $(\mathrm{SUPG})_{82}$ terms in a fashion similar to how it was done in [73] for the advection–diffusion–reaction equation.

Remark 2.5 The discontinuity-capturing parameter "CAU DC," reported in 1996 [74], is just an extension of δ_{91} from its form based on the steady-state residual to its form based on the time-dependent residual.

2.2 Pressure-Stabilizing/Petrov-Galerkin (PSPG) Method, 1990

The name "PSPG," coined in [1, 7], refers to the specific pressure stabilization method introduced in 1990 [1, 7] and the descendants of that method given in [9, 13–15, 24]. It does not refer to a general concept of pressure stabilization. An earlier version of the pressure stabilization, for Stokes flows, was introduced in [75]. The SUPG/PSPG ("SUPS") formulation [1, 7] is written as

$$\int_{\Omega} \mathbf{w}^h \cdot \rho \left(\frac{\partial \mathbf{u}^h}{\partial t} + \mathbf{u}^h \cdot \nabla \mathbf{u}^h - \mathbf{f}^h \right) \mathrm{d}\Omega + \int_{\Omega} \boldsymbol{\varepsilon}(\mathbf{w}^h) : \boldsymbol{\sigma}^h \mathrm{d}\Omega$$
$$- \int_{\Gamma_\mathrm{h}} \mathbf{w}^h \cdot \mathbf{h}^h \mathrm{d}\Gamma + \int_{\Omega} q^h \nabla \cdot \mathbf{u}^h \mathrm{d}\Omega$$
$$+ \sum_{e=1}^{(n_\mathrm{el})_n} \int_{\Omega^e} \frac{\tau_\mathrm{SUPS}}{\rho} \left(\rho \mathbf{u}^h \cdot \nabla \mathbf{w}^h + \nabla q^h \right) \cdot \mathbf{r}_\mathrm{M}^h \mathrm{d}\Omega$$
$$+ \sum_{e=1}^{(n_\mathrm{el})_n} \int_{\Omega^e} \nu_\mathrm{LSIC} \nabla \cdot \mathbf{w}^h \rho \nabla \cdot \mathbf{u}^h \mathrm{d}\Omega = 0, \tag{2.4}$$

where

$$\mathbf{r}_\mathrm{M}^h = \rho \left(\frac{\partial \mathbf{u}^h}{\partial t} + \mathbf{u}^h \cdot \nabla \mathbf{u}^h - \mathbf{f}^h \right) - \nabla \cdot \boldsymbol{\sigma}^h \tag{2.5}$$

is the residual of the momentum equation. Here, ρ, $\mathbf{u}$, $\mathbf{f}$, and $\mathbf{h}$ are the density, velocity, body force, and the stress specified at the boundary. The stress tensor is defined as $\boldsymbol{\sigma} = -p\mathbf{I} + 2\mu\boldsymbol{\varepsilon}(\mathbf{u})$, where p is the pressure, $\mathbf{I}$ is the identity tensor, $\mu = \rho\nu$ is the viscosity, ν is the kinematic viscosity, and $\boldsymbol{\varepsilon}(\mathbf{u}) = \left(\nabla\mathbf{u} + (\nabla\mathbf{u})^\mathrm{T}\right)/2$ is the strain-rate tensor. The test functions associated with the $\mathbf{u}$ and p are $\mathbf{w}$ and q. The last integral, which is the LSIC stabilization, was added in 1992 [76]. The stabilization parameters τ_SUPS and ν_LSIC, in some of their newer forms, can be found in [77, 78].

Chapter 3
Methods and Solutions in 1990–1994

3.1 Space–Time Computational Flow Analysis (STCFA), 1990

3.1.1 STCFA Concept

The STCFA, as truly flow analysis, was introduced in 1990 [1–3]. In the STCFA, the formulation is written over a sequence of N ST slices Q_n, where Q_n is the slice of the ST domain between the time levels n and $n+1$ (see Fig. 3.1). The lateral boundary P_n will have complementary subsets where essential and natural boundary conditions are enforced, just like how it is with Γ_t. The symbol Ω_n represents the spatial domain at time level n. At each time step, the integrations are performed over Q_n. The functions are discontinuous in time at each time level, and the superscripts "−" and "+" indicate the values of the functions just below and above the time level. The trial solution and test function spaces are defined over Q_n by using ST basis functions. A superscript "h" will indicate that the function is coming from a finite-dimensional space. Each Q_n is decomposed into elements Q_n^e, where $e = 1, 2, \ldots, (n_{\mathrm{el}})_n$. The subscript n used with n_{el} is for the general case where the number of ST elements may change from one ST slice to another. For more on the general concepts of the ST methods, see [24].

3.1.2 Basis Functions and Their Derivatives in Space and Time

The ST basis functions are written as a product of their parts in space and time:

$$N_a^\alpha = T^\alpha(\theta) N_a(\boldsymbol{\xi}), \qquad a = 1, 2, \ldots, n_{\mathrm{ens}}, \qquad \alpha = 1, 2, \ldots, n_{\mathrm{ent}}, \tag{3.1}$$

T. E. Tezduyar, K. Takizawa, *Space–Time Computational Flow Analysis*,
Modeling and Simulation in Science, Engineering and Technology,
https://doi.org/10.1007/978-3-031-88727-7_3

Fig. 3.1 ST slice

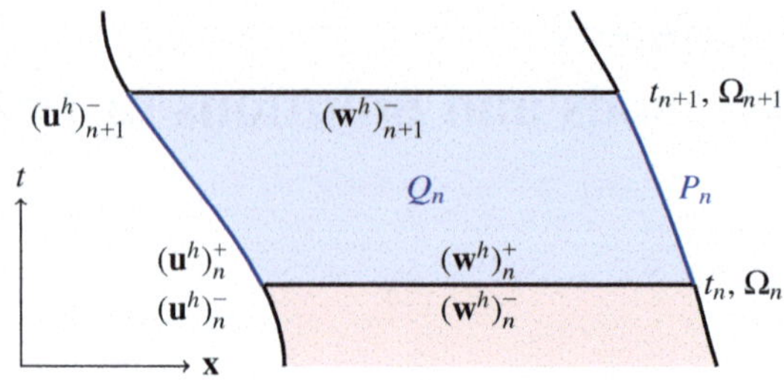

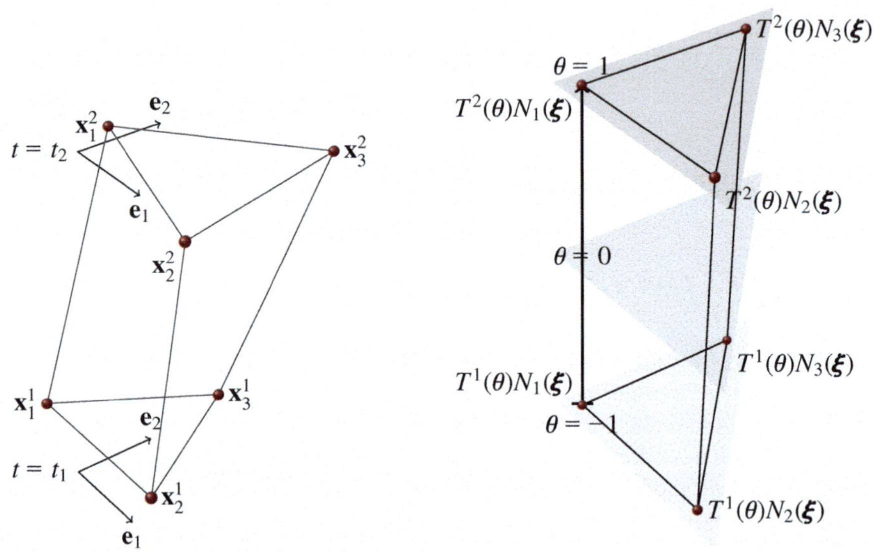

Fig. 3.2 A physical ST element and its parent element. The example is for an element that is triangular in space and linear in time ($n_{\text{ens}} = 3$, $n_{\text{ent}} = 2$). The spatial basis vectors $\mathbf{e}_1$ and $\mathbf{e}_2$ are replicated so that they are depicted at both time levels t_1 and t_2

where $\boldsymbol{\xi}$ and $\theta \in [-1, 1]$ are the parametric coordinates in space and time, and n_{ens} and n_{ent} are the number of element nodes in space and time. Figure 3.2 shows, as an example in 2D space, a physical ST element and its parent element. With the basis functions, position and time are represented as

$$\mathbf{x}^h = \sum_{\alpha=1}^{n_{\text{ent}}} \sum_{a=1}^{n_{\text{ens}}} \mathbf{x}_a^\alpha T^\alpha(\theta) N_a(\boldsymbol{\xi}), \tag{3.2}$$

$$t^h = \sum_{\alpha=1}^{n_{\text{ent}}} \sum_{a=1}^{n_{\text{ens}}} t_a^\alpha T^\alpha(\theta) N_a(\boldsymbol{\xi}). \tag{3.3}$$

The ST Jacobian tensor is defined as

$$\mathbf{Q}^{\mathrm{ST}} = \begin{bmatrix} \frac{\partial t}{\partial \theta} & \frac{\partial t}{\partial \boldsymbol{\xi}} \\ \frac{\partial \mathbf{x}}{\partial \theta} & \mathbf{Q} \end{bmatrix}, \tag{3.4}$$

with the space-only Jacobian tensor being

$$\mathbf{Q} = \frac{\partial \mathbf{x}^h}{\partial \xi^\alpha} \mathbf{e}_\alpha . \tag{3.5}$$

With that, the basis-function derivatives in space and time can be expressed as

$$\begin{bmatrix} \frac{\partial N_a^\alpha}{\partial t} \\ \frac{\partial N_a^\alpha}{\partial \mathbf{x}} \end{bmatrix} = \left(\mathbf{Q}^{\mathrm{ST}}\right)^{-\mathsf{T}} \begin{bmatrix} N_a \frac{\mathrm{d}T^\alpha}{\mathrm{d}\theta} \\ T^\alpha \frac{\partial N_a}{\partial \boldsymbol{\xi}} \end{bmatrix}. \tag{3.6}$$

If time has no dependence on position, $t_a^\alpha = t^\alpha$, and we can write it as

$$t^h = \sum_{\alpha=1}^{n_{\mathrm{ent}}} t^\alpha T^\alpha(\theta) \tag{3.7}$$

and express $\mathbf{Q}^{\mathrm{ST}}$ as

$$\mathbf{Q}^{\mathrm{ST}} = \begin{bmatrix} \frac{\mathrm{d}t^h}{\mathrm{d}\theta} & \mathbf{0}^{\mathsf{T}} \\ \mathbf{v}^h \frac{\mathrm{d}t^h}{\mathrm{d}\theta} & \mathbf{Q} \end{bmatrix}, \tag{3.8}$$

where

$$\mathbf{v}^h = \left.\frac{\partial \mathbf{x}^h}{\partial t}\right|_{\boldsymbol{\xi}} \tag{3.9}$$

is the mesh velocity. Then $\left(\mathbf{Q}^{\mathrm{ST}}\right)^{-\mathsf{T}}$ can be expressed as

$$\left(\mathbf{Q}^{\mathrm{ST}}\right)^{-\mathsf{T}} = \begin{bmatrix} \left(\frac{\mathrm{d}t^h}{\mathrm{d}\theta}\right)^{-1} & -\mathbf{v}^h \cdot \mathbf{Q}^{-\mathsf{T}} \\ \mathbf{0} & \mathbf{Q}^{-\mathsf{T}} \end{bmatrix}. \tag{3.10}$$

3.2 Deforming-Spatial-Domain/Stabilized Space–Time (DSD/SST) ("ST-SUPS") Method, 1990

The ST-SUPS was introduced in 1990 [1–3]. It is written as

$$
\begin{aligned}
&\int_{Q_n} \mathbf{w}^h \cdot \rho \left(\frac{\partial \mathbf{u}^h}{\partial t} + \mathbf{u}^h \cdot \nabla \mathbf{u}^h - \mathbf{f}^h \right) \mathrm{d}Q + \int_{Q_n} \boldsymbol{\varepsilon}(\mathbf{w}^h) : \boldsymbol{\sigma}^h \mathrm{d}Q \\
&- \int_{(P_n)_\mathrm{h}} \mathbf{w}^h \cdot \mathbf{h}^h \mathrm{d}P + \int_{Q_n} q^h \nabla \cdot \mathbf{u}^h \mathrm{d}Q \\
&+ \int_{\Omega_n} (\mathbf{w}^h)_n^+ \cdot \rho \left((\mathbf{u}^h)_n^+ - (\mathbf{u}^h)_n^- \right) \mathrm{d}\Omega \\
&+ \sum_{e=1}^{(n_\mathrm{el})_n} \int_{Q_n^e} \frac{\tau_\mathrm{SUPS}}{\rho} \left(\rho \left(\frac{\partial \mathbf{w}^h}{\partial t} + \mathbf{u}^h \cdot \nabla \mathbf{w}^h \right) + \nabla q^h \right) \cdot \mathbf{r}_\mathrm{M}^h \mathrm{d}Q \\
&+ \sum_{e=1}^{(n_\mathrm{el})_n} \int_{Q_n^e} \nu_\mathrm{LSIC} \nabla \cdot \mathbf{w}^h \rho \nabla \cdot \mathbf{u}^h \mathrm{d}Q = 0.
\end{aligned} \tag{3.11}
$$

As mentioned in Sect. 2.2, the LSIC stabilization was added in 1992.

3.3 Block-Iterative Coupling in ST-SUPS Computations, 1990

Full discretization in methods for FSI and fluid–particle interaction leads to a coupled, nonlinear equation system that needs to be solved at every time step. In a conceptual form that is partitioned with respect to the models represented, such a nonlinear equation system can be written as

$$\mathbf{N}_1 \left(\mathbf{d}_1, \mathbf{d}_2, \mathbf{d}_3 \right) = \mathbf{F}_1, \tag{3.12}$$

$$\mathbf{N}_2 \left(\mathbf{d}_1, \mathbf{d}_2, \mathbf{d}_3 \right) = \mathbf{F}_2, \tag{3.13}$$

$$\mathbf{N}_3 \left(\mathbf{d}_1, \mathbf{d}_2, \mathbf{d}_3 \right) = \mathbf{F}_3, \tag{3.14}$$

where $\mathbf{d}_1$, $\mathbf{d}_2$, and $\mathbf{d}_3$ are the vectors of nodal unknowns representing the fluid, structure (or particle), and mesh unknowns, respectively. For the ST-SUPS, $\mathbf{d}_1$ represents the unknowns associated with the ST slice Q_n.

The block-iterative coupling method in ST-SUPS computations was introduced in 1990 [1, 3] in the context of fluid–particle interactions and was written in a more comprehensive way in [12, 13]. We summarize the method from [13]. In an iteration step taking us from iterative solution i to $i + 1$, the following three blocks of equations are solved:

$$\left.\frac{\partial \mathbf{N}_1}{\partial \mathbf{d}_1}\right|_{(\mathbf{d}_1^i,\mathbf{d}_2^i,\mathbf{d}_3^i)} \Delta \mathbf{d}_1^i = \mathbf{F}_1 - \mathbf{N}_1\left(\mathbf{d}_1^i, \mathbf{d}_2^i, \mathbf{d}_3^i\right), \tag{3.15}$$

$$\mathbf{d}_1^{i+1} = \mathbf{d}_1^i + \Delta \mathbf{d}_1^i, \tag{3.16}$$

$$\left.\frac{\partial \mathbf{N}_2}{\partial \mathbf{d}_2}\right|_{\left(\mathbf{d}_1^{i+1},\mathbf{d}_2^i,\mathbf{d}_3^i\right)} \Delta \mathbf{d}_2^i = \mathbf{F}_2 - \mathbf{N}_2\left(\mathbf{d}_1^{i+1}, \mathbf{d}_2^i, \mathbf{d}_3^i\right), \tag{3.17}$$

$$\mathbf{d}_2^{i+1} = \mathbf{d}_2^i + \Delta \mathbf{d}_2^i, \tag{3.18}$$

$$\left.\frac{\partial \mathbf{N}_3}{\partial \mathbf{d}_3}\right|_{\left(\mathbf{d}_1^{i+1},\mathbf{d}_2^{i+1},\mathbf{d}_3^i\right)} \Delta \mathbf{d}_3^i = \mathbf{F}_3 - \mathbf{N}_3\left(\mathbf{d}_1^{i+1}, \mathbf{d}_2^{i+1}, \mathbf{d}_3^i\right), \tag{3.19}$$

$$\mathbf{d}_3^{i+1} = \mathbf{d}_3^i + \Delta \mathbf{d}_3^i. \tag{3.20}$$

3.4 First 2D Solutions with the ST-SUPS, 1990

We summarize three of the 2D computations reported in 1990 [1–3] when the ST-SUPS was introduced. All three were computed with special-purpose mesh moving methods and the block-iterative coupling method.

3.4.1 *Cylinder Drifting in Shear Flow*

The shear flow profile at the inflow boundary is linear, with higher velocity at the upper wall. The Reynolds number based on the average inflow velocity and the cylinder diameter is 50. The cylinder density is twice the fluid density. The block-iterative coupling method is used in solving the equations governing the fluid mechanics and cylinder dynamics. As the cylinder drifts downstream, it also moves downward and rotates. Figure 3.3 shows the mesh and vorticity.

3.4.2 *Pulsating Drop*

The drop initially has an elliptical shape, with the axial dimensions being 1.25 and 0.80. We are working with nondimensional numbers. The density, viscosity, and surface tension coefficient are 1.0, 0.001, and 0.001. The block-iterative coupling

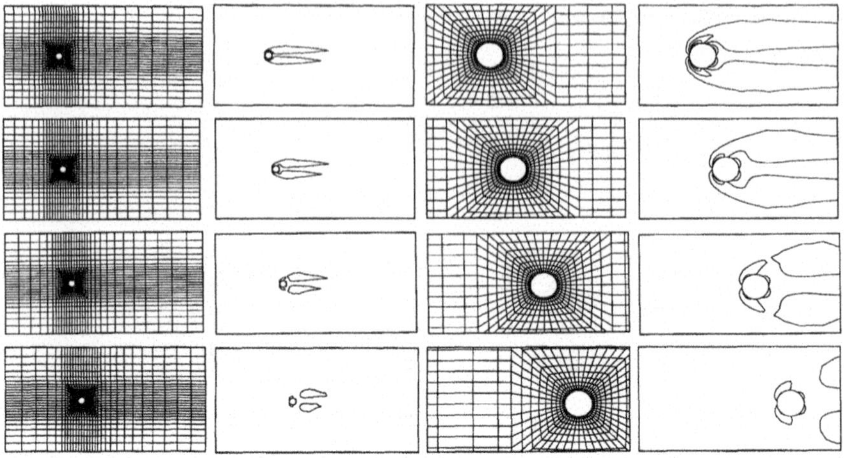

Fig. 3.3 Cylinder drifting in shear flow. Global and near-cylinder views of the mesh and vorticity at uniformly spaced instants

method is used in solving the equations governing the fluid mechanics and free-surface-node positions. As the drop pulsates, the magnitude of the oscillations around the circular shape gradually decreases.

3.4.3 Large-Amplitude Sloshing in a Rectangular Tank

The fluid is initially stationary, occupying a 2.667×1.0 rectangular region. We are working with nondimensional numbers. The density and viscosity are 1.0 and 0.002. The gravity is 1.0 and the surface tension is neglected. The sloshing is driven by a horizontal body force of $0.01 \sin(0.978t)$. The Reynolds number based on the gravity and fluid height is 514. Slip condition is used on the walls. The block-iterative coupling method is used in solving the equations governing the fluid mechanics and free-surface-node positions. After a long computing duration, the sloshing reaches a fully periodic pattern.

3.5 First 2D FSI Computations with the ST-SUPS: Vortex-Induced Vertical Oscillations of a Cylinder, 1991

The first 2D FSI computation with the ST-SUPS was in 1991 [79] for vortex-induced vertical oscillations of a cylinder at Reynolds number 324. At this Reynolds number, the nondimensional natural frequency of the cylinder spring–mass system and the Strouhal number for flow past a cylinder are very close. In 1992 [76], the computations were extended to a range of Reynolds numbers, from 290 to 360. All computations reported in 1991 and 1992 were carried out with a special-purpose mesh moving method and the block-iterative coupling method. The mesh structure was the same as in Fig. 3.3, though with much higher refinement.

3.6 First 2D ST-SUPS Computation of Flow Past an Airfoil with Prescribed Oscillations, 1992

The first 2D ST-SUPS computation of flow past an airfoil with prescribed oscillations was in 1992 [76], for a NACA0012 airfoil at Reynolds number 1000. The nondimensional frequency of oscillations is 1.0 and the angle of attack varies between 10° and 30°. The computation was with a special-purpose mesh moving method. The global mesh structure is the same as in Fig. 3.3, with the near-airfoil mesh shown in Fig. 3.4.

3.7 First 2D ST-SUPS Computation of the Free-Fall Aerodynamics of an Airfoil, 1992

The first 2D ST-SUPS computation of the free-fall aerodynamics of an airfoil was in 1992 [10], with a more extensive study in the same year [80]. A NACA0012 airfoil is dropped at a 10° angle from the line of gravity. Two cases were computed. In the first case [10, 80], the density of the airfoil is uniform and is 50 times the fluid density. In the second case [80], in the region extending from the leading edge to 10% of the chord the density is 500 times the fluid density, and elsewhere 50 times. The computations were with a special-purpose mesh moving method. The block-iterative coupling method is used in solving the equations governing the fluid mechanics and airfoil dynamics. In the uniform-density case, the airfoil quickly reaches time-periodic oscillations between approximately 45° and 125° from the line of gravity, with an average Reynolds number of approximately 225. In the nonuniform-density case, the airfoil eventually reaches a nearly-gliding motion at approximately 80° from the line of gravity, with a Reynolds number of approximately 500. Both cases involve periodic vortex shedding. Figure 3.5 shows the flow patterns for the two cases.

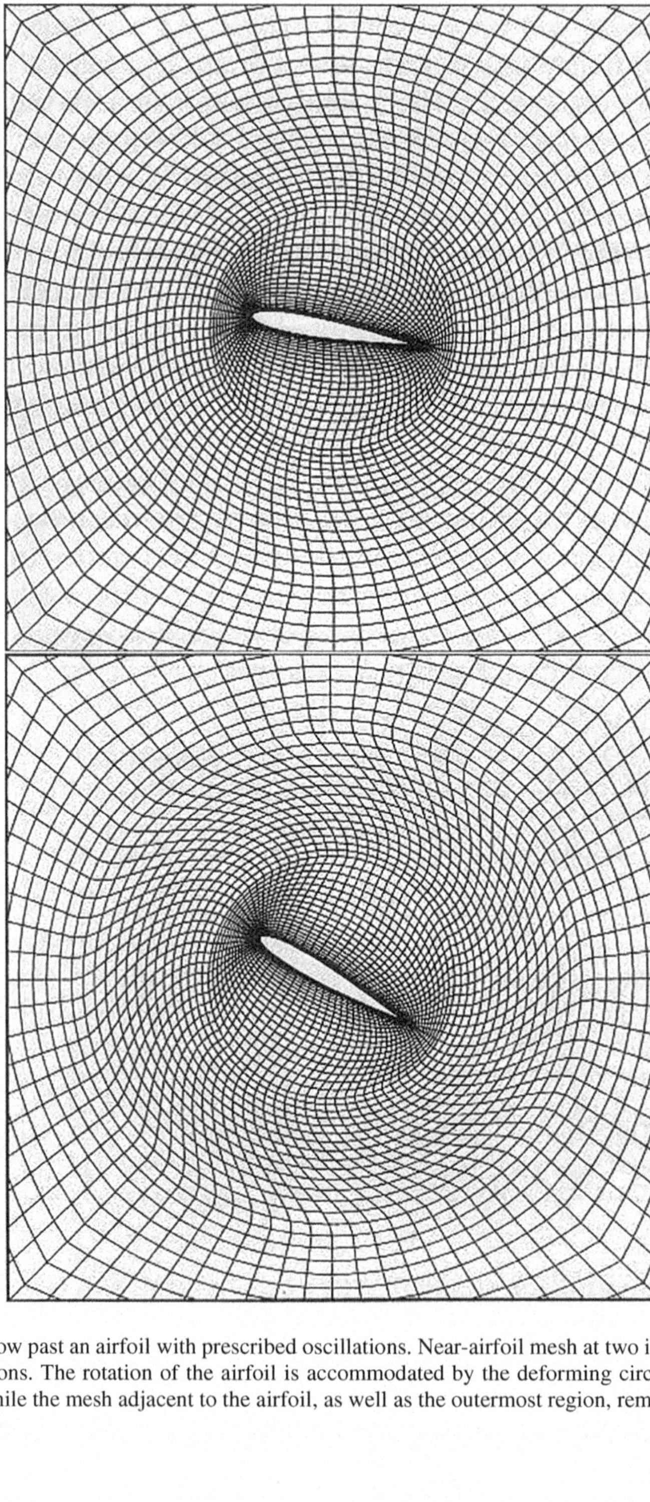

Fig. 3.4 Flow past an airfoil with prescribed oscillations. Near-airfoil mesh at two instants during the oscillations. The rotation of the airfoil is accommodated by the deforming circular region in the mesh while the mesh adjacent to the airfoil, as well as the outermost region, remain rigid

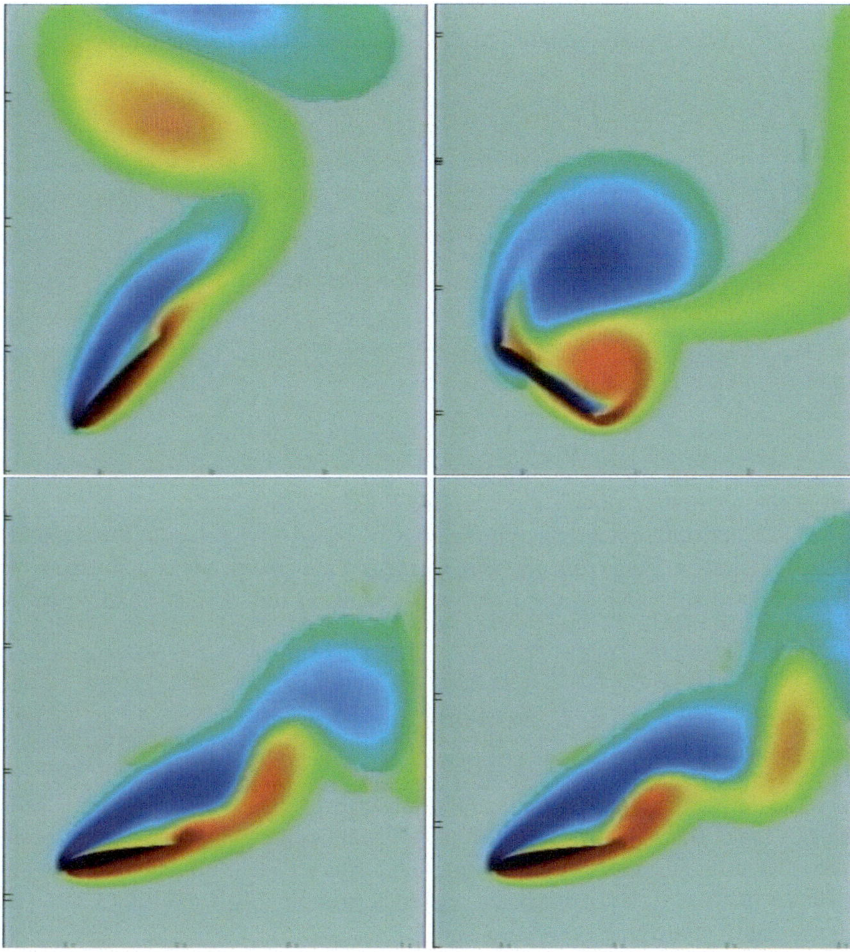

Fig. 3.5 Free-fall aerodynamics of an airfoil. Vorticity. Two instants during the airfoil-oscillation cycle in the uniform-density case (*top*) and two instants during the vortex-shedding cycle in the nonuniform-density case (*bottom*)

3.8 Linear-Elasticity Mesh Moving Method with Mesh-Jacobian-Based Stiffening (MJBS), 1992

The linear-elasticity mesh moving method with MJBS was introduced in 1992 [10]. The early description of the method can also be found in journal articles in 1993 [11] and 1994 [81]. The overview here is from a 2020 journal article [82].

The motion of the internal nodes is determined by solving the equations of linear elasticity. In specifying the boundary conditions, the normal component of the mesh velocity at the boundaries and interfaces need to match the normal component of the fluid velocity. With the notation of [24], the formulation can be written as

$$\int_{\Omega_{\tilde{t}}} \boldsymbol{\varepsilon}(\mathbf{w}^h) : \boldsymbol{\sigma}(\hat{\mathbf{y}}^h - \hat{\mathbf{y}}^h_{\tilde{t}}) \mathrm{d}\Omega = 0. \tag{3.21}$$

Here $\hat{\mathbf{y}}^h$ is the displacement from the reference configuration $\mathbf{X}^h_{\mathrm{REF}}$ to the current configuration $\mathbf{x}^h = \mathbf{X}^h_{\mathrm{REF}} + \hat{\mathbf{y}}^h$, and $\hat{\mathbf{y}}^h_{\tilde{t}}$ is the displacement from $\mathbf{X}^h_{\mathrm{REF}}$ to the $\tilde{t}$ configuration $\mathbf{x}^h_{\tilde{t}} = \mathbf{X}^h_{\mathrm{REF}} + \hat{\mathbf{y}}^h_{\tilde{t}}$ we compute the mesh motion from. The most common selection is $\tilde{t} = t_n$ when we are computing the mesh at t_{n+1}. The stress tensor is given as

$$\boldsymbol{\sigma}(\mathbf{y}) = \frac{E}{1+\nu} \left(\frac{\nu}{1-2\nu} \mathrm{tr}\left(\boldsymbol{\varepsilon}(\mathbf{y})\right) \mathbf{I} + \boldsymbol{\varepsilon}(\mathbf{y}) \right), \tag{3.22}$$

where E and ν are the Young's modulus and Poisson's ratio.

The mesh deformation is dealt with selectively based on the sizes of the elements. Selective treatment based on element sizes is attained by altering the way we account for the Jacobian of the transformation from the element domain to the physical domain:

$$J_{\mathrm{M}} = \det \left(\frac{\partial \mathbf{x}^h_{\tilde{t}}}{\partial \boldsymbol{\xi}} \right), \tag{3.23}$$

where $\boldsymbol{\xi}$ is the element coordinate, and the subscript "M" is to clarify that this is the mesh Jacobian. The objective is to stiffen the smaller elements, which are typically placed near solid surfaces, more than the larger ones. When this method was first introduced in [10, 11, 81], it consisted of simply dropping J_{M} from Eq. (3.21). That is equivalent to modifying Eq. (3.21) as

$$\int_{\Omega_{\tilde{t}}} \boldsymbol{\varepsilon}(\mathbf{w}^h) : \boldsymbol{\sigma}(\hat{\mathbf{y}}^h - \hat{\mathbf{y}}^h_{\tilde{t}}) \left(\frac{J_{\mathrm{M}}}{(J_{\mathrm{M}})_0} \right)^{-1} \mathrm{d}\Omega = 0, \tag{3.24}$$

where $(J_{\mathrm{M}})_0$ was originally intended to be a global scaling value and is now seen as a free parameter. The modification results in the smaller elements being stiffened more than the larger ones, and the element stiffening factor is $\frac{(J_{\mathrm{M}})_0}{J_{\mathrm{M}}}$. The method was augmented in [83] to a more extensive kind by introducing a stiffening power χ that determines the degree by which the smaller elements are rendered stiffer than the larger ones:

$$\int_{\Omega_{\tilde{t}}} \boldsymbol{\varepsilon}(\mathbf{w}^h) : \boldsymbol{\sigma}(\hat{\mathbf{y}}^h - \hat{\mathbf{y}}^h_{\tilde{t}}) \left(\frac{J_{\mathrm{M}}}{(J_{\mathrm{M}})_0} \right)^{-\chi} \mathrm{d}\Omega = 0. \tag{3.25}$$

This approach, when $\chi = 1$, would be identical to the method first introduced in [10], and most of the time $\chi = 1$.

Figure 3.6 shows, from [83], an example of how the MJBS protects thin layers of elements near solid surfaces.

From earlier tests with different $\tilde{t}$ settings, we know that setting $\tilde{t} = t_n$ gives reasonably good performance for most problems. We believe that is partly because when we compute the mesh motion from a configuration close to the current configuration, J_{M} provides the method feedback on which elements to protect.

We note that in practical computations the element stiffening method proposed in [84] in 1996 is approximately equivalent to the MJBS method with $\chi = 1$. To show that, we first approximate $\frac{(J_{\mathrm{M}})_0}{J_{\mathrm{M}}}$ as

$$\frac{(J_{\mathrm{M}})_0}{J_{\mathrm{M}}} \approx \frac{V_0}{V}, \tag{3.26}$$

where V is the element volume and V_0 is a global scaling value. The approximation becomes exact for simplex elements. In [84], the stiffening factor, represented by "$1 + \tau_{\mathrm{m}}$", is given as

$$1 + \tau_{\mathrm{m}} = 1 + \frac{1 - \frac{V_{\min}}{V_{\max}}}{\frac{V}{V_{\max}}}, \tag{3.27}$$

where $V_{\min}$ and $V_{\max}$ are the minimum and maximum element volumes. This can also be written as

$$1 + \tau_{\mathrm{m}} = 1 + \frac{V_{\max} - V_{\min}}{V}. \tag{3.28}$$

In practical computations, $V_{\max} >> V_{\min}$, and for most of the elements that matter,

$$\frac{V_{\max} - V_{\min}}{V} >> 1. \tag{3.29}$$

Therefore

$$1 + \tau_{\mathrm{m}} \approx \frac{V_{\max} - V_{\min}}{V}. \tag{3.30}$$

Because $V_{\max} - V_{\min}$ is a global scaling value just like $(J_{\mathrm{M}})_0$ is, the stiffening factor given by Eq. (3.30) is equivalent to the one given by Eq. (3.26).

Remark 3.1 We note that a stiffening-factor expression in terms of the mesh Jacobian gives us the option of evaluating the factor at the integration points, without the need for any precalculation. We also note that such an expression is naturally applicable to isogeometric discretization.

Fig. 3.6 Linear-elasticity mesh moving method with MJBS. Mesh deformation tests. Without and with MJBS. We note how the method protects the thin layer of elements on each side of the zero-thickness structure

3.9 First Axisymmetric ST-SUPS Computation of a Viscous Drop Falling in a Viscous Fluid, 1992

The first axisymmetric ST-SUPS computation of a viscous drop falling in a viscous fluid was in 1992 [10], with a detailed report in [81]. A drop made of a heavier viscous fluid is falling in another viscous fluid, with the surface tension accounted for. The drop, initially spherical, changes shape until a terminal shape and speed are reached. The computation was with the linear-elasticity mesh moving method with MJBS. The block-iterative coupling method is used in solving the equations governing the fluid mechanics and interface-node positions. At terminal speed, the Reynolds and Weber numbers are 12 and 6.24, and the drag coefficient is 1.42. Figure 3.7 shows the drop at different instants.

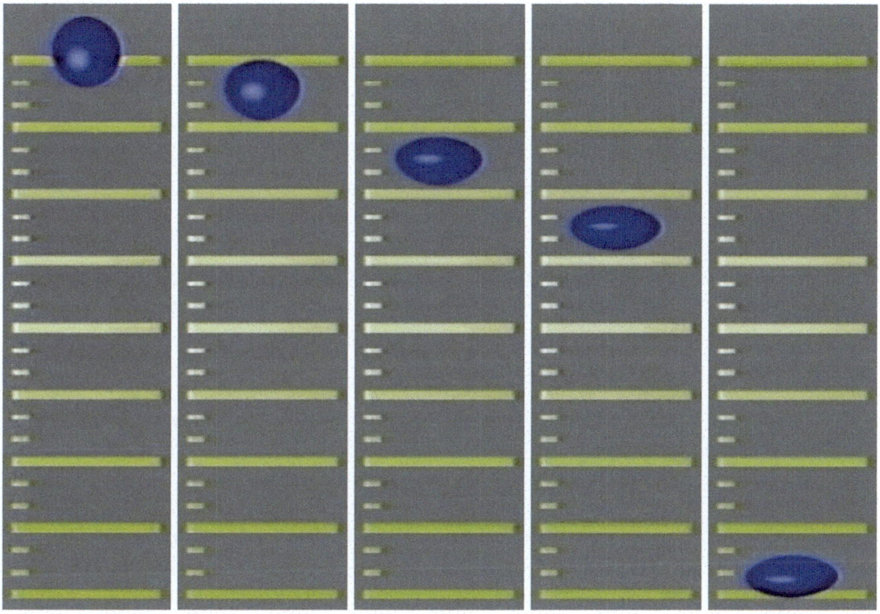

Fig. 3.7 Viscous drop falling in a viscous fluid

3.10 Compressible-Flow DSD/SST ("Compressible-Flow ST-SUPG") Method, 1992

The compressible-flow ST-SUPG, first formulated in [85], is a straightforward mixture of the DSD/SST concept and the compressible-flow SUPG. The compressible-flow ST-SUPG formulation [85] is written as

$$\begin{aligned}
&\int_{Q_n} \mathbf{W}^h \cdot \left(\frac{\partial \mathbf{U}^h}{\partial t} + \mathbf{A}_i^h \cdot \frac{\partial \mathbf{U}^h}{\partial x_i} - \mathbf{S}^h \right) \mathrm{d}Q \\
&+ \int_{Q_n} \frac{\partial \mathbf{W}^h}{\partial x_i} \cdot \mathbf{K}_{ij}^h \cdot \frac{\partial \mathbf{U}^h}{\partial x_j} \mathrm{d}Q - \int_{(P_n)_\mathrm{H}} \mathbf{W}^h \cdot \mathbf{H}^h \mathrm{d}P \\
&+ \int_{\Omega_n} (\mathbf{W}^h)_n^+ \cdot \left((\mathbf{U}^h)_n^+ - (\mathbf{U}^h)_n^- \right) \mathrm{d}\Omega \\
&+ \sum_{e=1}^{(n_\mathrm{el})_n} \int_{Q_n^e} \left(\boldsymbol{\tau}_\mathrm{SUPG} \cdot \left(\frac{\partial \mathbf{W}^h}{\partial t} + \frac{\partial \mathbf{W}^h}{\partial x_k} \cdot \mathbf{A}_k^h \right) \right) \cdot \mathbf{R}^h \mathrm{d}Q \\
&+ \sum_{e=1}^{(n_\mathrm{el})_n} \int_{Q_n^e} \nu_\mathrm{SHOC} \frac{\partial \mathbf{W}^h}{\partial x_i} \cdot \frac{\partial \mathbf{U}^h}{\partial x_i} \mathrm{d}Q = 0.
\end{aligned} \tag{3.31}$$

Remark 3.2 The advection–diffusion ST-SUPG, which can easily be inferred from the compressible-flow ST-SUPG or the ST-SUPS, is given in Appendix B.

3.11 First 2D Solutions with the Compressible-Flow ST-SUPG, 1992

The first 2D computations with the compressible-flow ST-SUPG, reported in 1992 [85], were for Mach 0.2 flow past an oscillating NACA0012 airfoil at Reynolds number 1000 and Mach 1.4 flow between two wedge-shaped solid surfaces in relative motion at Reynolds number 7500. The airfoil angle of attack varies between 10° and 30°, with the aerodynamic coefficients reaching time-periodicity. The Mach 1.4 flow involves moving shocks and their interaction with the boundary layers. Both computations were with special-purpose mesh moving methods.

3.12 First 3D Solution with the ST-SUPS, 1992

The first 3D computation with the ST-SUPS, reported in 1992 [86] and also in 1993 [11], was for sloshing in a partially-filled tank subjected to vertical vibration. A special-purpose mesh moving method was used in the computation, with a mesh having 6000 trilinear hexahedral elements and with the free-surface nodes moving with the normal component of the fluid velocity at the surface. The block-iterative coupling method is used in solving the equations governing the fluid mechanics and free-surface-node positions. The boundary conditions on the tank walls and bottom are slip-surface conditions. The simulation is continued for a large number of sloshing cycles to see that the sloshing direction slowly rotates.

3.13 First ST-SUPS Computation of the Taylor–Couette Flow, 1993

The first ST-SUPS computation of the Taylor–Couette flow was reported in 1993 [11], at Reynolds number 150, 250, and 1498, and also in early 1994 [87], at Reynolds number 150 and 250. The mesh is made of 38,400 trilinear hexahedral elements, with 6, 32, and 200 elements in the radial, circumferential, and axial directions. The boundary conditions are no-slip on the inner and outer surfaces and slip-surface conditions at the two axial ends. At Reynolds number 150, we see an axisymmetric Taylor vortex flow, and at 250 and 1498, a wavy Taylor vortex flow, with four and three waves traveling in the circumferential direction.

3.14 First ST-SUPS Computation of Flow Around a Submarine, 1993

The first ST-SUPS computation of flow around a submarine was in 1993 [87], for a Los Angeles-class submarine, at Reynolds number 10^9. The mesh is made of 71,035 nodes and 345,129 linear tetrahedral elements. The images from the computation were featured on the cover page of the *Slide Book of the ARPA High Performance Computing Software PI Meeting, San Diego, California, 27–29 September 1993* (see Figure 5 in [87]), following a cover-page competition.

3.15 First Axisymmetric Computations with the Compressible-Flow ST-SUPG, 1994

The first axisymmetric computations with the compressible-flow ST-SUPG were in 1994 for the air intake of a jet engine with adjustable spool [87] and a regenerative liquid propellant gun [88]. The flow in the air intake is at Reynolds number 0.8×10^6 and Mach 2. The motion of the spool involves moving shocks and their interaction with the boundary layers. The computation was with a special-purpose mesh moving method. The regenerative liquid propellant gun involves two pistons in relative motion. The fluid is barotropic, with the pressure being only a function of density. The linear-elasticity mesh moving method with MJBS was used in dealing with the moving solid surfaces.

3.16 First 3D FSI Computations with the ST-SUPS: Flow-Induced Vibrations of a Flexible Pipe, 1994

The first 3D FSI computations with the ST-SUPS were in 1994 [87] for flow-induced vibrations of a cantilevered flexible pipe, with details provided also in 1994 [89]. The inflow profile is parabolic, the Reynolds number based on the flow velocity at the pipe centerline is 1000, and the pipe length is 20 times its diameter. The mesh is made of 18,720 trilinear hexahedral elements. A special-purpose mesh moving method was used in dealing with the moving solid surfaces. The block-iterative coupling method is used in solving the coupled FSI equations. The ratio of the fluid and structure masses per unit length of the pipe is treated as one of the parameters varied in the study, together with the scaled average flow speed. Depending on the combinations of these two parameters, the pipe vibrations damp out or reach a time-periodic pattern.

3.17 First 3D ST-SUPS Computation of Flapping-Wing Aerodynamics, 1994

The first 3D ST-SUPS computation of flapping-wing aerodynamics was in 1994 [89], for a pair of wings, with the wing cross-section being a NACA0012 airfoil. The wings are each 1.5 long in the spanwise direction, have a chord length of 1.0 at the root, and taper off quadratically to 0.5 at the tips. We are working with nondimensional numbers. The wing motion is based on the description given by Lighthill [90] of the flight of birds and includes a variation of the angle of attack between 4° and 20°. The Reynolds number based on the flight speed and the chord length at the root is 1000. The nondimensional flapping frequency is 0.5. The mesh is made of 137,280 trilinear hexahedral elements. A special-purpose mesh moving method was used in dealing with the moving solid surfaces. The lift and drag coefficients reach a time-periodic pattern, showing a net lift and thrust.

3.18 Shear–Slip Mesh Update Method (SSMUM), 1994

The SSMUM was introduced, though not under the name "SSMUM," in 1994 (see Remark 3.3) in a conference proceedings paper [91] that was also included as a chapter [91] in a book published by the conference organizers. It was presented in a journal article first time in 1996 [92]. It increases the scope of interface-tracking (moving-mesh) techniques by making them applicable to MBI problems that would otherwise be too taxing for general mesh moving and remeshing methods. What motivated the creation of the SSMUM was computing the aerodynamics of two high-speed trains passing each other in a tunnel [91, 92] (see Remark 3.3). The challenge was to accurately and efficiently update the meshes used in ST computations involving two solid surfaces or objects in fast, linear relative motion, with the meshes covering the subdomains containing the objects moving with the objects. The idea behind the SSMUM is to restrict the mesh moving and remeshing to a thin layer of elements at the interfaces between the subdomains. The mesh update at each time step is done by a "shear" deformation of the elements in this layer, followed by a "slip" in node connectivities. The slip in the node connectivities, to an extent, un-does the deformation of the elements and results in elements with better shapes than those that were shear-deformed. Because the remeshing consists of just redefining the node connectivities, both the projection errors and the mesh generation cost are reduced.

Remark 3.3 The SSMUM was actually created in 1993 by Tezduyar, to be implemented by Masayuki Kaiho (Hitachi, Japan), who was a Visiting Scholar at that time, and to be used in computing the aerodynamics of two high-speed trains passing each other in a tunnel. The implementation, at least most of it, was completed by Kaiho before his return to Japan in 1993. That was unpublished work.

Remark 3.4 In computations involving objects in linear relative motion, such as trains passing each other in a tunnel, in addition to the SSMUM process, the elements are periodically transplanted from the mesh regions ahead of the objects to the regions behind. Otherwise, the elements in those regions may become too compressed or too stretched. One does not face that challenge in computations involving solid surfaces or objects in rotational relative motion.

3.19 First 3D Computation with the SSMUM: ST-SUPS Computation of the Aerodynamics of Two High-Speed Trains Passing Each Other in a Tunnel, 1994

The first 3D computation with the SSMUM was reported in 1994 [91] for the aerodynamics of two high-speed trains passing each other in a tunnel. The trains are moving at 100 m/s as they pass each other. The mesh is made of 51,584 trilinear hexahedral elements. In addition to the SSMUM process, the elements are

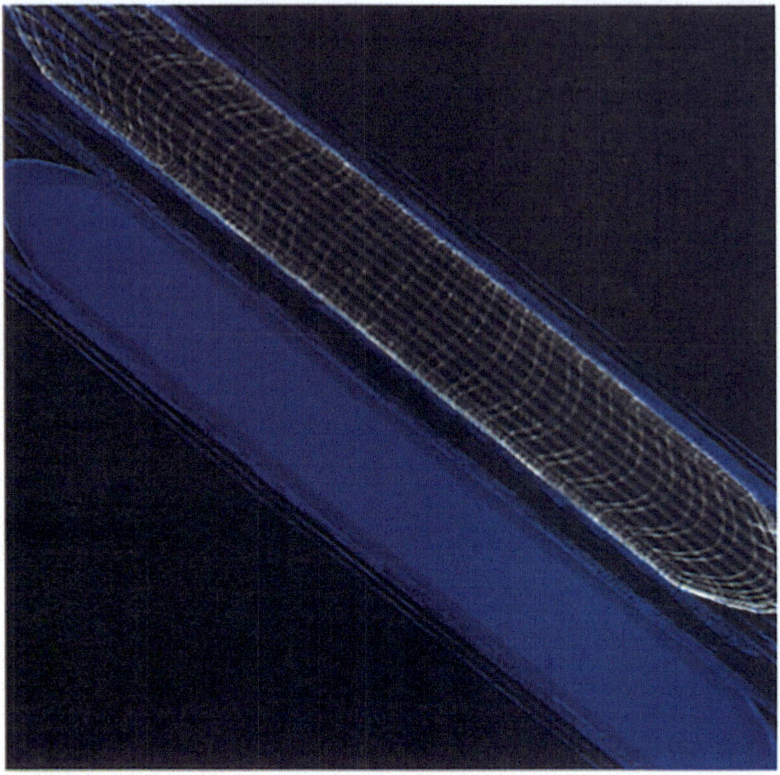

Fig. 3.8 Incompressible-flow aerodynamics of two high-speed trains passing each other in a tunnel, computed in 1994 with the ST-SUPS and SSMUM. Surface pressure and mesh

periodically transplanted from the mesh regions ahead of the trains to the regions behind. In this first computation, the flow was treated as incompressible, and the ST-SUPS was used. Figure 3.8 shows the trains from the computation reported in [91].

Chapter 4
Methods and Solutions in 1994–1999

4.1 First in 3D Direct Numerical Simulation of Fluid–Particle Interactions: ST-SUPS, 1995

The ST-SUPS was the first in 3D direct numerical simulation of fluid–particle interactions, reported in 1995 [57]. Four cases of spherical particles falling in a liquid-filled tube were computed, one with two spheres, one with three spheres, and two with five spheres. Collisions between the spheres are modeled as near contact (with near topology change in the domain), as described in [57], and the nearness is sufficiently near for the purpose of solving this class of problems. The meshes are made of linear tetrahedral elements, with an element count of roughly 140,000 in the two-sphere case, 190,000 in the three-sphere case, and 276,000 and 321,000 in the two five-sphere cases. The computations were with the linear-elasticity mesh moving method with MJBS and remeshing as needed. The frequency of remeshing was, on the average, every 75 time steps in the two-sphere case, every 415 in the three-sphere case, and every 125 and 500 in the two five-sphere cases. The block-iterative coupling method was used in solving the equations governing the fluid mechanics and sphere dynamics. In the two-sphere case, the spheres are initially in a staggered configuration. As expected, the trailing sphere gets attracted to the low-pressure wake of the leading sphere, they collide, and fall side by side. That is the equilibrium configuration for that case. The spheres in the other three cases also start in a nonequilibrium configuration and after a while reach a corresponding equilibrium configuration. The Reynolds number associated with the sphere terminal speed varies from 72 to 93 in the four cases. In all four cases, the solution quantitatively match the experimental and theoretical data.

T. E. Tezduyar, K. Takizawa, *Space–Time Computational Flow Analysis*,
Modeling and Simulation in Science, Engineering and Technology,
https://doi.org/10.1007/978-3-031-88727-7_4

4.2 First 3D Computation with the Compressible-Flow ST-SUPG and First 3D Compressible-Flow Computation with the SSMUM: Aerodynamics of Two High-Speed Trains Passing Each Other in a Tunnel, 1996

The first 3D computation with the compressible-flow ST-SUPG, which was also the first 3D compressible-flow computation with the SSMUM, was reported in 1996 [92] for the aerodynamics of two high-speed trains passing each other in a tunnel. The trains are moving at 100 m/s as they pass each other. The mesh is made of 101,888 trilinear hexahedral elements. In addition to the SSMUM process, the elements are periodically transplanted from the mesh regions ahead of the trains to the regions behind. Figure 4.1 shows the trains from the computation reported in [92].

Fig. 4.1 Compressible-flow aerodynamics of two high-speed trains passing each other in a tunnel, computed in 1996 with the compressible-flow ST-SUPG and SSMUM. Surface pressure and mesh

Remark 4.1 Extending the SSMUM to 3D computations involving solid surfaces in rotational relative motion (see Remark 3.4) was reported in 1998 [93] for a propeller and in 1999 [94] for a helicopter.

Remark 4.2 The SSMUM involves a complicated, special-purpose implementation, and therefore has high startup cost in each analysis. It is a method that is neither simple to use nor sufficiently general. A number of options were proposed in 2000 [95] to increase its generality. Those options can also be found in [96, 97].

4.3 Shallow-Water-Flow DSD/SST ("Shallow-Water-Flow ST-SUPG") Method, 1996

The shallow-water-flow ST-SUPG was first formulated in 1996 [98]. The shallow-water equations (SWE) is a special case of free-surface flow modeling, where the problem scale in the vertical direction is much smaller compared to the problem scale in the other two directions. It is an effective mathematical model for practical computer simulation of tidal flows and floods, including those caused by tsunamis and storm surges. It is known that the SWE can be represented by the strong form of the ST-SUPG formulation given in Sect. 3.10, with appropriate expressions for $\mathbf{U}$, $\mathbf{A}_i$, $\mathbf{K}_{ij}$, and $\mathbf{S}$ (see, for example, [95]). The shallow-water-flow ST-SUPG was presented also in [95], with details on the expressions for $\mathbf{U}$, $\mathbf{F}_i$, $\mathbf{E}_i$, $\mathbf{A}_i$, $\mathbf{K}_{ij}$, and $\mathbf{S}$. Those expressions are given in Appendix C. With the SWE expressions for $\mathbf{U}$, $\mathbf{A}_i$, $\mathbf{K}_{ij}$, and $\mathbf{S}$, the ST-SUPG formulation given in Sect. 3.10 becomes the shallow-water-flow ST-SUPG. The first shallow-water-flow ST-SUPG computation was reported in 1996 [98], in the same article where the shallow-water-flow ST-SUPG was formulated. The computation was for the tidal flow in Tokyo Bay.

The SWE model takes care of the free-surface aspect of the moving boundaries. The moving-shoreline aspect requires a method that can handle 2D computational domains with changing shapes. It was pointed out in [95] that the shallow-water-flow ST-SUPG can automatically take into account the motion of the water–land interfaces. It was also pointed out in [95] that all techniques introduced in conjunction with the DSD/SST method, such as the mesh moving methods, can also be used with the shallow-water-flow ST-SUPG. The shallow-water-flow ST-SUPG was used in 2011 [99] in computations with moving shorelines. The computations were with the linear-elasticity mesh moving method with MJBS.

4.4 First in 3D Direct Numerical Simulation of Fluid–Particle Interactions Reaching 100 Particles: ST-SUPS, 1996

The ST-SUPS was the first in 3D direct numerical simulation of fluid–particle interactions reaching 100 particles, reported in 1996 [100]. Two cases of 101 spherical particles falling in a liquid-filled tube were computed, one with random-sized spheres and one with uniform-sized. The initial distribution of the spheres is random in both cases. Collisions between the spheres are modeled as near contact, as described in [57], and the nearness is sufficiently near for the purpose of solving this class of problems. The meshes are made of linear tetrahedral elements, with an element count of roughly 1.2 M in both cases. The computations were with the linear-elasticity mesh moving method with MJBS and remeshing as needed. The frequency of remeshing was, on the average, every 8 time steps in the random-sized-spheres case and every 9 in the uniform-sized-spheres case. The block-iterative coupling method was used in solving the equations governing the fluid mechanics and sphere dynamics. In the random-sized-spheres case, as we expect based on what we know from basic fluid mechanics, larger spheres fall faster, expanding the column of particles in the tube. In the uniform-sized-spheres case, on the other hand, the column height does not change much. In both cases, we see many instances of a sphere getting attracted to the wake of a leading sphere, the two colliding, and falling side by side (see Sect. 4.1). Figure 4.2 shows, from the computations reported in [100], the spheres for the random-sized-spheres case. In the random-sized-spheres case, the group-averaged Reynolds number corresponding to the terminal speed is 23 for the smallest one third of the spheres, 40 for the middle one third, and 57 for the largest one third. In the uniform-sized-spheres case, the average Reynolds number is 38.

4.5 First in 3D Direct Numerical Simulation of Fluid–Particle Interactions Reaching 1000 Particles: ST-SUPS, 1998

The ST-SUPS was the first in 3D direct numerical simulation of fluid–particle interactions reaching 1000 particles, reported in 1998 [101, 102]. The simulation was for 1000 uniform-sized spherical particles falling in a liquid-filled tube, with a random initial distribution. Collisions between the spheres are modeled as near contact, as described in [57], and the nearness is sufficiently near for the purpose of solving this class of problems. The mesh is made of linear tetrahedral elements, with an element count of roughly 2.5 M. The computations were with the linear-elasticity mesh moving method with MJBS and remeshing as needed. The frequency of remeshing was, on the average, every 14 time steps. The block-iterative coupling

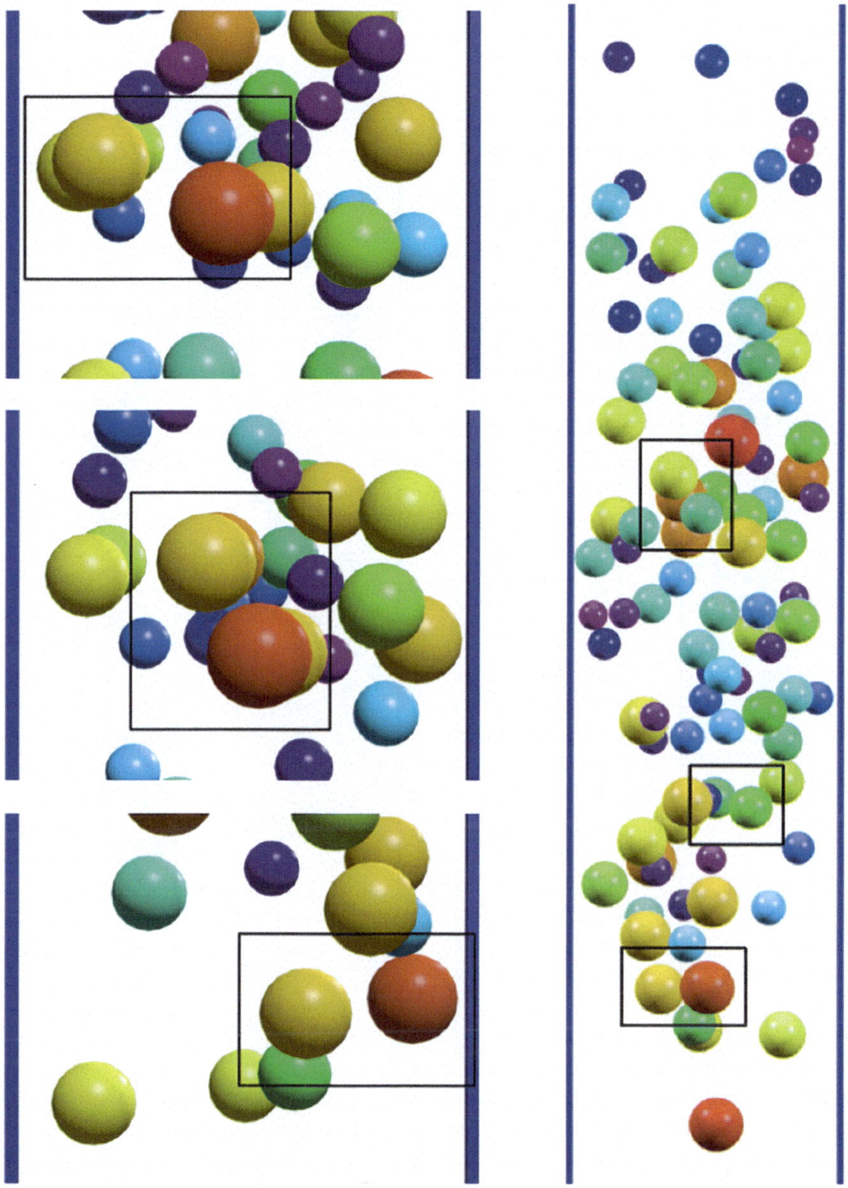

Fig. 4.2 3D direct numerical simulation of fluid–particle interactions reaching 100 particles. Random-sized-spheres case. The *left* pictures show how a pair of spheres collide and fall side by side. The *right* picture shows three pairs of spheres as they are colliding and falling side by side

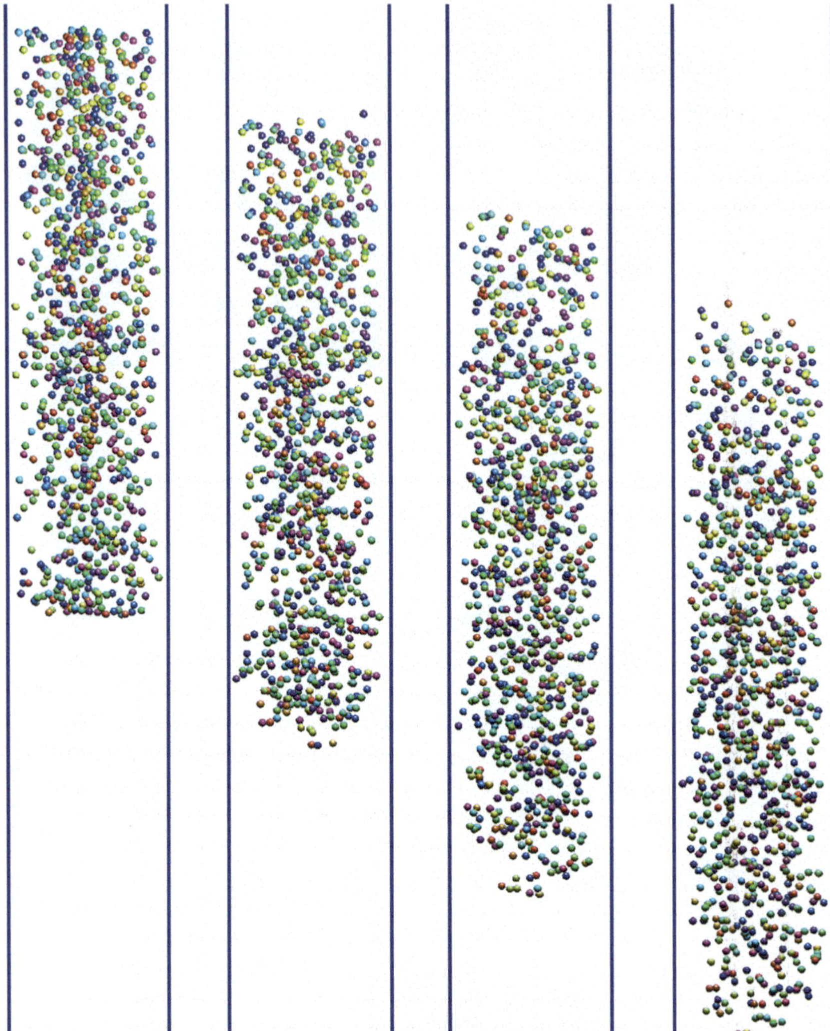

Fig. 4.3 3D direct numerical simulation of fluid–particle interactions reaching 1000 particles. The first picture shows the initial distribution of the spheres

method was used in solving the equations governing the fluid mechanics and sphere dynamics. The spheres near the center of the tube fall with higher speeds compared to those near the tube walls, and as they exit the particle column at the bottom, they spread out towards the tube walls. This increases the column height. The average Reynolds number is around 8. Figure 4.3 shows, from the simulation reported in [101, 102], the spheres at four instants.

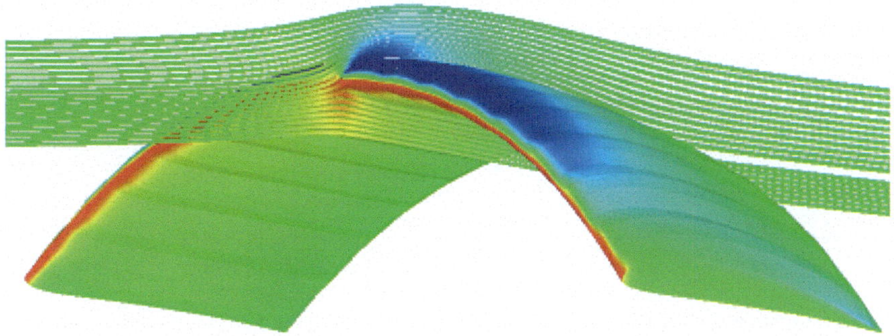

Fig. 4.4 3D parachute FSI for a ram-air parachute. Equilibrium configuration. Surface pressure and stream tubes

4.6 First in 3D Parachute FSI: ST-SUPS, 1998

The ST-SUPS was the first in 3D parachute FSI, reported in 1998 [103]. The computation was for a ram-air parachute. The structure mesh is made of 9454 two-node cable elements and 4796 four-node membrane elements. The fluid mechanics mesh is made of 236,688 trilinear hexahedral elements. The computations were with the linear-elasticity mesh moving method with MJBS. The block-iterative coupling method was used in solving the coupled FSI equations. Figure 4.4 shows, from the computation reported in [103], the parachute in equilibrium configuration. More 3D parachute FSI computations with the ST-SUPS were reported in 1998 (see [104, 105]); they were for a round parachute.

4.7 Multidomain Method (MDM), 1998

The MDM was introduced in 1998 [106], originally intended for cost-effective computation of long-wake flows and later extended in scope to cover additional classes of flow problems. In computation of long-wake flows, the wake computation alone might be the final objective, or the ultimate objective might be the computation of the wake influence on a secondary object placed far downstream. The MDM computation is based on a sequence of subdomains with slight overlap between them. The first one covers the object producing the wake, and that would be the primary object if there is a secondary object downstream. The secondary object would be in the last subdomain. The velocity specified at the inflow boundary of the first subdomain is the free-stream velocity. The inflow velocity for each subsequent subdomain is extracted from the preceding subdomain. If the outflow boundary of a subsequent subdomain is also in the preceding subdomain, then the stress at the outflow boundary is also extracted from the preceding subdomain.

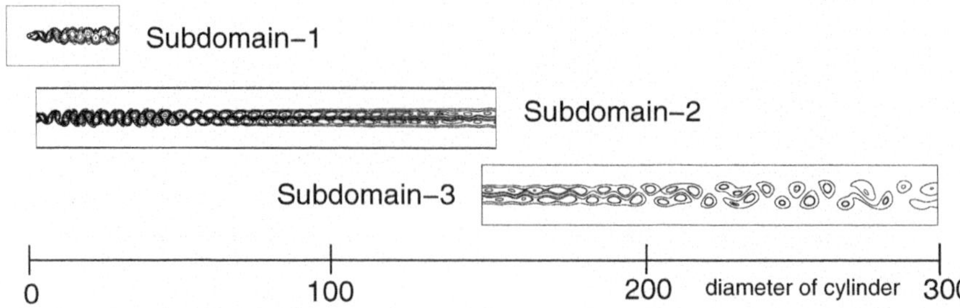

Fig. 4.5 3D MDM computation of long-wake flow for a cylinder at Reynolds number 140

The first computations with the MDM were reported also in 1998 [106]. They were 2D long-wake flow for a circular cylinder, which was a verification study, and 3D wake influence on a wing placed downstream of a larger wing. Other 3D computations conducted soon after that were a cylinder wake extending 300 diameters downstream (1998) [107–109], aerodynamics of a parachute crossing the far wake of an aircraft (1999) [108], and FSI of that parachute as it is crossing the wake (2000) [110]. Figure 4.5 shows, from the 3D computation reported in [107–109], long-wake flow for a cylinder at Reynolds number 140. The meshes for the three subdomains are made of 201,600, 4,308,480, and 4,224,000 trilinear hexahedral elements. The secondary vortex street, appearing far downstream and known from laboratory experiments, is successfully captured.

4.8 First in 3D Direct Numerical Simulation of Spatially Triperiodic Fluid–Particle Interactions Reaching 125 Particles: ST-SUPS, 1999

The ST-SUPS was the first in 3D direct numerical simulation of spatially triperiodic fluid–particle interactions reaching 125 particles, reported in 1999 [111]. Special versions of the ST-SUPS method for spatially uni-periodic, bi-periodic, and triperiodic flows were also introduced in [111]. Two sequences of triperiodic computations with uniform-sized spherical particles falling in a liquid-filled tube were reported in [111]. In the first sequence, for a given particle density, the number of spheres in the periodic cell is 1, 8, 27, 64, and 125. In the second sequence, the number of spheres in the periodic cell is 27, and the particle density is set to four different values. The flow parameters used in all simulations are set to give, for a single sphere, a terminal-speed Reynolds number of approximately 10. Collisions between the spheres are modeled as near contact, as described in [57], and the nearness is sufficiently near for the purpose of solving this class of problems. All meshes are made of linear tetrahedral elements. The computations were with the linear-elasticity mesh moving method with MJBS, with remeshing, on the average, every 7 time steps. The block-

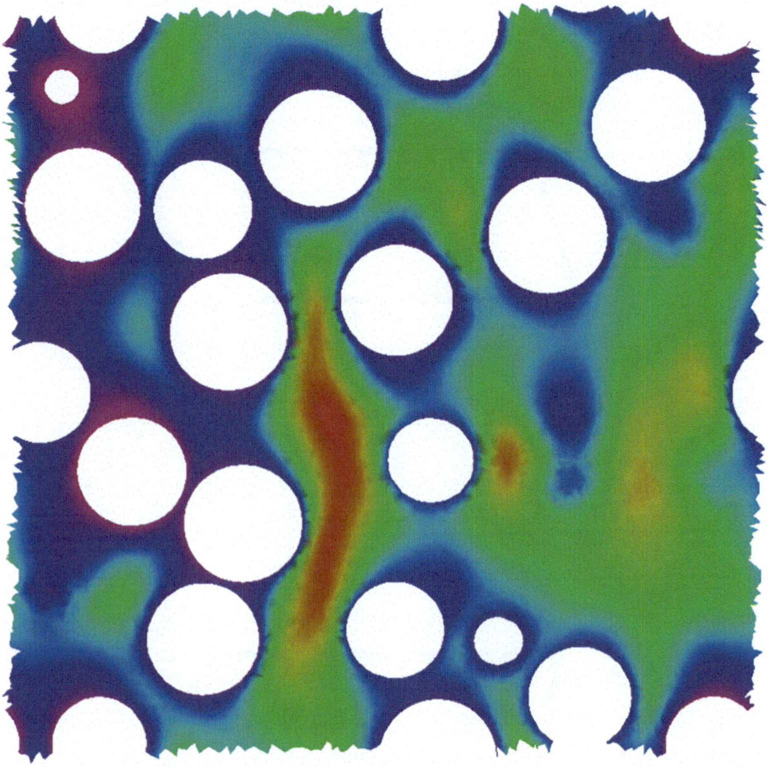

Fig. 4.6 3D direct numerical simulation of spatially triperiodic fluid–particle interactions with 125 particles. Vertical component of the flow velocity at a cross-section. *Red* and *blue* indicate zones of rising and descending fluid

iterative coupling method was used in solving the equations governing the fluid mechanics and sphere dynamics. As a sample, we report from [111] computation of the case with 125 spheres. The number of elements is roughly 1.8 M. Figure 4.6 shows the flow velocity.

4.9 Space–Time Contact Technique (STCT), 1999

The STCT was introduced in 1990 [108], recognizing that the ST framework, combined with the STCT, provides a natural mechanism to handle time-dependent flow problems with contacting and de-contacting surfaces. The technique was described in the context of liquid surfaces contacting and de-contacting solid walls.

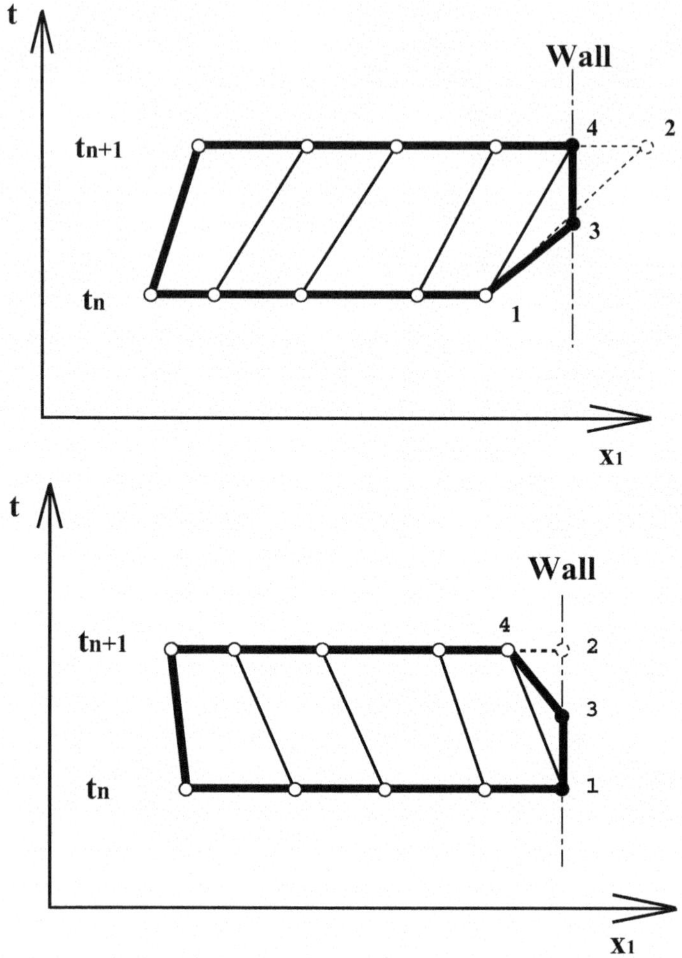

Fig. 4.7 STCT concept in 1D. Fluid contacting (*top*) and de-contacting (*bottom*) the wall

Figures 4.7 and 4.8 illustrate the STCT concept in 1D and 2D (for details, see [95, 96, 108]). Because the STCT targeted representing the contact and de-contact times with a precision finer than the time-step size, it required unstructured ST mesh generation, though only partial, still not encouragingly practical.

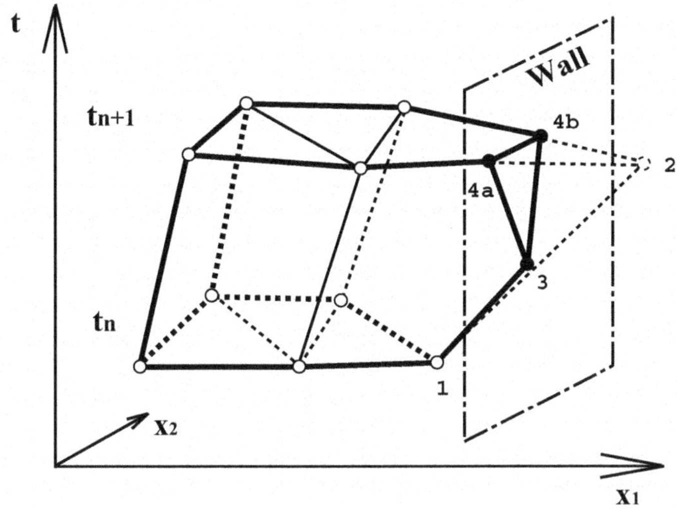

Fig. 4.8 STCT concept in 2D. Fluid contacting the wall

Chapter 5
Methods and Solutions in 2000–2004

5.1 First in 3D FSI Computation of a Parachute Crossing the Far Wake of an Aircraft: ST-SUPS, 2000

The ST-SUPS, in combination with the MDM, was the first in 3D FSI computation of a parachute crossing the far wake of an aircraft, reported in 2000 [110]. The objective is to predict the aerodynamic forces the parachute canopy is subjected to as it crosses the unsteady wake and how that influences, through the FSI, the performance of the parachute. Figure 5.1 shows how the MDM process works in this case. The computations for Subdomain-1 and Subdomain-2, carried out with the SUPS method, were reported in [108], and the data from Subdomain-2 is used here to specify the boundary conditions for Subdomain-3. The mesh for Subdomain-1 is made of 1,286,063 linear tetrahedral elements and 212,177 six-node prismatic elements in the one-element-thick layer surrounding the domain. Figure 5.2 shows the parachute and the flow patterns.

5.2 Mixed Interface-Tracking/Interface-Capturing Technique (MITICT), 2000

In computations involving fluid–solid interfaces, an interface-tracking (moving-mesh) technique, where the fluid mesh moves to track the interface, would enable us to control the mesh resolution near the interface and have high-resolution representation of the boundary layers. In computations involving fluid–fluid interfaces where the interface is too complex to track while keeping the remeshing frequency under control (such as free-surface and two-fluid flows with splashing), an interface-capturing (nonmoving-mesh) technique would enable us to make the problem computable. Sometimes the computation may involve both fluid–solid interfaces

T. E. Tezduyar, K. Takizawa, *Space–Time Computational Flow Analysis*,
Modeling and Simulation in Science, Engineering and Technology,
https://doi.org/10.1007/978-3-031-88727-7_5

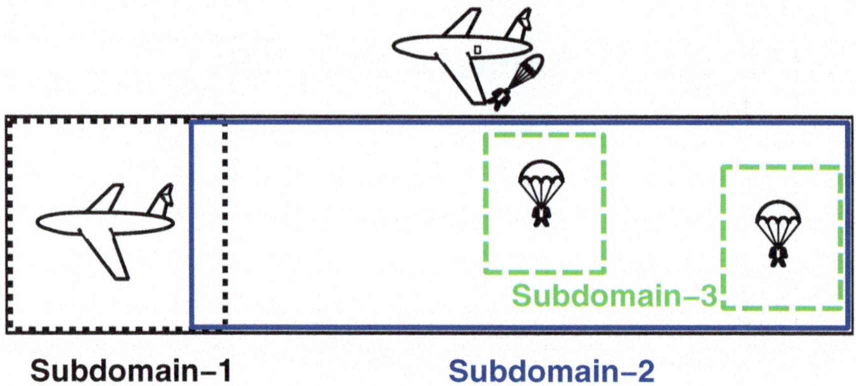

Fig. 5.1 3D FSI computation of a parachute crossing the far wake of an aircraft. MDM subdomains

and very complex fluid–fluid interfaces. The MITICT was introduced in 2000 [95], initially intended for fluid–object and fluid–particle interactions with multiple fluids. The classes of applications targeted were fluid–particle–gas interactions and free-surface flow of fluid–particle mixtures.

In the MITICT, the fluid–solid interfaces are tracked by a moving-mesh method, with a mesh that follows those interfaces, enabling mesh resolution control in the boundary layers. The fluid–fluid interfaces and free surfaces, on the other hand, are captured over that moving mesh. Because the mesh motion is not driven by the motion of the interfaces that are captured, the method used in computing those interfaces is, in essence, a nonmoving-mesh method. In a more general sense, in the MITICT, interfaces that can be tracked are tracked with a moving mesh, and interfaces that cannot be tracked are captured over that moving mesh. That comprehensiveness of the MITICT was emphasized in 2000 [95] with the following statement: "In more general cases, the MITICT can be used for classes of problems that involve both interfaces that can be accurately tracked with a moving-mesh method and interfaces that are too complex or unsteady to be tracked and therefore require an interface-capturing technique."

Remark 5.1 The intent in the creation of the MITICT was to use as its interface-tracking component a moving-mesh method so that we can have mesh resolution control in the boundary layers near the interfaces that are tracked. What was meant by "interface-tracking" and "moving-mesh" was clearly declared as early as in 2002 [9] with the following statement: "A consequence of the mesh not moving to 'track' the interface is that for fluid–solid interfaces, independent of how well the interface geometry is represented, the resolution of the boundary layer will be limited by the resolution of the fluid mesh where the interface is." Clearly, that statement is referring to moving a fluid mechanics volume mesh.

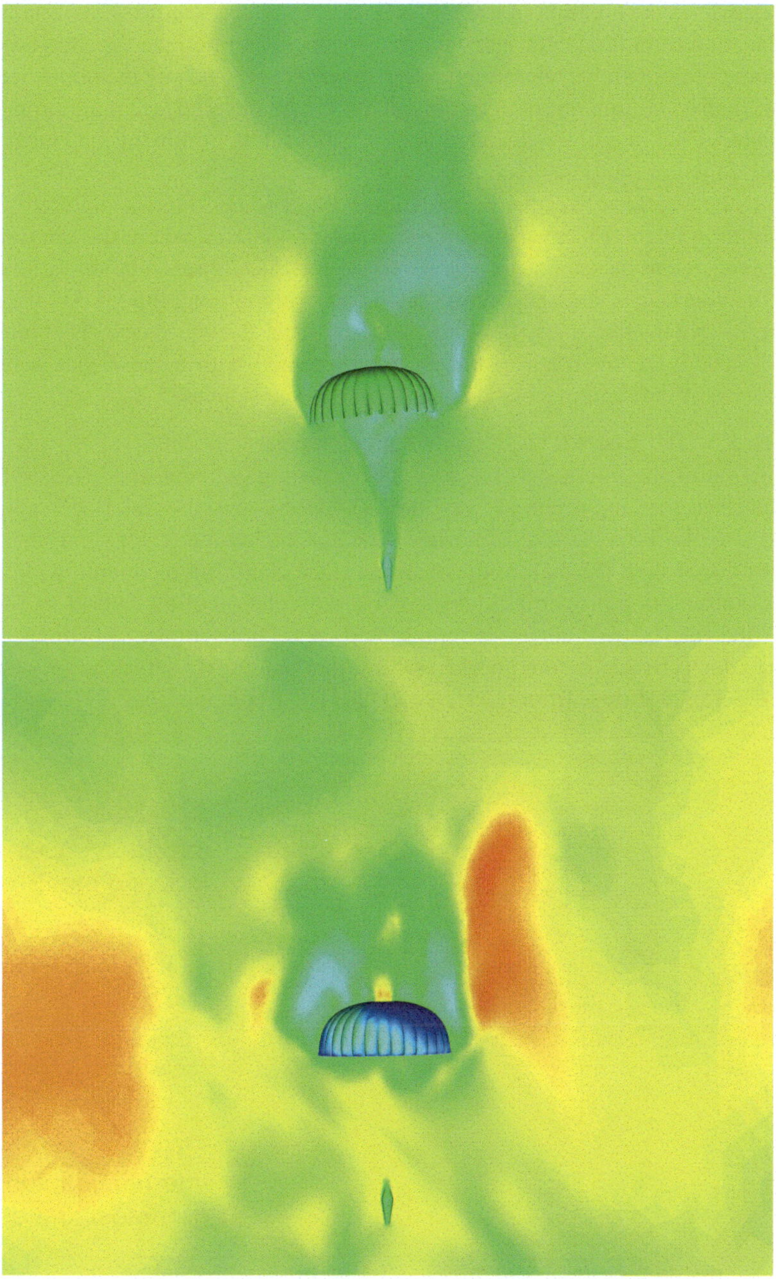

Fig. 5.2 3D FSI computation of a parachute crossing the far wake of an aircraft. Vertical velocity component and the pressure on the parachute and payload surfaces. Parachute in uniform flow and in the unsteady wake

Remark 5.2 In the creation of the MITICT, the "interface-capturing" was meant to be a general terminology, not just the methods using an interface function (such as the volume of fluid and level set methods). It was meant to be any method where the fluid mechanics volume mesh does not move to track the interface that is somehow represented. That intent was declared in 2002 [9] with the following statement: "An interface function, marking the location of the interface, needs to be computed to 'capture' the interface over the nonmoving mesh. The interface is captured within the resolution of the finite element mesh covering the area where the interface is. This approach can be seen as a special case of interface representation techniques, where the interface is somehow represented over a nonmoving fluid mesh, the main point being that the fluid mesh does not move to 'track' the interfaces." That includes methods where the interface geometry is represented with a mesh that is not an integral part of the fluid mechanics volume mesh (see Remark 5.3).

Remark 5.3 The first MITICT computation, reported in 2005 [112], was a 2D computation of the interaction between an oscillating cylinder and a free surface. The interface-tracking method was the ST-SUPS, and the interface-capturing method was the Edge-Tracked Interface Locator Technique (ETILT). The ETILT was introduced in 2000 [95] "to have an interface-capturing technique with better volume conservation properties and sharper representation of the interfaces." In the ETILT, the interface geometry is represented by a "collection of positions along element edges crossed by the interfaces," and that practically serves as an interface mesh. Figure 5.3 shows, from [112], the 2D MITICT computation.

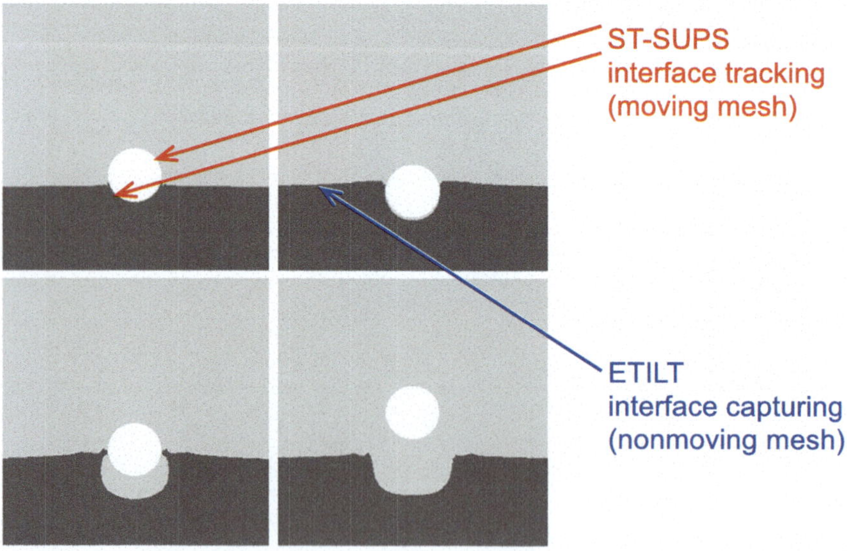

Fig. 5.3 2D MITICT computation of an oscillating cylinder interacting with a free surface

Remark 5.4 The MITICT was used in 2012 [113] in 3D computation of ship hydrodynamics. In that case the interface-tracking technique was the ALE-VMS method, and the interface-capturing technique was a level set method.

Remark 5.5 We recognize, at least in the context of the MITICT and MITICT-like techniques, the need for separately classifying the interface-capturing techniques with interface function and interface mesh (see Remarks 5.2 and 5.3), especially if the two versions are used in a mixed setting. We introduce the classification "interface-meshing technique" to identify the techniques with an interface mesh. With that, we introduce the classifications "Mixed Interface-Tracking/Interface-Meshing Technique (MITIMT)" and "Mixed Interface-Meshing/Interface-Capturing Technique (MIMICT)." The MITIMT is a special case of the MITICT, and the MIMICT is a MITICT-like technique.

5.3 Temporally Multiscale ST Method: Enhanced-Discretization Space–Time Technique (EDSTT), 2001

The EDSTT was introduced in 2001 [114]. The method was also described in 2003 in a journal article [115]. It increases the scope of interface-tracking (moving-mesh) techniques by enabling time-accurate computation of FSI problems where the fluid and structural mechanics parts have different time scales and therefore require very different time-step sizes in temporal discretization. In general, the EDSTT can be used in time-accurate computations where we require smaller time steps in certain parts of the fluid domain. There are two ways of formulating the EDSTT.

5.3.1 EDSTT-Single-Mesh (EDSTT-SM)

In the EDSTT-SM [114, 115], a single ST mesh, unstructured both in space and time, would be used to enhance the time discretization in regions of the fluid domain requiring smaller time steps. This concept is illustrated in Fig. 5.4 with a fictitious 2D ST mesh for FSI computation. The EDSTT-SM was successfully tested in 2003 [115] with 1D and 2D computation of time-dependent pure advection problems. The EDSTT-SM enables in such computations local adjustment of the Courant number to obtain accurate solutions as the profiles of a passive scalar were advected. Application of the EDSTT-SM to a spatially 3D problem would in general require an unstructured 4D ST mesh generation.

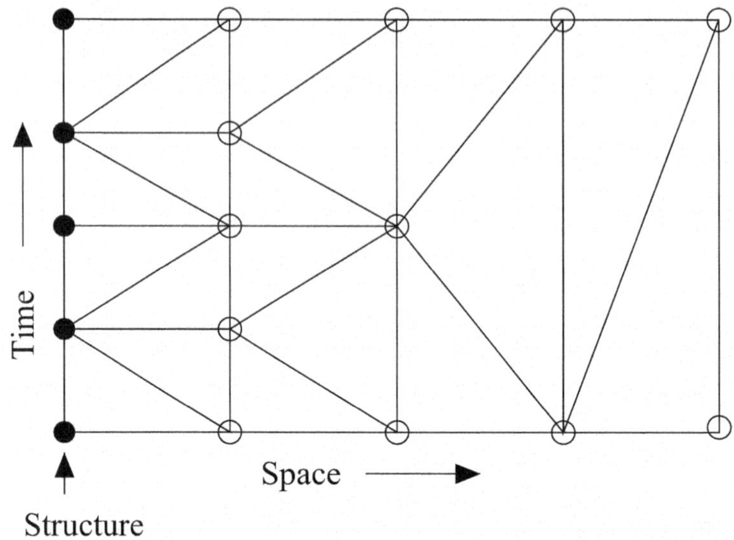

Fig. 5.4 EDSTT-SM concept. A fictitious 2D ST mesh for FSI computation. Spatially, the fluid and structural mechanics parts are 1D and 0D. The time-step size for the fluid mechanics part is four times the time-step size for the structural mechanics part. The fluid-only nodes are denoted by empty circles, while the nodes shared by the fluid and structure are denoted by solid circles

5.3.2 EDSTT-Multi-Mesh (EDSTT-MM)

In the EDSTT-MM, multiple ST meshes, all structured in time, would be used to enhance the time-discretization in regions of the fluid domain requiring smaller time steps. This concept is illustrated in Fig. 5.5 with fictitious multiple 2D ST meshes for FSI computation. The EDSTT-MM was successfully tested in 2003 [115] in FSI computation of 2D flow past a thin flexible 1D structure modeled as Bernoulli–Euler beam. The time-step size for the fluid mechanics part was 20 times the time-step size for the structural mechanics part.

5.4 Mesh-Resolution Control in Boundary Layers Near Deforming Solid Surfaces: Solid-Extension Mesh Moving Technique (SEMMT), 2001

The SEMMT was introduced in 2001 [114]. It was also described in 2003 in a journal article [116]. In mesh motion determined by solving the elasticity equations, it enables better mesh-resolution control in boundary layers near deforming solid surfaces. That in turn results in more accurate representation of those boundary layers. In the SEMMT, the thin layers of elements placed the near solid surfaces

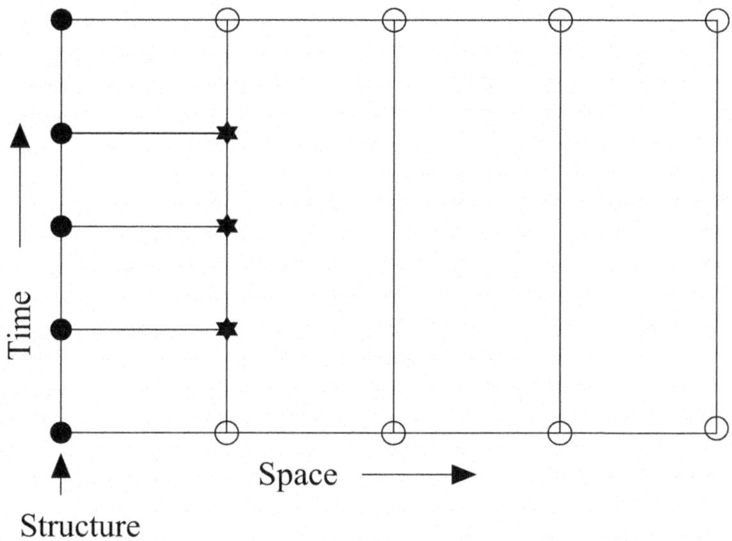

Fig. 5.5 EDSTT-MM concept. Fictitious multiple 2D ST meshes for FSI computation. Spatially, the fluid and structural mechanics parts are 1D and 0D. The time-step size for the fluid mechanics part is four times the time-step size for the structural mechanics part. The structural mechanics part has a single mesh, and the fluid mechanics part has two ST meshes. The two components of the fluid mechanics function space associated with the two meshes are based on the nodes denoted by the empty and solid circles. Neither component is based on the nodes denoted by the stars. The solid circles also denote the structural mechanics nodes

are treated almost like an extension of the solid elements. In solving the elasticity equations governing the mesh motion, higher stiffness is assigned to the thin layers of elements compared to the other elements in the mesh. Two ways of accomplishing this were proposed in [114]: solving the elasticity equations for the nodes connected to the thin layers of elements separate from the elasticity equations for the other nodes, or together. If they are solved separately, for the thin layers of elements, as boundary conditions at the interface with the other elements, traction-free conditions would be used. The separate-solution option is referred to as "SEMMT – Multiple Domain (SEMMT-MD)," and the unified-solution option as "SEMMT – Single Domain (SEMMT-SD)." Figure 5.6 shows, from [116], how the SEMMT protects the thin layers of elements placed near the near solid surfaces.

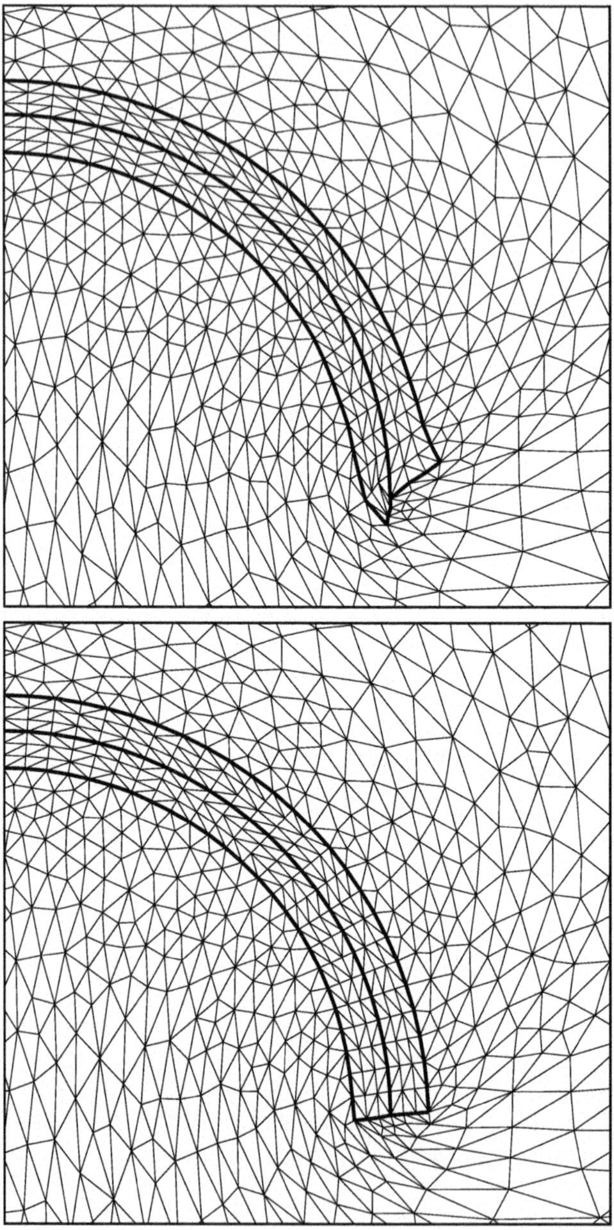

Fig. 5.6 SEMMT. Mesh deformation tests with the linear-elasticity mesh moving method with MJBS. Standard method and SEMMT-MD. We note how the SEMMT protects the thin layer of elements on each side of the zero-thickness structure

5.5 Variationally Consistent FSI Interface Projection for Nonmatching Fluid and Structure Meshes in the Context of the ST-SUPS, 2004

A variationally consistent FSI interface projection method for nonmatching fluid and structure meshes in the context of the ST-SUPS was introduced in 2004 [12]. The formulation is written as

$$\begin{aligned}
&\int_{Q_n} \mathbf{w}_{1\mathrm{E}}^h \cdot \rho \left(\frac{\partial \mathbf{u}^h}{\partial t} + \mathbf{u}^h \cdot \boldsymbol{\nabla}\mathbf{u}^h - \mathbf{f}^h \right) \mathrm{d}Q \\
&+ \int_{Q_n} \boldsymbol{\varepsilon}(\mathbf{w}_{1\mathrm{E}}^h) : \boldsymbol{\sigma}^h \mathrm{d}Q - \int_{(P_n)_\mathrm{h}} \mathbf{w}_{1\mathrm{E}}^h \cdot \mathbf{h}_{1\mathrm{E}}^h \mathrm{d}P \\
&+ \int_{Q_n} q_{1\mathrm{E}}^h \boldsymbol{\nabla} \cdot \mathbf{u}^h \mathrm{d}Q + \int_{\Omega_n} (\mathbf{w}_{1\mathrm{E}}^h)_n^+ \cdot \rho \left((\mathbf{u}^h)_n^+ - (\mathbf{u}^h)_n^- \right) \mathrm{d}\Omega \\
&+ \sum_{e=1}^{(n_\mathrm{el})_n} \int_{Q_n^e} \frac{\tau_\mathrm{SUPS}}{\rho} \left(\rho \left(\frac{\partial \mathbf{w}_{1\mathrm{E}}^h}{\partial t} + \mathbf{u}^h \cdot \boldsymbol{\nabla}\mathbf{w}_{1\mathrm{E}}^h \right) + \boldsymbol{\nabla} q_{1\mathrm{E}}^h \right) \cdot \mathbf{r}_\mathrm{M}^h \mathrm{d}Q \\
&+ \sum_{e=1}^{(n_\mathrm{el})_n} \int_{Q_n^e} \nu_\mathrm{LSIC} \boldsymbol{\nabla} \cdot \mathbf{w}_{1\mathrm{E}}^h \rho \boldsymbol{\nabla} \cdot \mathbf{u}^h \mathrm{d}Q = 0,
\end{aligned} \tag{5.1}$$

$$\int_{Q_n} q_{1\mathrm{I}}^h \boldsymbol{\nabla} \cdot \mathbf{u}^h \mathrm{d}Q + \sum_{e=1}^{(n_\mathrm{el})_n} \int_{Q_n^e} \frac{\tau_\mathrm{SUPS}}{\rho} \boldsymbol{\nabla} q_{1\mathrm{I}}^h \cdot \mathbf{r}_\mathrm{M}^h \mathrm{d}Q = 0, \tag{5.2}$$

$$\int_{(\Gamma_{1\mathrm{I}})_{n+1}} (\mathbf{w}_{1\mathrm{I}}^h)_{n+1}^- \cdot \left((\mathbf{u}_{1\mathrm{I}}^h)_{n+1}^- - \mathbf{u}_{2\mathrm{I}}^h \right) \mathrm{d}\Gamma = 0, \tag{5.3}$$

$$\begin{aligned}
&\int_{Q_n} (\mathbf{w}_{1\mathrm{I}}^h)_{n+1}^- \cdot \rho \left(\frac{\partial \mathbf{u}^h}{\partial t} + \mathbf{u}^h \cdot \boldsymbol{\nabla}\mathbf{u}^h - \mathbf{f}^h \right) \mathrm{d}Q \\
&+ \int_{Q_n} \boldsymbol{\varepsilon}((\mathbf{w}_{1\mathrm{I}}^h)_{n+1}^-) : \boldsymbol{\sigma}^h \mathrm{d}Q - \int_{(P_n)_\mathrm{h}} (\mathbf{w}_{1\mathrm{I}}^h)_{n+1}^- \cdot \mathbf{h}_{1\mathrm{I}}^h \mathrm{d}P \\
&+ \sum_{e=1}^{(n_\mathrm{el})_n} \int_{Q_n^e} \frac{\tau_\mathrm{SUPS}}{\rho} \left(\rho \left(\frac{\partial (\mathbf{w}_{1\mathrm{I}}^h)_{n+1}^-}{\partial t} + \mathbf{u}^h \cdot \boldsymbol{\nabla} (\mathbf{w}_{1\mathrm{I}}^h)_{n+1}^- \right) \right) \cdot \mathbf{r}_\mathrm{M}^h \mathrm{d}Q \\
&+ \sum_{e=1}^{(n_\mathrm{el})_n} \int_{Q_n^e} \nu_\mathrm{LSIC} \boldsymbol{\nabla} \cdot (\mathbf{w}_{1\mathrm{I}}^h)_{n+1}^- \rho \boldsymbol{\nabla} \cdot \mathbf{u}^h \mathrm{d}Q = 0.
\end{aligned} \tag{5.4}$$

Here subscripts 1 and 2 refer to fluid and structure, subscript I refers to the fluid–structure interface, subscript E refers to "elsewhere" in the fluid and structure domain and boundaries, and $\mathbf{u}_{2I}^h$ is the structural displacement rate at the interface.

5.6 Quasi-Direct Coupling, 2004

The quasi-direct coupling method was introduced in 2004 [12] and was written in a more comprehensive way in [13]. We summarize the method from [13]. In an iteration step taking us from iterative solution i to $i + 1$, the following two blocks of equations are solved:

$$\left.\frac{\partial \mathbf{N}_1}{\partial \mathbf{d}_1}\right|_{(\mathbf{d}_1^i,\mathbf{d}_2^i,\mathbf{d}_3^i)} \Delta\mathbf{d}_1^i + \left.\frac{\partial \mathbf{N}_1}{\partial \mathbf{d}_2}\right|_{(\mathbf{d}_1^i,\mathbf{d}_2^i,\mathbf{d}_3^i)} \Delta\mathbf{d}_2^i = \mathbf{F}_1 - \mathbf{N}_1\left(\mathbf{d}_1^i, \mathbf{d}_2^i, \mathbf{d}_2^i\right), \tag{5.5}$$

$$\left.\frac{\partial \mathbf{N}_2}{\partial \mathbf{d}_1}\right|_{(\mathbf{d}_1^i,\mathbf{d}_2^i,\mathbf{d}_3^i)} \Delta\mathbf{d}_1^i + \left.\frac{\partial \mathbf{N}_2}{\partial \mathbf{d}_2}\right|_{(\mathbf{d}_1^i,\mathbf{d}_2^i,\mathbf{d}_3^i)} \Delta\mathbf{d}_2^i = \mathbf{F}_2 - \mathbf{N}_2\left(\mathbf{d}_1^i, \mathbf{d}_2^i, \mathbf{d}_2^i\right), \tag{5.6}$$

$$\mathbf{d}_1^{i+1} = \mathbf{d}_1^i + \Delta\mathbf{d}_1^i, \tag{5.7}$$

$$\mathbf{d}_2^{i+1} = \mathbf{d}_2^i + \Delta\mathbf{d}_2^i, \tag{5.8}$$

$$\left.\frac{\partial \mathbf{N}_3}{\partial \mathbf{d}_3}\right|_{\left(\mathbf{d}_1^{i+1},\mathbf{d}_2^{i+1},\mathbf{d}_3^i\right)} \Delta\mathbf{d}_3^i = \mathbf{F}_3 - \mathbf{N}_3\left(\mathbf{d}_1^{i+1}, \mathbf{d}_2^{i+1}, \mathbf{d}_2^i\right), \tag{5.9}$$

$$\mathbf{d}_3^{i+1} = \mathbf{d}_3^i + \Delta\mathbf{d}_3^i. \tag{5.10}$$

5.7 First 3D FSI Computation with the ST-SUPS and Quasi-Direct Coupling: Flow Past a Flag, 2006

The first 3D FSI computation with the ST-SUPS and quasi-direct coupling, reported in 2004 [12], was for flow past a flag. The Reynolds number was artificially low, at around 250. The computation was with the linear-elasticity mesh moving method with MJBS. The quasi-direct coupling method was used in solving the coupled FSI equations. Figure 5.7 shows, from the computation reported in [12], the flag and

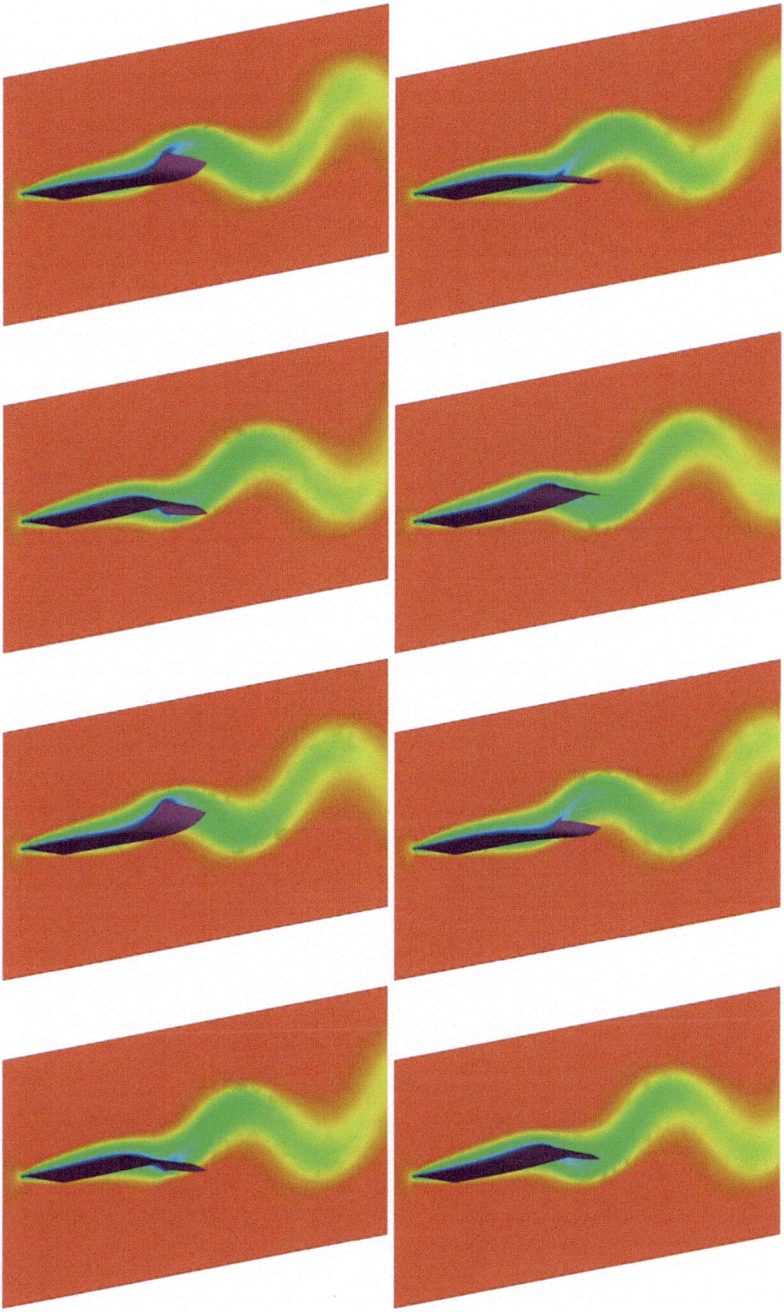

Fig. 5.7 3D FSI computation with the ST-SUPS and quasi-direct coupling. Flow past a flag. The flag and the horizontal component of the velocity on a cross-sectional plane

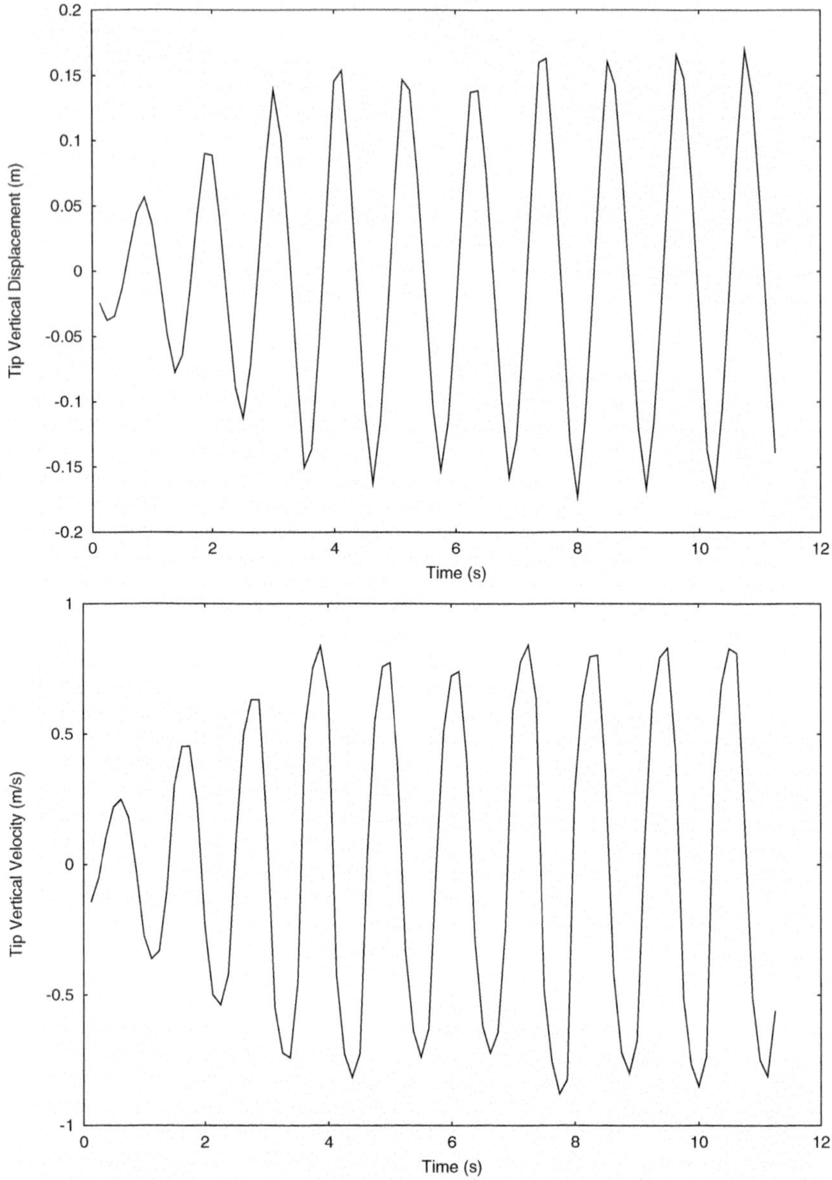

Fig. 5.8 3D FSI computation with the ST-SUPS and quasi-direct coupling. Flow past a flag. Vertical displacement and velocity for the midpoint of the free edge of the flag

the flow patterns, at eight instants during the FSI computation. Figure 5.8 shows the vertical displacement and velocity for the midpoint of the free edge of the flag.

5.8 First 3D Patient-Specific Arterial FSI Computation with the ST-SUPS: Internal Carotid Artery, 2004

The first 3D patient-specific arterial FSI computation with the ST-SUPS, reported in 2004 [117], was for a segment of the internal carotid artery. Both the fluid and structure meshes are made of trilinear hexahedral elements, with an element count of 23,708 for the fluid and 7808 for the structure. The computation was with the linear-elasticity mesh moving method with MJBS. The block-iterative coupling method was used in solving the coupled FSI equations. Figure 5.9 shows, from [117], the deformation of the arterial wall at the peak systole.

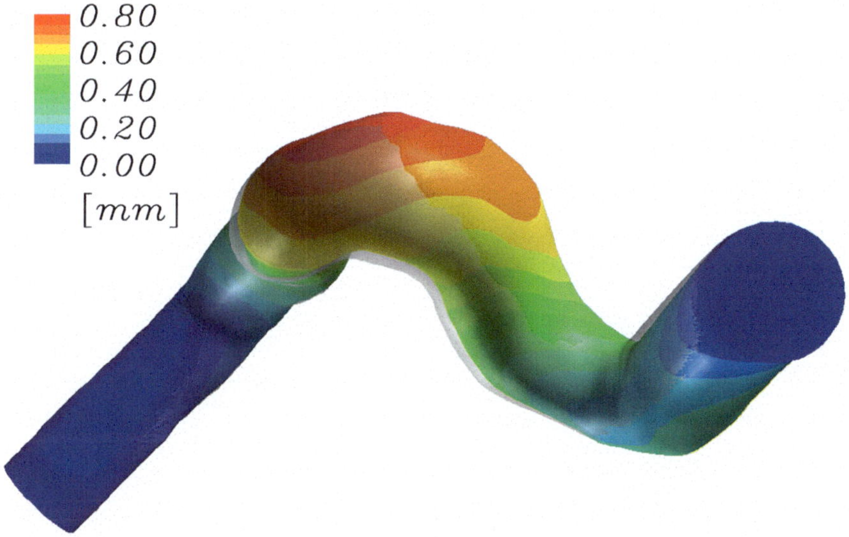

Fig. 5.9 3D patient-specific arterial FSI computation for a segment of the internal carotid artery. Arterial-wall deformation at the peak systole, with the transparent geometry showing the configuration at the beginning of the systole

Chapter 6
Methods and Solutions in 2005–2009

6.1 Surface-Edge-Node Contact Tracking (SENCT) Method, 2006

The SENCT method was introduced in 2006 [13], in connections with the ST-SUPS, to increase the scope of interface-tracking (moving-mesh) techniques by enabling computation of an FSI problem even when the solid surfaces are in near contact and the nearness is sufficiently near for the purpose of solving the problem. In a contact algorithm we need for FSI computations where we model the contact as near contact, the objective is to prevent the structural surfaces from coming closer than a predetermined minimum distance we would like to maintain to protect the quality of the fluid mechanics mesh between the solid surfaces. The SENCT serves that purpose.

Two versions of the SENCT were proposed when it was introduced in 2006 [13]. In the SENCT-Force (SENCT-F) method, the contacted node is subjected to penalty forces that are inversely proportional to the projection distances to the contacting surfaces, edges, and nodes. In the SENCT-Displacement (SENCT-D) method, the displacement of the contacted node is adjusted to correlate with the motion of the contacting surfaces, edges, and nodes. Several 3D FSI computations with the SENCT-F and SENCT-D were reported in 2008 [118]. For FSI problems with nonmatching fluid and structure meshes at the interface, it was proposed in 2008 [59] to formulate the contact model based on the fluid mechanics mesh at the interface. This version of the SENCT is denoted with the option key "-M1."

The SENCT-FC method, used in the spacecraft parachute cluster computations first reported in 2010 [119], was formally introduced, and described in detail, in 2011 [120]. It has some features in common with the SENCT-F but is more robust. Also, compared to the SENCT-F, the forces are applied in a conservative fashion, and the letter "C" in "FC" stands for "conservative." Comments on this aspect of the differences between the SENCT-FC and SENCT-F can be found in [24]. The

T. E. Tezduyar, K. Takizawa, *Space–Time Computational Flow Analysis*,
Modeling and Simulation in Science, Engineering and Technology,
https://doi.org/10.1007/978-3-031-88727-7_6

SENCT-FC method is used as SENCT-FC-M1 in the computations. It can be seen as having three parts: contact detection, force representation, and solving the contact force equations. Description of those parts can be found in [24, 120].

6.2 First 3D ST-SUPS FSI Computation with Contact: A Cloth Piece Falling Over a Rigid Rod, 2006

The first 3D ST-SUPS FSI computation with contact, reported in 2006 [13], was for a cloth piece falling over a rigid rod. The contact modeling was with the SENCT-D. The structure mesh is made of 5616 three-node membrane elements. The fluid mechanics mesh is made of 335,000 linear tetrahedral elements. The computation was with the linear-elasticity mesh moving method with MJBS. The quasi-direct coupling method was used in solving the coupled FSI equations. Figure 6.1 shows, from [13], the cloth at eight instants.

6.3 First 3D ST-SUPS FSI Computation with Quasi-Direct Coupling and Nonmatching Fluid and Structure Interface Meshes: Flow Through and Around a Windsock, 2006

The first 3D ST-SUPS FSI computation with quasi-direct coupling and nonmatching fluid and structure interface meshes, reported in 2006 [13], was for flow through and around a windsock. The structure mesh is made of 1920 three-node membrane elements. The fluid mechanics mesh is made of 113,245 linear tetrahedral elements. The computation was with the linear-elasticity mesh moving method with MJBS. The quasi-direct coupling method was used in solving the coupled FSI equations. Figure 6.2 shows, from the computation reported in [13], the flow patterns.

6.4 First 3D Patient-Specific Arterial FSI Computations with the ST-SUPS and Quasi-Direct Coupling, 2006

The first 3D patient-specific arterial FSI computations with the ST-SUPS and quasi-direct coupling were reported in 2006 [121]. Many cases were computed with the arterial geometries being close approximation to the patient-specific image-based geometries. The arterial walls were mostly modeled with membrane elements, but also continuum elements. The computations were with the linear-elasticity mesh moving method with MJBS. We summarize here the computation for a segment of the middle cerebral artery, where the arterial geometry is a close approximation

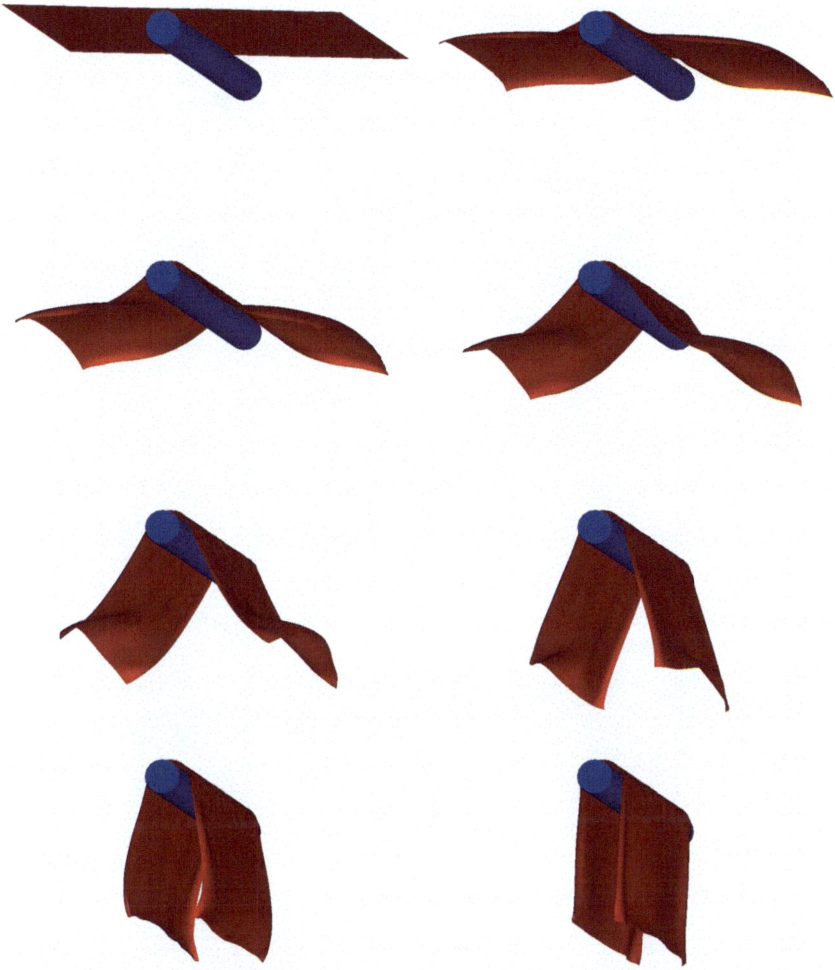

Fig. 6.1 A cloth piece falling over a rigid rod

to the patient-specific image-based geometry used in [122] and the arterial wall is modeled with continuum elements made of hyperelastic (Mooney–Rivlin) material. Both the fluid and structure meshes are made of linear tetrahedral elements, with an element count of 43,960 for the fluid and 52,944 for the structure. Figure 6.3 shows, from [121], the flow patterns.

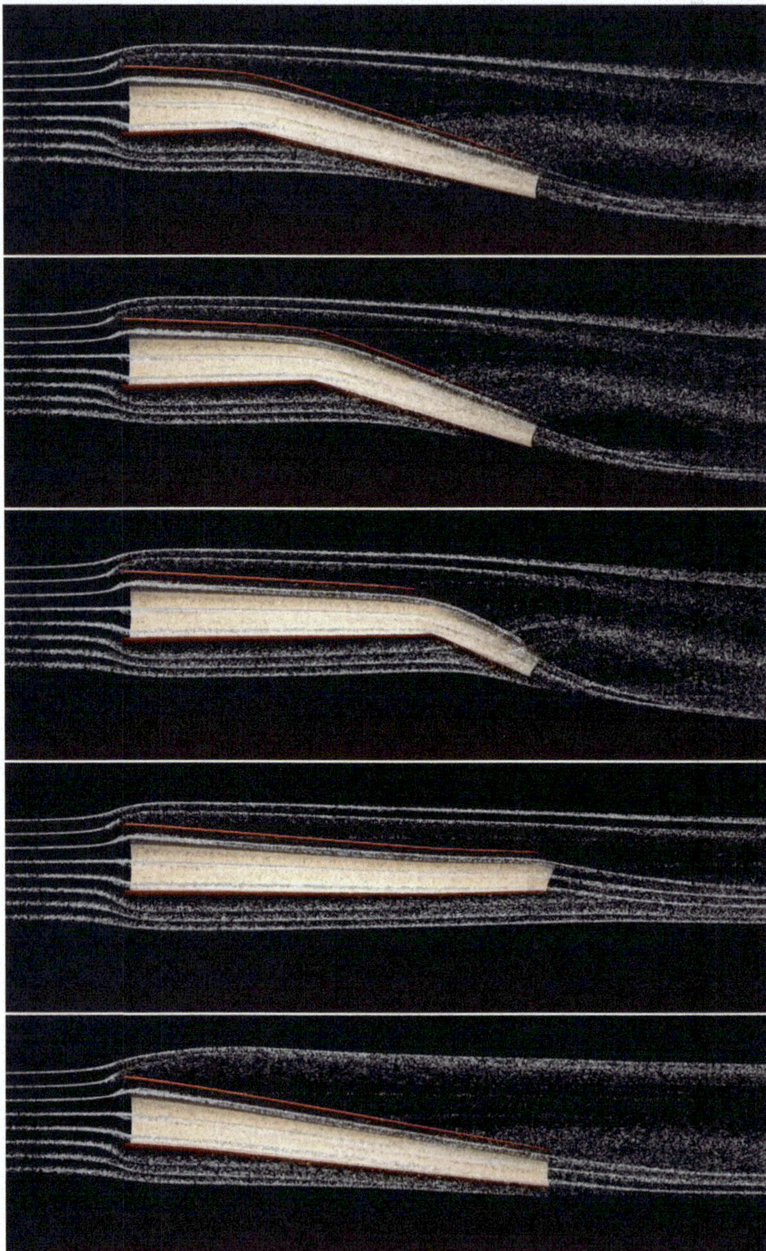

Fig. 6.2 Flow through and around a windsock

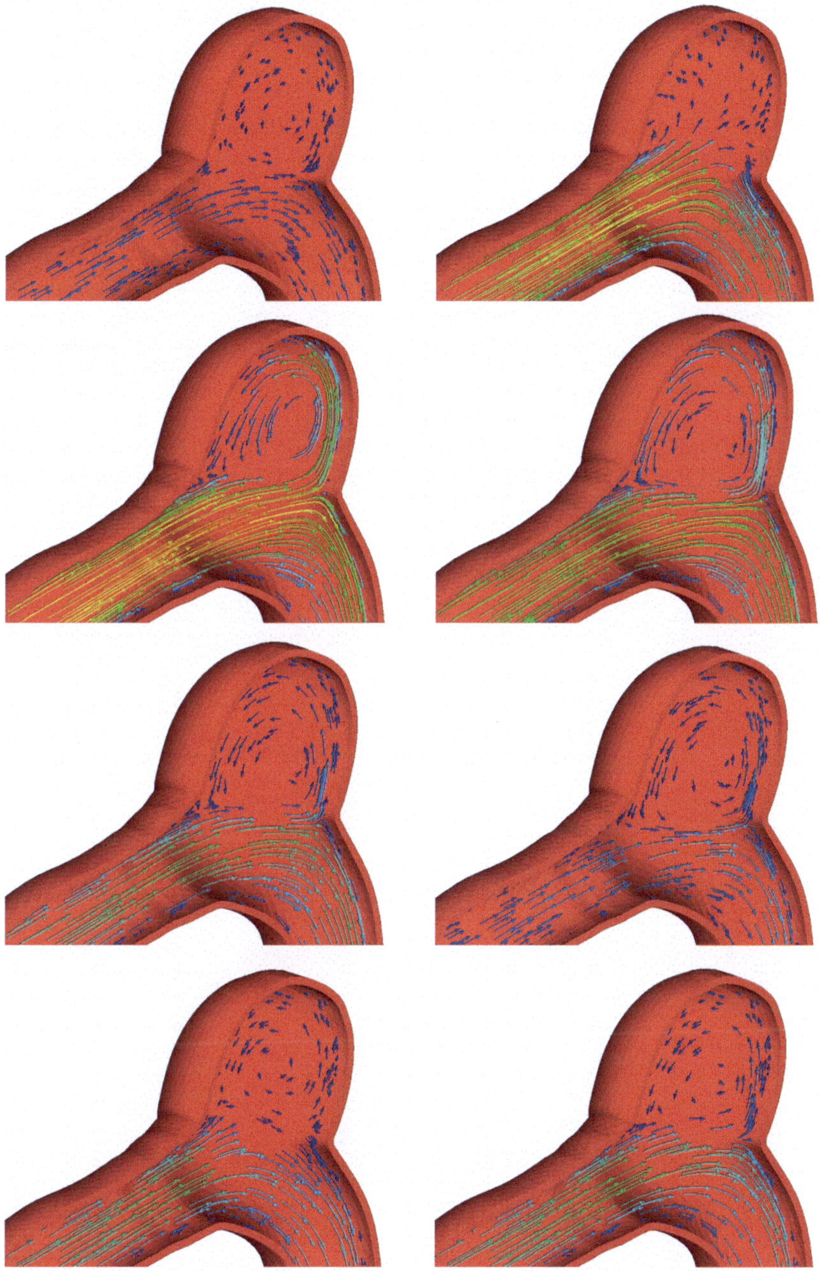

Fig. 6.3 3D patient-specific arterial FSI computation for a segment of the middle cerebral artery. Flow patterns at eight instants in the cardiac cycle. Velocity vectors colored by magnitude

6.5 Homogenized Modeling of Geometric Porosity (HMGP), 2007

The HMGP was introduced in 2007 [58] to address the geometric-porosity computational challenge in FSI computation of spacecraft parachutes. The method was also described in 2008 in a journal article [59]. Geometric porosity is a consequence of the ringsail construction of spacecraft parachutes, which are made of a large number of gores, where a gore is the slice of the parachute canopy between two radial reinforcement cables running from the vent to the skirt (see Fig. 6.4 from [123]). Ringsail parachute gores are constructed from rings and sails, resulting in a canopy with hundreds of ring gaps and sail slits (see Fig. 6.5 from [123]). The complexity created by the geometric porosity makes the FSI computation inherently challenging.

With the HMGP, and its 2010 version, "HMGP-FG" [124], the intractable complexities of the geometric porosity are bypassed in the fluid mechanics computation by approximating the geometric porosity with an equivalent, locally-varying homogenized porosity. The fluid mechanics computation sees a parachute canopy with no gaps or slits. The structural mechanics computation retains the actual canopy geometry, with all the gaps and slits. In the HMGP-FG, the normal velocity component crossing the parachute canopy under a pressure differential Δp is modeled as

$$u_n = -(k_{\mathrm{F}})_J \frac{A_{\mathrm{F}}}{A_1} \Delta p - (k_{\mathrm{G}})_J \frac{A_{\mathrm{G}}}{A_1} \mathrm{sgn}(\Delta p) \sqrt{\frac{|\Delta p|}{\rho}}, \tag{6.1}$$

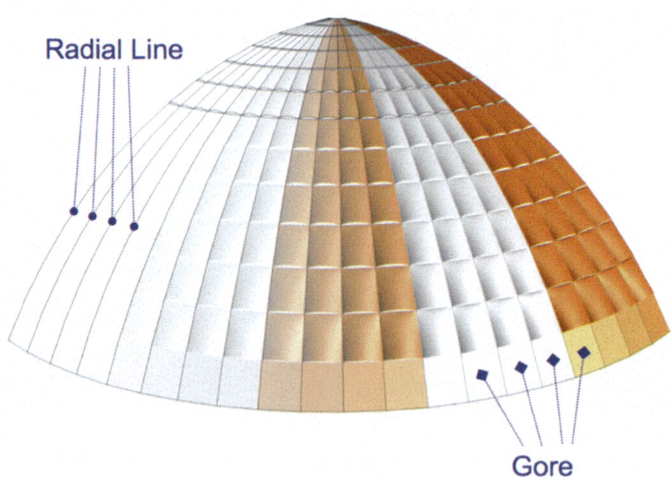

Fig. 6.4 Spacecraft parachute radial lines and gores

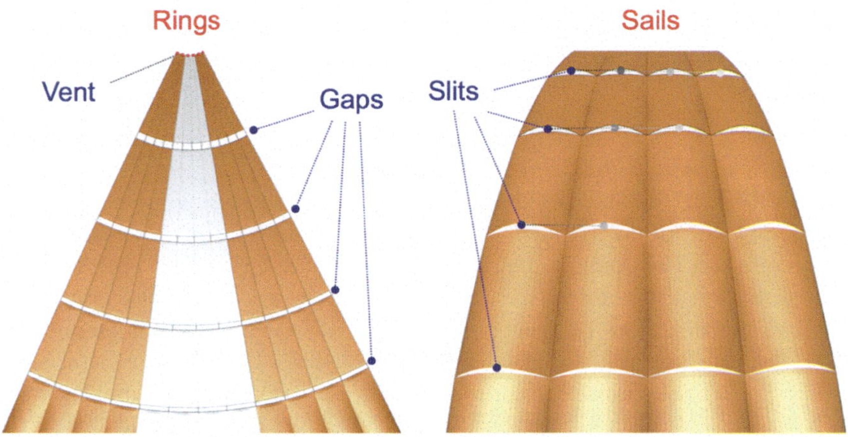

Fig. 6.5 Spacecraft parachute rings, sails, ring gaps, and sail slits

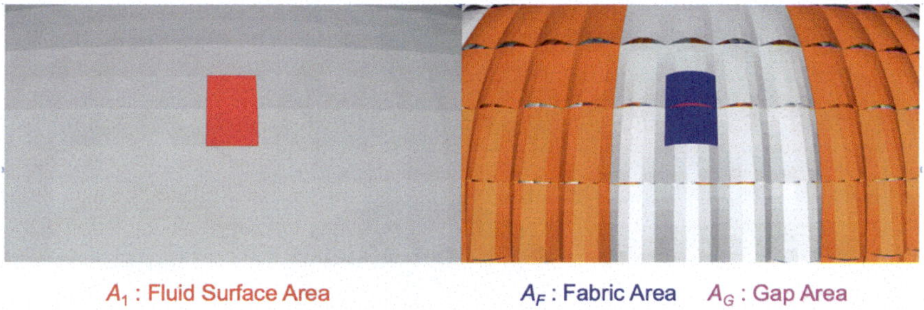

Fig. 6.6 HMGP-FG area definitions

where A_1, A_F, and A_G are defined in Fig. 6.6 (from [123]), and $(k_F)_J$ and $(k_G)_J$ are the homogenized porosity coefficients for each patch J, calculated in a one-time fluid mechanics computation with an n-gore slice of the parachute canopy (see Fig. 6.7 from [123]).

Even in fully open configuration, the parachute canopy goes through a periodic breathing motion where the diameter varies between its minimum and maximum values. The shapes and areas of the gaps and slits vary significantly during the breathing (see Fig. 6.8 from [123]). Still, the porosity coefficients have good invariance properties with respect to these shape and area changes, and this can be seen in Fig. 6.9 (from [123]).

Fig. 6.7 HMGP-FG one-time fluid mechanics computation with an n-gore slice of the parachute canopy, where the flow through all the gaps and slits is resolved. The two porosity coefficients for each patch are calculated from this one-time computation. Except for the first and last patches, each patch contains a gap or a slit. See [59, 124] for details

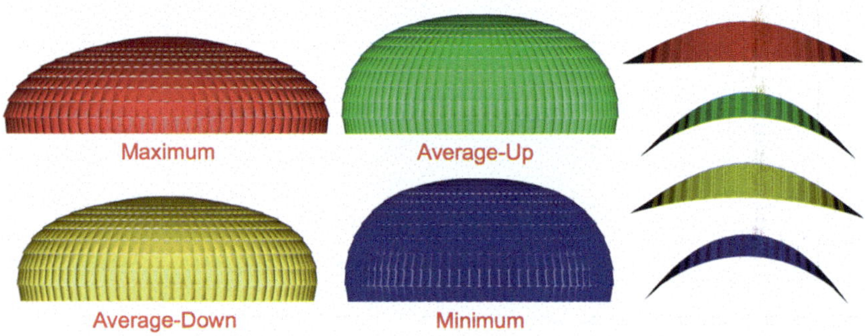

Fig. 6.8 Spacecraft parachute canopy breathing motion. The shapes and the areas of the slits vary significantly during the breathing motion

6.6 First in 3D Spacecraft Parachute FSI: ST-SUPS, 2007

The ST-SUPS, in combination with the HMGP, was the first in 3D spacecraft parachute FSI, reported in 2007 [58]. The computation, also described in two 2008 journal articles [59, 60], was for an earlier design of the Orion spacecraft main (i.e. landing) parachute. The main parachute has a ringsail construction and 80 gores (see Sect. 6.5 for the terminology). This earlier design had 760 gaps and slits. The computation was for the process of dropping the spacecraft heat shield, which had a weight approximately 13.5% of the spacecraft weight. The structure mesh consists of 28,642 nodes and 48,160 three-node membrane elements, 12,761 two-node cable elements, and 1 payload point mass. The fluid volume mesh consists of 121,370 nodes and 745,937 four-node tetrahedral elements. The membrane part of the structure forms the structure interface and has 27,120 nodes

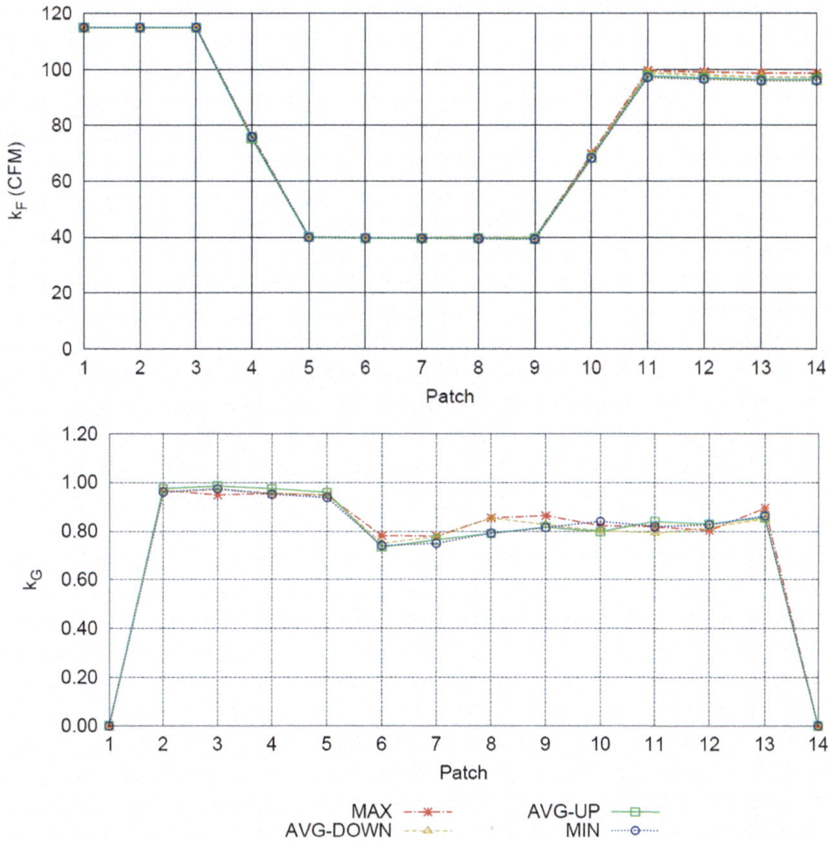

Fig. 6.9 HMGP-FG porosity coefficients $(k_{\mathrm{F}})_J$ and $(k_{\mathrm{G}})_J$ for each patch J, for different canopy shapes during the breathing motion. The plots show good invariance for these coefficients with respect to the shape changes

and 48,160 elements. The homogenized fluid interface has 2320 nodes and 4520 elements. The computation was with the linear-elasticity mesh moving method with MJBS. The quasi-direct coupling method was used in solving the coupled FSI equations. Figures 6.10 and 6.11 show, from the computation reported in [58–60], the parachute before and after the heat shield is dropped.

6.7 Sequentially-Coupled FSI (SCFSI) Method, 2007

The SCFSI method was introduced in 2007 [125, 126] as an approximate FSI method in the context of arterial FSI, with the name "Sequentially-Coupled Arterial FSI (SCAFSI)." In the SCAFSI, first we compute a "reference" (i.e. "base") arterial

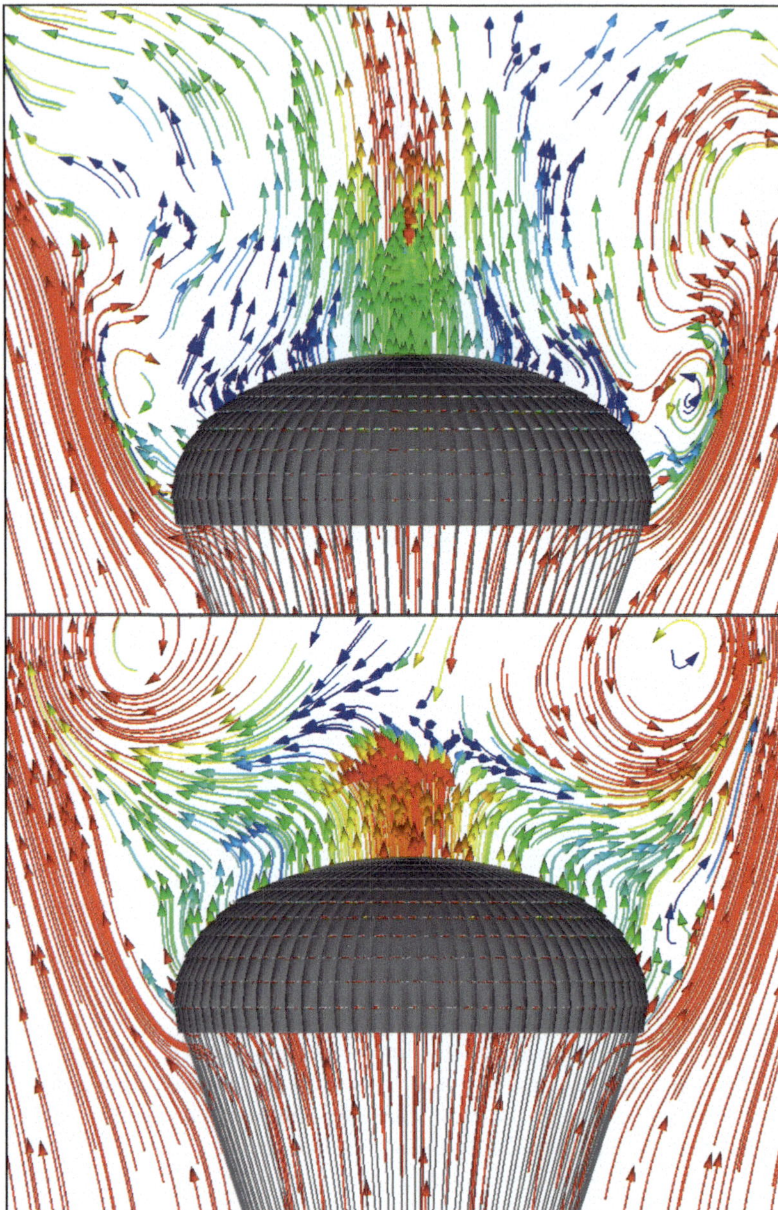

Fig. 6.10 3D spacecraft parachute FSI for an earlier design of the Orion spacecraft main parachute. Flow field before and about 6 s after the heat shield is dropped. The velocity vectors are colored by magnitude

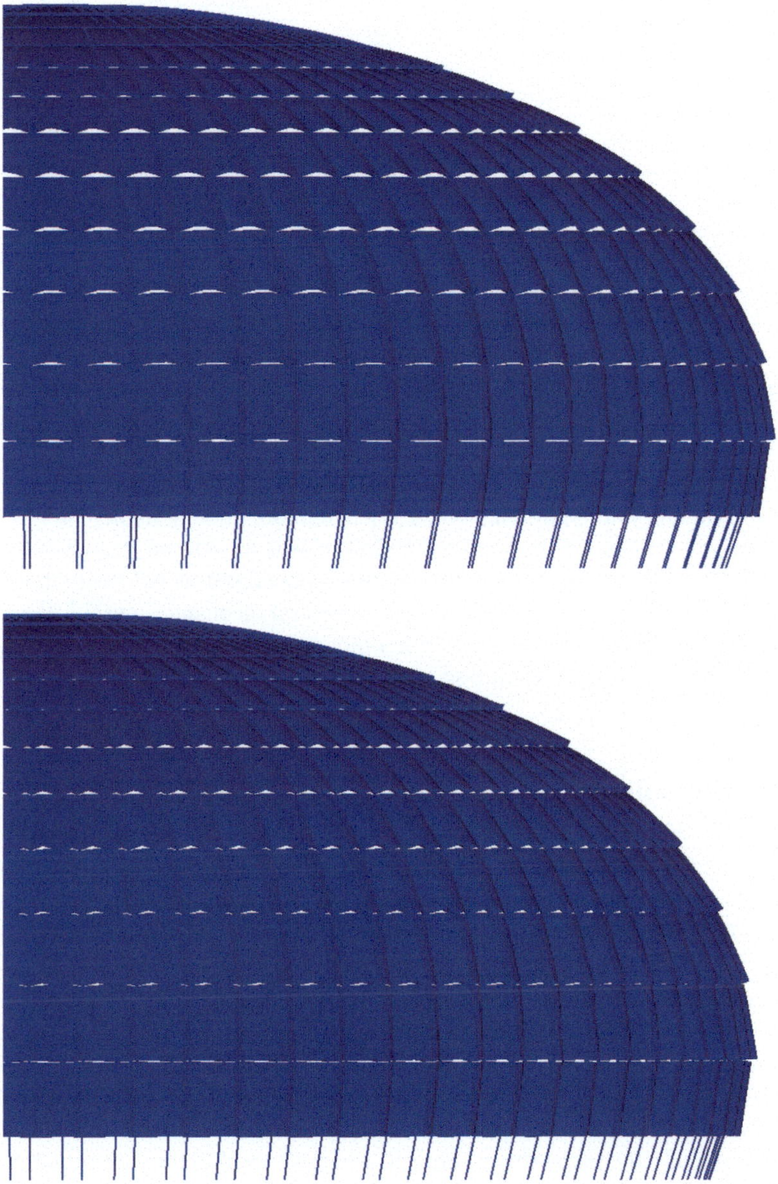

Fig. 6.11 3D spacecraft parachute FSI for an earlier design of the Orion spacecraft main parachute. Parachute shape before and about 6 s after the heat shield is dropped

deformation as a function of time, driven only by the blood pressure, which is given as a function of time in a cardiac cycle. Then we compute a sequence of updates involving mesh motion, fluid dynamics calculation, and recomputing the arterial deformation.

Temporally and spatially multiscale versions of the SCAFSI were introduced in 2008 [127], and test computations were presented for the temporally multiscale version, using different time-step sizes for the fluid and structural mechanics parts. In the spatially multiscale versions, fluid mechanics meshes with different refinement levels are used at different stages of the FSI computation. We use a relatively coarse mesh at the early stages and reserve the highly-refined mesh for the stage where we do the high-fidelity fluid mechanics computation. The spatially multiscale versions introduced in 2008 [127] included the SCAFSI M1C. In that version, we first compute the time-dependent structure shape with the (fully) coupled FSI method and a relatively coarse fluid mechanics mesh, followed by mesh motion and fluid mechanics computation with a more refined mesh, obtaining a higher-fidelity fluid mechanics solution. Test computations with the spatially multiscale versions were reported in 2009, first in a book chapter [128] and a conference paper [129], and then in journal articles [48, 130]. The more general name "SCFSI" was introduced in [48, 128, 129], and the SCFSI M2C was introduced in 2009 [48, 129]. In the SCFSI M2C, we first compute the time-dependent flow field with the (fully) coupled FSI method and a relatively coarse structural mechanics mesh, followed by a structural mechanics computation with a more refined mesh, obtaining a higher-fidelity structural mechanics solution.

6.8 Fluid–Solid Interface-Tracking/Interface-Capturing Technique (FSITICT), 2009

The FSITICT was introduced in 2009 [48]. It is a fluid–solid interaction version of the MITICT. It is for FSI and other flow problems with contact between the solid surfaces, where the nearness that can be modeled with a moving-mesh method without an actual contact between the surfaces (without an actual TC in the domain) is not near enough for the purpose of solving the problem. A good example is FSI computation of heart valves, where the flow has to be completely blocked at contact. In the FSITICT, we track the interfaces wherever and whenever we can with a moving mesh, and capture over that moving mesh the interfaces we cannot track, specifically the interfaces where and when we need to have an actual contact between the solid surfaces. As commented on in [53], essentially, the FSITICT is based on giving up on the interface-tracking accuracy in the parts of the domain where and when we expect an actual contact. Figure 6.12 illustrates the FSITICT concept.

Two versions of the FSITICT were proposed when it was introduced in 2009 [48], one with fixed-partitioning (FSITICT-FP), and one with adaptive-partitioning

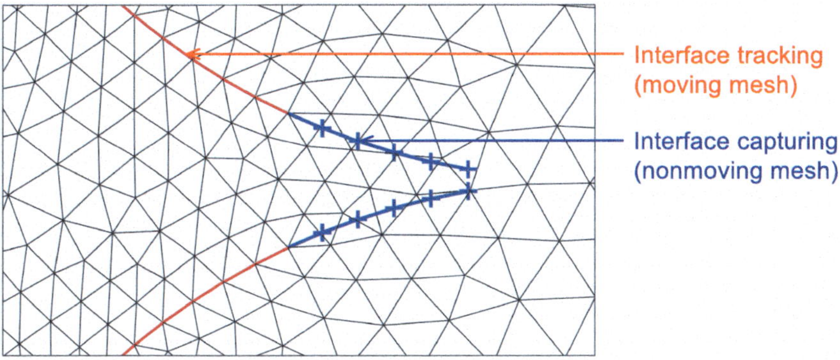

Fig. 6.12 FSITICT concept. The *red* parts of the fluid–solid interface can be tracked and the mesh will move to track them. The *blue* parts would be difficult to track because, in this example, the two solid surfaces are too close, leaving a space in between that is too tight for a boundary-conforming fluid mechanics mesh and will become zero at contact. The *blue* parts will just be captured over the fluid mechanics mesh that is moving to track the *red* parts

(FSITICT-AP). In the FSITICT-FP, the tracked/captured partitioning of the fluid–solid interface will be fixed during the computation. In the FSITICT-AP, the fluid mechanics mesh will stop tracking the parts that have become too difficult to track, leaving them just to be captured. This does not require remeshing. If any of the parts that are being captured become suitable for tracking, a new mesh that is boundary-conforming for those parts will be generated and the tracking of those parts will start. The FSITICT-FP was successfully extended in 2014 [131] to 3D FSI computation of a bioprosthetic heart valve. In that case the interface-tracking technique was the ALE-VMS method, and the interface-capturing technique was a variational immersed-boundary method.

6.9 First 3D ST-SUPS Computation of Fluid–Cable Interaction with One-Way-Dependence: Flow-Driven Dynamics of Parachute Suspension Lines, 2009

First 3D ST-SUPS computation of fluid–cable interaction with one-way-dependence was reported in 2009 [48], for flow-driven dynamics of spacecraft parachute suspension lines. Because a cable is a very thin structure, its influence on the flow is very small. With the one-way-dependence model, we compute the influence of the flow on the cable dynamics, while avoiding the formidable task of computing the influence of the cable on the flow. Using the fluid–cable interactions of parachute suspension lines as a context, we explain the one-way-dependence model.

The line drag is generated mainly in the cross-flow direction, and therefore the flow parallel to the line creates negligible drag (see [132]). The relative velocity is

$$\mathbf{u}_{\mathrm{R}} = \mathbf{u}_{W} - \dot{\mathbf{y}}, \tag{6.2}$$

where $\mathbf{u}_{\mathrm{W}}$ is the wind velocity and $\dot{\mathbf{y}}$ is the structural displacement rate. The velocity $\mathbf{u}_{\mathrm{W}}$ is evaluated at the centroid of each cable group, where a group consists of one or more cable elements. We can define, for example, each suspension line as a group or all the suspension lines as a single group or have a single cable element in each group. The relative velocity perpendicular to the line is

$$\mathbf{u}_{\mathrm{RP}} = \mathbf{u}_{\mathrm{R}} - (\mathbf{u}_{\mathrm{R}} \cdot \mathbf{s}_{\mathrm{LINE}})\, \mathbf{s}_{\mathrm{LINE}}, \tag{6.3}$$

where

$$\mathbf{s}_{\mathrm{LINE}} = \frac{\mathbf{y}_2 - \mathbf{y}_1}{\|\mathbf{y}_2 - \mathbf{y}_1\|}, \tag{6.4}$$

with 1 and 2 representing the cable-element nodes, as seen in Fig. 6.13. For a cable with circular cross-section and radius R, the drag force (per unit length) is calculated as

$$\mathbf{f}_{\mathrm{D}} = \frac{1}{2} C_{\mathrm{D}} \rho \|\mathbf{u}_{\mathrm{RP}}\|^2 (2R) \frac{\mathbf{u}_{\mathrm{RP}}}{\|\mathbf{u}_{\mathrm{RP}}\|}. \tag{6.5}$$

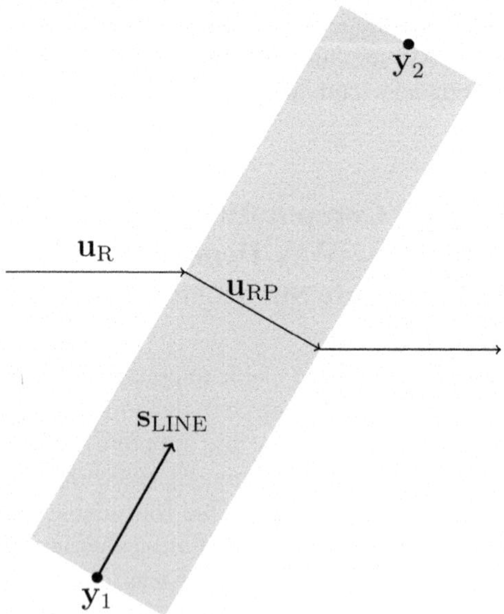

Fig. 6.13 3D ST-SUPS computation of fluid–cable interaction with one-way-dependence. A cable element and relative flow directions

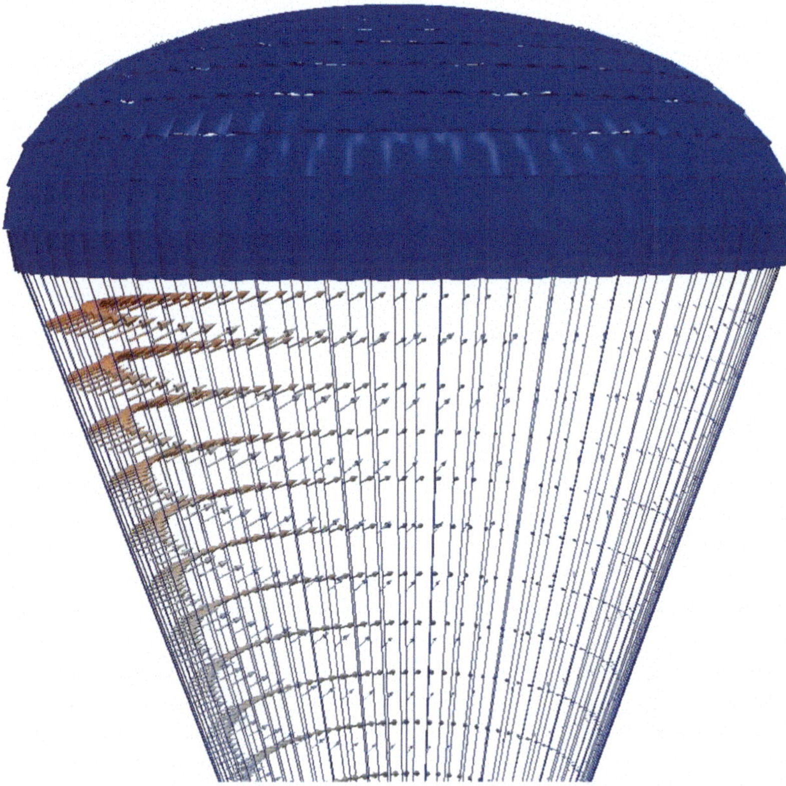

Fig. 6.14 3D ST-SUPS computation of fluid–cable interaction with one-way-dependence. Orion spacecraft parachute FSI with line drag. Parachute shape and line drag vectors

Here C_D is the drag coefficient, determined from tabulated experimental data for flow past a circular cylinder, based on the Reynolds number

$$\mathrm{Re} = \frac{\|\mathbf{u}_{\mathrm{RP}}\| \, (2R)}{\nu}. \tag{6.6}$$

Using nearly the same conditions used for the spacecraft parachute side-wind case reported in [60], FSI computations with and without line drag were carried out in [48] to compare the results. Each cable group consists of a single element. The information on the parachute and suspension line geometries, side-wind speed, descent speed, fluid and structural mechanics meshes, and other computational conditions can be found in [48]. Figure 6.14 shows, from [48], the parachute shape and line drag vectors. It was reported in [48] that the difference in drag when we account for the line drag in addition to the canopy drag is less than 1% in the vertical direction and about 10% in the wind direction.

Chapter 7
Methods and Solutions in 2010–2015

7.1 First in Orion Spacecraft Parachute 3D FSI-Based Design Studies: ST-SUPS, 2010

The ST-SUPS, in combination with the HMGP, was the first in Orion spacecraft parachute 3D FSI-based design studies. These studies, which led to the final parachute design, were first reported in 2010 [124] and then in 2011 [133]. The 2010 studies were with design variations based on the modified geometric porosity, and the 2011 studies were with design variations based on the suspension line length, canopy loading, and the length of the overinflation control line (OICL).

In the studies based on the modified geometric porosity, the earlier canopy configuration described in Sect. 6.6 ("PA") and two alternate canopy configurations ("PM5" and "PM11") were investigated. The configurations PM5 and PM11 are "missing" Sail 5 and Sail 11, respectively, when numbered from the top including the rings. In the studies based on the suspension line length, six different values of the suspension line length ratio S_L/D_o were investigated, where D_o is the parachute nominal diameter. The six values are S_L/D_o = 1.00, 1.15, 1.30, 1.44, 1.60, and 1.76. The canopy loading is defined as W_p/S_o, where W_p is the payload weight and S_o is the parachute nominal area. In the studies based on the canopy loading, six different values of W_p/S_o were investigated. The six values are W_p/S_o = 0.527, 0.639, 0.790, 0.912, 1.060, and 1.230 lbs/ft^2. In the studies based on the OICL, eight different OICL values were applied to various values of S_L/D_o. The eight values are OICL = 76, 78, 80, 81, 82, 83, 84, and 85 ft. In all studies, the computations were with the linear-elasticity mesh moving method with MJBS and the quasi-direct coupling method was used in solving the coupled FSI equations. The structural mechanics meshes are made of four-node membrane elements, two-node cable elements, and payload point mass. The fluid mechanics

T. E. Tezduyar, K. Takizawa, *Space–Time Computational Flow Analysis*,
Modeling and Simulation in Science, Engineering and Technology,
https://doi.org/10.1007/978-3-031-88727-7_7

meshes are made of four-node tetrahedral elements. We summarize the studies based on the modified geometric porosity and suspension line length.

7.1.1 Studies Based on the Modified Geometric Porosity

While the geometric porosity of the gaps and slits are still modeled with the HMGP, the added geometric porosity of a missing sail is resolved in the flow computation. Figure 7.1 shows the parachute shape and structure mesh for PA, PM5, and PM11. The number of nodes and elements for the fluid and structure meshes are given in Table 7.1. All computations were completed without any remeshing. Figure 7.2 shows, from the computations reported in [124], the parachute and flow field for each configuration when the tilt angle, as measured between the vertical axis and a line connecting the payload and vent, is at a maximum.

7.1.2 Studies Based on the Suspension Line Length

The structural mechanics mesh is made of 30,722 nodes and 26,000 membrane elements, 12,521 cable elements, and 1 payload point mass. The fluid mechanics mesh is made of 178,270 nodes and 1,101,643 elements. The fluid interface has 2140 nodes and 4180 elements. Figure 7.3 shows the parachute configurations for the six values of S_L/D_o. All computations were completed without any remeshing. Figures 7.4 and 7.5 show the canopy projected area (S_p) and C_D for the six values of S_L/D_o. Table 7.2 shows the time-averaged values of S_p, C_D, and horizontal velocity (V_H). The averaging is from the time when the full dynamics of the parachute have developed to the end of the time range in Figs. 7.4 and 7.5.

7.2 First in 3D Spacecraft Parachute Cluster FSI: ST-SUPS, 2010

The ST-SUPS, in combination with the HMGP and SENCT-FC, was the first in 3D spacecraft parachute cluster FSI, reported in 2010 [119]. The computations were for the 2-parachute and 3-parachute clusters of the earlier design of the Orion spacecraft main parachute described in Sect. 6.6. The SENCT-FC was used for dealing with the contact between the parachute canopies, handled as a near contact. The fluid mechanics volume mesh consists of four-node tetrahedral elements, and the membrane elements used in the parachute canopy structure are four-node quadrilaterals. The number of nodes and elements for the fluid and structure meshes are given in Table 7.3. The computations were with the linear-elasticity mesh

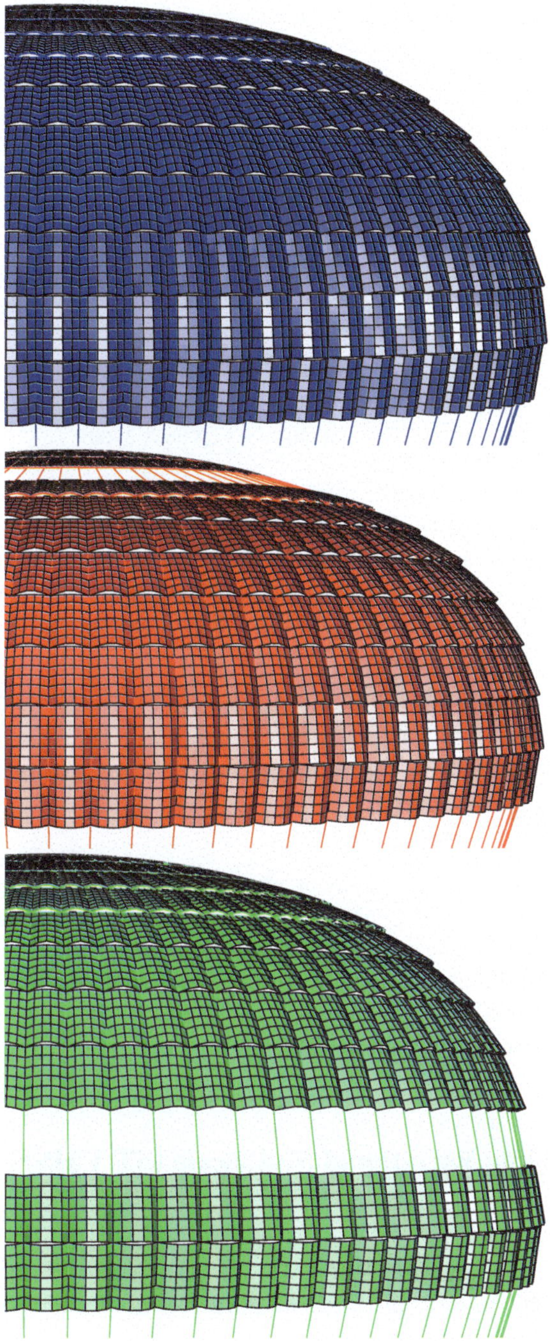

Fig. 7.1 Orion spacecraft parachute 3D FSI-based design studies based on the modified geometric porosity. Parachute shape and structure mesh for PA, PM5, and PM11

Table 7.1 Orion spacecraft parachute 3D FSI-based design studies based on the modified geometric porosity. Number of nodes (*nn*) and elements (*ne*) for the parachutes

		PA	PM5	PM11
Structure				
	nn	30,722	28,642	28,082
Membrane	*ne*	26,000	24,080	23,600
Cable	*ne*	12,521	11,401	12,121
Payload	*ne*	1	1	1
Interface	*nn*	29,200	27,120	26,560
Fluid				
Volume	*nn*	178,270	192,412	180,917
	ne	1,101,643	1,192,488	1,119,142
Interface	*nn*	2140	2060	2060
	ne	4180	3860	3860

moving method with MJBS, with only one resmesh for the 2-parachute cluster and no remesh for the 3-parachute cluster. The quasi-direct coupling method was used in solving the coupled FSI equations. Figure 7.6 shows, from the computations reported in [119], the 2-parachute and 3-parachute clusters at an instant during the simulation.

7.3 First 3D ST-SUPS Computation of Wind Turbine Rotor Aerodynamics, 2010

The first 3D ST-SUPS computation of wind turbine aerodynamics was reported in 2010 [134], for the NREL 5MW offshore baseline wind turbine rotor. The simulation was performed at realistic wind velocity and rotor speed conditions. Because the model did not include the tower, only a slice of 120° was used as the computational domain, with rotational-periodicity representation of the flow field. The mesh was not sufficiently refined and the computation was preliminary.

7.4 Space–Time Variational Multiscale (ST-VMS) Method, 2010

The ST-VMS was introduced in 2010 [14]. The method was first identified with the acronym "DSD/SST-VMST" in [14], with the acronym "ST-VMS" coined in 2011 [15]. The ST-VMS is written as

$$\int_{Q_n} \mathbf{w}^h \cdot \rho \left(\frac{\partial \mathbf{u}^h}{\partial t} + \mathbf{u}^h \cdot \boldsymbol{\nabla} \mathbf{u}^h - \mathbf{f}^h \right) \mathrm{d}Q$$

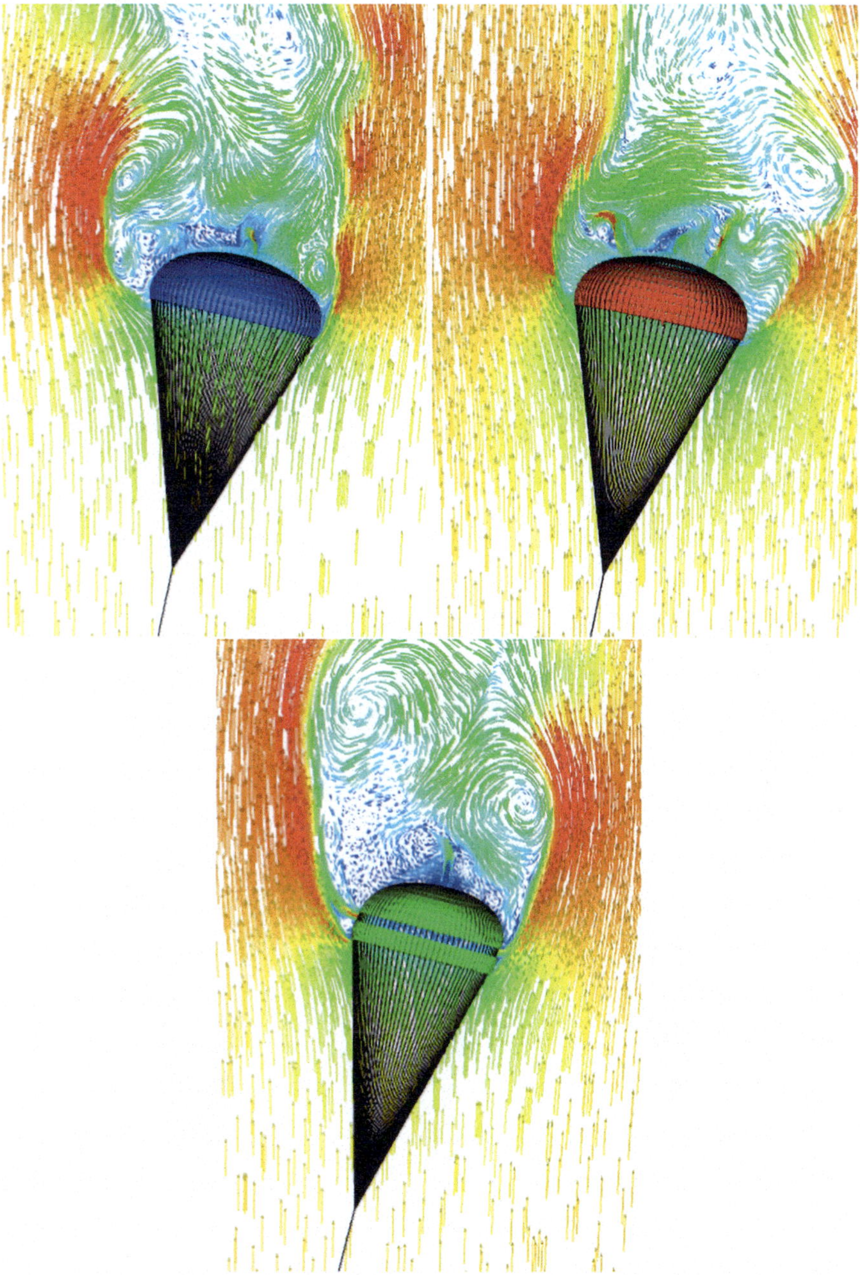

Fig. 7.2 Orion spacecraft parachute 3D FSI-based design studies based on the modified geometric porosity. Parachute and flow field for PA, PM5, and PM11 when the tilt angle is at a maximum

Fig. 7.3 Orion spacecraft parachute 3D FSI-based design studies based on the suspension line length. Parachute configuration for $S_L/D_o = 1.00, 1.15, 1.30, 1.44, 1.60$, and 1.76

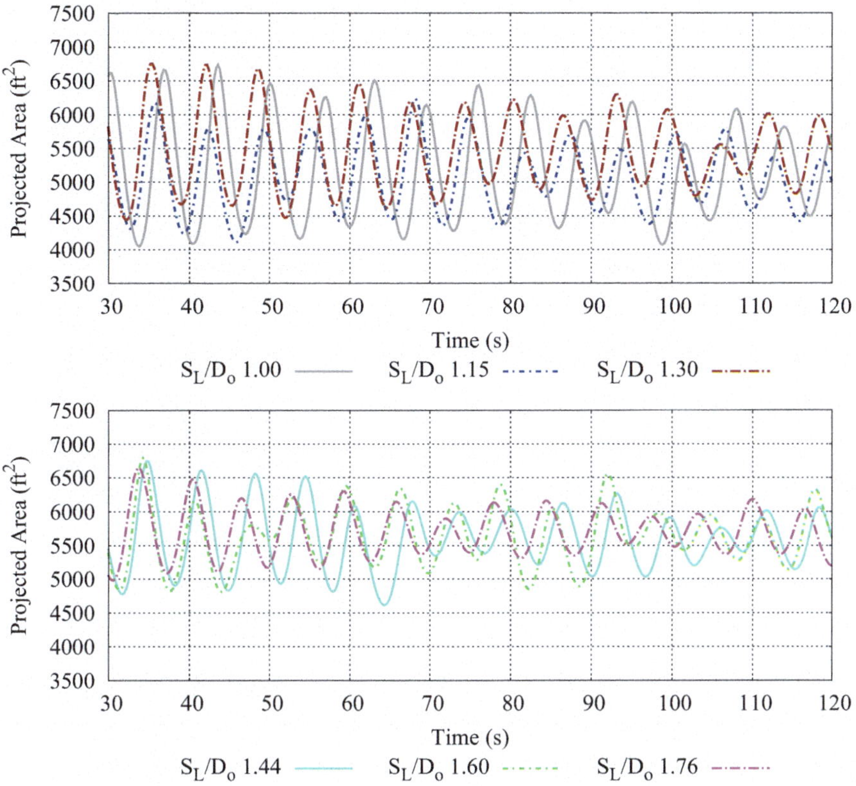

Fig. 7.4 Orion spacecraft parachute 3D FSI-based design studies based on the suspension line length. S_{p}

$$
\begin{aligned}
&+\int_{Q_n} \boldsymbol{\varepsilon}(\mathbf{w}^h) : \boldsymbol{\sigma}^h \mathrm{d}Q - \int_{(P_n)_\mathrm{h}} \mathbf{w}^h \cdot \mathbf{h}^h \mathrm{d}P \\
&+\int_{Q_n} q^h \boldsymbol{\nabla} \cdot \mathbf{u}^h \mathrm{d}Q + \int_{\Omega_n} (\mathbf{w}^h)_n^+ \cdot \rho \left((\mathbf{u}^h)_n^+ - (\mathbf{u}^h)_n^- \right) \mathrm{d}\Omega \\
&+\sum_{e=1}^{(n_\mathrm{el})_n} \int_{Q_n^e} \frac{\tau_\mathrm{SUPS}}{\rho} \left(\rho \left(\frac{\partial \mathbf{w}^h}{\partial t} + \mathbf{u}^h \cdot \boldsymbol{\nabla} \mathbf{w}^h \right) + \boldsymbol{\nabla} q^h \right) \cdot \mathbf{r}_\mathrm{M}^h \mathrm{d}Q \\
&+\sum_{e=1}^{(n_\mathrm{el})_n} \int_{Q_n^e} \nu_\mathrm{LSIC} \boldsymbol{\nabla} \cdot \mathbf{w}^h \rho \boldsymbol{\nabla} \cdot \mathbf{u}^h \mathrm{d}Q \\
&-\sum_{e=1}^{(n_\mathrm{el})_n} \int_{Q_n^e} \tau_\mathrm{SUPS} \mathbf{w}^h \cdot \left(\mathbf{r}_\mathrm{M}^h \cdot \boldsymbol{\nabla} \mathbf{u}^h \right) \mathrm{d}Q
\end{aligned}
$$

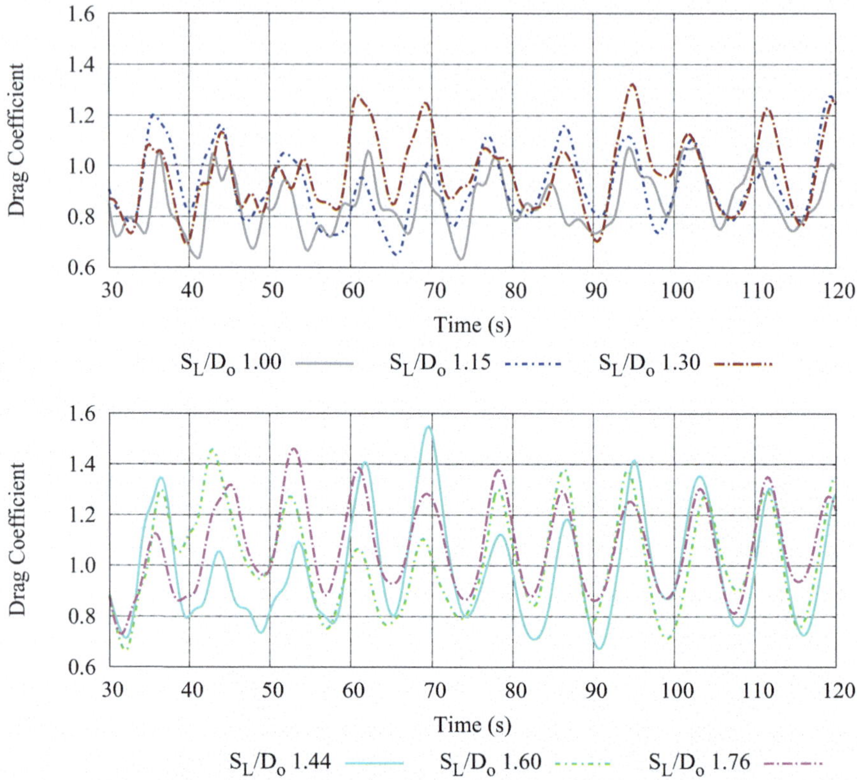

Fig. 7.5 Orion spacecraft parachute 3D FSI-based design studies based on the suspension line length. C_{D}

$$
-\sum_{e=1}^{(n_{\mathrm{el}})_n} \int_{Q_n^e} \frac{\tau_{\mathrm{SUPS}}^2}{\rho} \mathbf{r}_{\mathrm{M}}^h \cdot \left(\boldsymbol{\nabla} \mathbf{w}^h \right) \cdot \mathbf{r}_{\mathrm{M}}^h \mathrm{d}Q = 0. \tag{7.1}
$$

Remark 7.1 While the ST-VMS formulation is the outcome of the ST version [14] of the space-only derivation of the VMS method, at the end, the ST-VMS is equivalent to adding to the ST-SUPS the eighth and ninth integrals.

7.5 First 2D Computations with the ST-VMS, 2010

The first 2D computations with the ST-VMS were reported in 2010 [14], for flow past a NACA 64-618 airfoil at Reynolds number 6×10^6. Both finite element and IGA discretizations were tested, with rather coarse meshes for such a high Reynolds number. We highlight the computations from [14]. The finite element discretization

Table 7.2 Orion spacecraft parachute 3D FSI-based design studies based on the suspension line length. Time-averaged values of S_p, C_D, and V_H

S_L/D_o	S_p (ft^2)	C_D	V_H (ft/s)
1.00	4897	0.85	7.7
1.15	5075	0.92	9.0
1.30	5241	0.97	9.4
1.44	5389	1.00	8.2
1.60	5473	1.03	8.1
1.76	5558	1.08	11.1

Table 7.3 3D spacecraft parachute cluster FSI for the earlier design of the Orion spacecraft main parachute described in Section 6.6. Number of nodes (*nn*) and elements (*ne*) for the parachute clusters

		2-parachute	3-parachute
Structure			
	nn	61,443	92,164
Membrane	*ne*	52,000	78,000
Cable	*ne*	25,042	37,563
Payload	*ne*	1	1
Interface	*nn*	58,400	87,600
Fluid			
Volume	*nn*	238,712	274,409
	ne	1,477,259	1,690,078
Interface	*nn*	4280	6420
	ne	8360	12,540

is with a mesh made of 2780 linear elements and with linear elements in time. Figure 7.7 shows the mesh and the velocity magnitude. The IGA discretization is with a mesh made of 1400 quadratic B-splines elements and with linear elements in time. Figure 7.8 shows the mesh and the velocity magnitude.

7.6 First 3D Computations with the ST-VMS: Wind Turbine Rotor Aerodynamics, 2011

The first 3D computations with the ST-VMS were reported in 2011 [135], for wind turbine aerodynamics with the NREL 5MW offshore baseline wind turbine rotor. As in the ST-SUPS computation outlined in Sect. 7.3, the simulation was performed at realistic wind velocity and rotor speed conditions. Again, because the model did not include the tower, only a slice of 120° was used as the computational domain, with rotational-periodicity representation of the flow field. The mesh in the ST-VMS computation, however, was significantly more refined near the rotor. Figure 7.9

Fig. 7.6 3D spacecraft parachute cluster FSI for the earlier design of the Orion spacecraft main parachute described in Sect. 6.6. 2-parachute and 3-parachute clusters at an instant during the FSI simulation

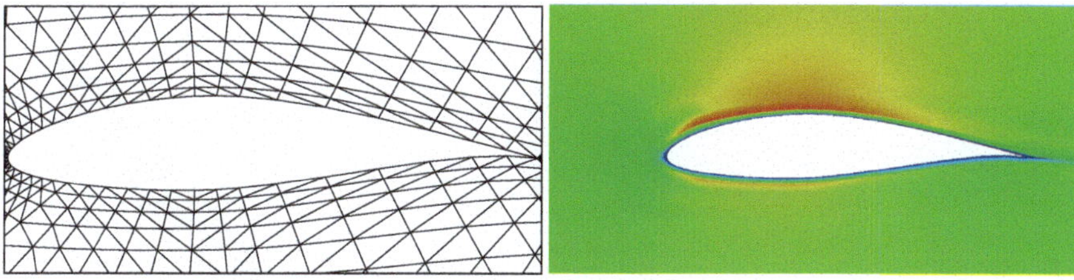

Fig. 7.7 Flow past a NACA 64-618 airfoil. Computed with the ST-VMS and finite element discretization. Mesh near the airfoil and the velocity magnitude

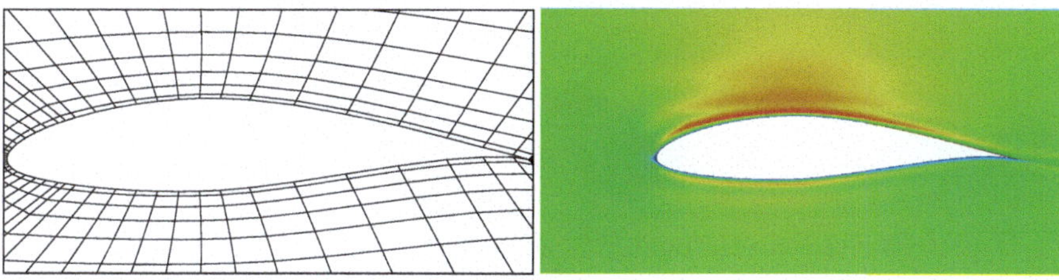

Fig. 7.8 Flow past a NACA 64-618 airfoil. Computed with the ST-VMS and IGA discretization. Mesh near the airfoil and the velocity magnitude

shows the surface mesh. The boundary layer mesh at $\frac{3}{4}R$, where R is the rotor radius, is shown in Fig. 7.10. The full mesh is made of 1,475,175 linear tetrahedral elements. Figure 7.11 shows the pressure coefficient at $0.65R$ and $0.90R$.

Remark 7.2 The ST-SUPS computation with this refined mesh was also reported in [135]. Compared to the ST-VMS computation, the torque is underpredicted by about 12%.

7.7 First in 3D FSI Computation of the Orion Spacecraft Parachute Final Design: ST-SUPS, 2012

The ST-SUPS, in combination with the HMGP-FG, was the first in 3D FSI computation of the Orion spacecraft parachute final design. The computation was reported in 2012 [136]. The parachute canopy design with the modified geometric porosity is obtained by starting with the earlier design described in Sect. 6.6 and removing every 5th gore of Sail 11 as well as the top 25% of Sail 6. With that, the canopy has 16 "windows" and a wider gap between Sail 5 and Sail 6. While the geometric porosity of the gaps and slits are still modeled with the HMGP, the added geometric porosity of the windows and wider gap is resolved in the flow computation. Figure 7.12 shows the canopy structure mesh and fluid mechanics interface mesh. The structure mesh consists of 29,074 nodes and 24,240 four-node

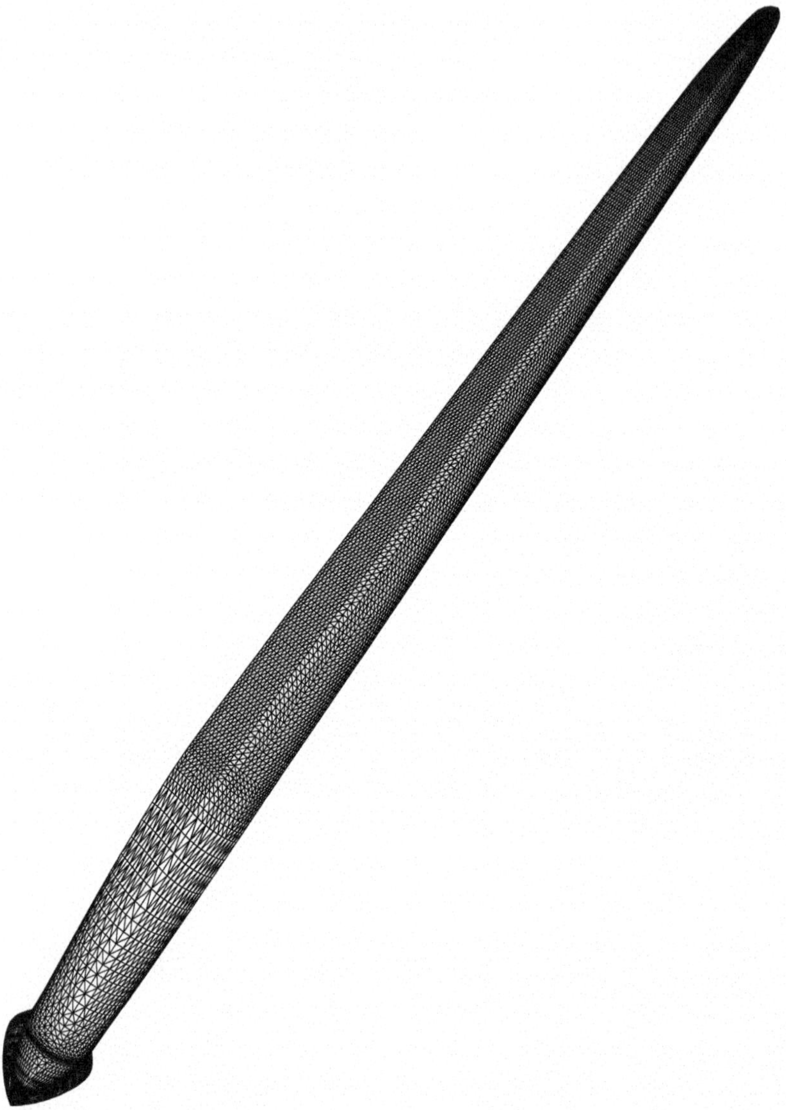

Fig. 7.9 3D ST-VMS computation of wind turbine rotor aerodynamics. Surface mesh

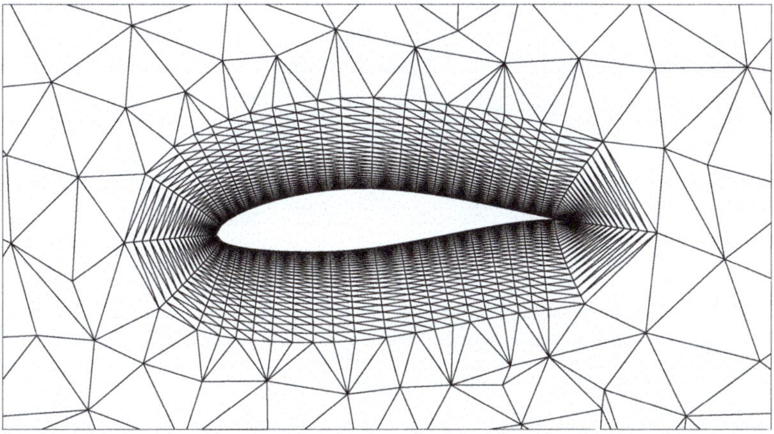

Fig. 7.10 3D ST-VMS computation of wind turbine rotor aerodynamics. Boundary layer mesh at $\frac{3}{4}R$

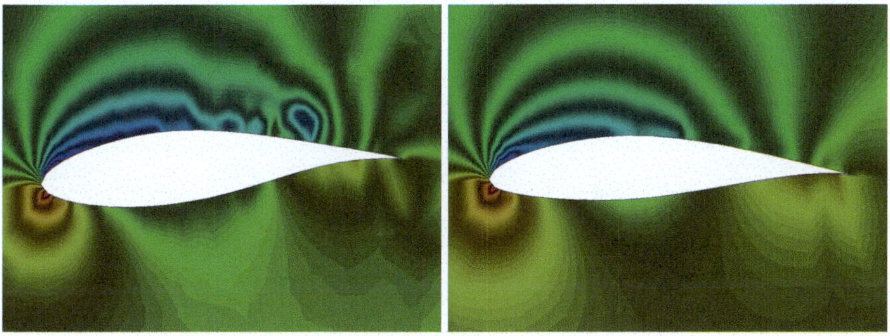

Fig. 7.11 3D ST-VMS computation of wind turbine rotor aerodynamics. Pressure coefficient at $0.65R$ and $0.90R$. Color range from *blue* to *red* corresponds to the pressure range from low to high

membrane elements, 12,321 two-node cable elements, and 1 payload point mass. The initial fluid volume mesh consists of 186,299 nodes and 1,150,150 four-node tetrahedral elements. The fluid interface has 2236 nodes and 4052 elements. The computation was with the linear-elasticity mesh moving method with MJBS. The quasi-direct coupling method was used in solving the coupled FSI equations. The computation involved one remesh, and the post-remesh fluid volume mesh consists of 197,787 nodes and 1,223,030 elements. Figure 7.13 shows, from the computation reported in [136], the velocity field.

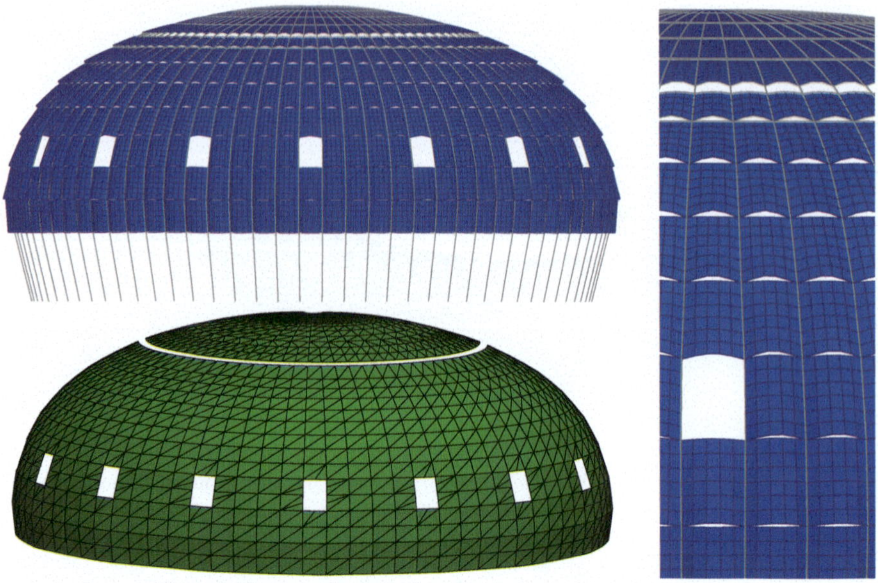

Fig. 7.12 3D FSI computation of the Orion spacecraft parachute final design. Canopy structure mesh (*blue*) and fluid mechanics interface mesh (*green*)

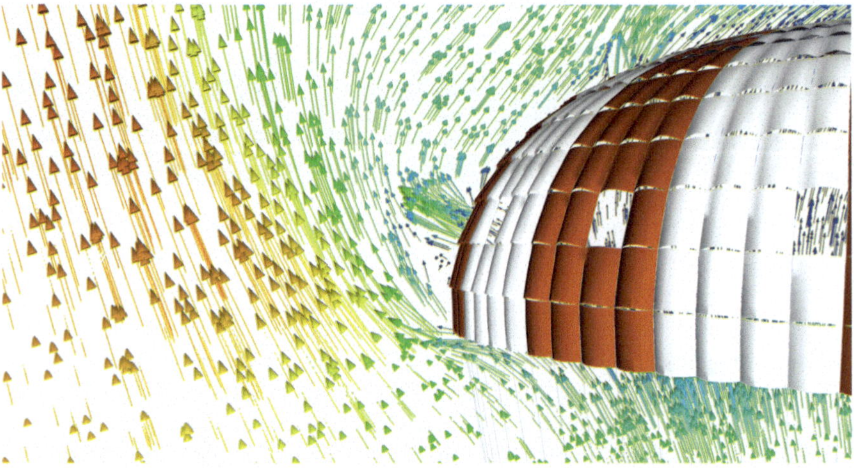

Fig. 7.13 3D FSI computation of the Orion spacecraft parachute final design. Velocity field. Note flow through added geometric porosity

7.8 First 3D ST-VMS Computation of Spacecraft Aerodynamics, 2012

The first 3D ST-VMS computation of spacecraft aerodynamics was reported in 2012 [137], for the Orion spacecraft. The flight condition used in the computation occurs within the spacecraft's flight envelope. Both the convective and conservative forms of the ST-VMS were used, together with the ST-SUPS, for comparison. The mesh is made of 2,617,941 linear tetrahedral elements, with six layers of constant-thickness elements around the spacecraft. Figures 7.14 and 7.15 show the mesh and the drag coefficient. Both the convective form of the ST-VMS and the ST-SUPS give drag coefficients matching the data from NASA within 5%. The conservative form of the ST-VMS yields an unrealistic drag coefficient.

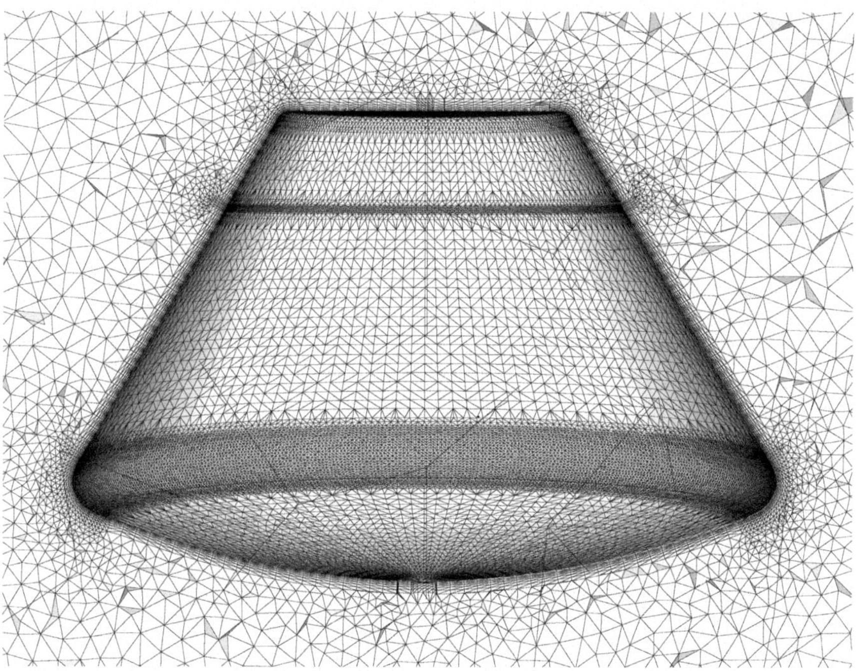

Fig. 7.14 3D ST-VMS computation of spacecraft aerodynamics for the Orion spacecraft. Mesh around the spacecraft

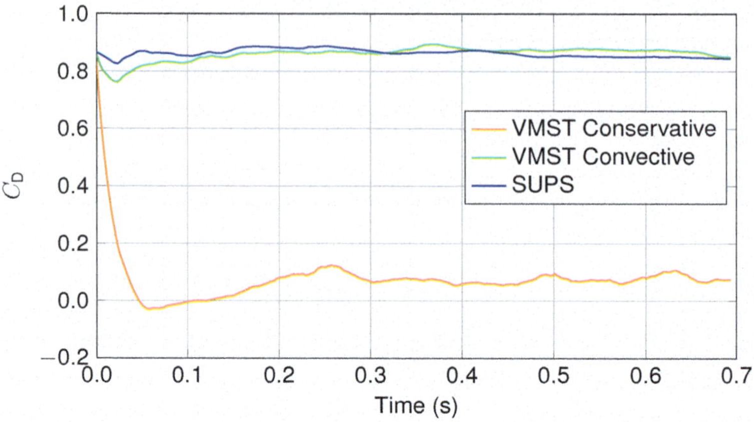

Fig. 7.15 3D ST-VMS computation of spacecraft aerodynamics for the Orion spacecraft. Drag coefficients from the conservative and convective forms of the ST-VMS ("VMST") and the ST-SUPS ("SUPS")

7.9 Variationally Consistent FSI Interface Projection for Nonmatching Fluid and Structure Meshes in the Context of the ST-VMS, 2012

A variationally consistent FSI interface projection method for nonmatching fluid and structure meshes in the context of the ST-VMS was introduced in 2012 [24]. The formulation is written as

$$
\begin{aligned}
&\int_{Q_n} \mathbf{w}_{1\mathrm{E}}^h \cdot \rho \left(\frac{\partial \mathbf{u}^h}{\partial t} + \mathbf{u}^h \cdot \boldsymbol{\nabla} \mathbf{u}^h - \mathbf{f}^h \right) \mathrm{d}Q \\
&+ \int_{Q_n} \boldsymbol{\varepsilon}(\mathbf{w}_{1\mathrm{E}}^h) : \boldsymbol{\sigma}^h \mathrm{d}Q - \int_{(P_n)_\mathrm{h}} \mathbf{w}_{1\mathrm{E}}^h \cdot \mathbf{h}_{1\mathrm{E}}^h \mathrm{d}P \\
&+ \int_{Q_n} q_{1\mathrm{E}}^h \boldsymbol{\nabla} \cdot \mathbf{u}^h \mathrm{d}Q + \int_{\Omega_n} (\mathbf{w}_{1\mathrm{E}}^h)_n^+ \cdot \rho \left((\mathbf{u}^h)_n^+ - (\mathbf{u}^h)_n^- \right) \mathrm{d}\Omega \\
&+ \sum_{e=1}^{(n_\mathrm{el})_n} \int_{Q_n^e} \frac{\tau_\mathrm{SUPS}}{\rho} \left(\rho \left(\frac{\partial \mathbf{w}_{1\mathrm{E}}^h}{\partial t} + \mathbf{u}^h \cdot \boldsymbol{\nabla} \mathbf{w}_{1\mathrm{E}}^h \right) + \boldsymbol{\nabla} q_{1\mathrm{E}}^h \right) \cdot \mathbf{r}_\mathrm{M}^h \mathrm{d}Q \\
&+ \sum_{e=1}^{(n_\mathrm{el})_n} \int_{Q_n^e} \nu_\mathrm{LSIC} \boldsymbol{\nabla} \cdot \mathbf{w}_{1\mathrm{E}}^h \rho \boldsymbol{\nabla} \cdot \mathbf{u}^h \mathrm{d}Q \\
&- \sum_{e=1}^{(n_\mathrm{el})_n} \int_{Q_n^e} \tau_\mathrm{SUPS} \mathbf{w}_{1\mathrm{E}}^h \cdot \left(\mathbf{r}_\mathrm{M}^h \cdot \boldsymbol{\nabla} \mathbf{u}^h \right) \mathrm{d}Q
\end{aligned}
$$

$$-\sum_{e=1}^{(n_{\text{el}})_n} \int_{Q_n^e} \frac{\tau_{\text{SUPS}}^2}{\rho} \mathbf{r}_{\text{M}}^h \cdot \left(\boldsymbol{\nabla} \mathbf{w}_{\text{1E}}^h\right) \cdot \mathbf{r}_{\text{M}}^h \mathrm{d}Q = 0, \tag{7.2}$$

$$\int_{Q_n} q_{\text{1I}}^h \boldsymbol{\nabla} \cdot \mathbf{u}^h \mathrm{d}Q + \sum_{e=1}^{(n_{\text{el}})_n} \int_{Q_n^e} \frac{\tau_{\text{SUPS}}}{\rho} \boldsymbol{\nabla} q_{\text{1I}}^h \cdot \mathbf{r}_{\text{M}}^h \mathrm{d}Q = 0, \tag{7.3}$$

$$\int_{(\Gamma_{\text{1I}})_{n+1}} (\mathbf{w}_{\text{1I}}^h)_{n+1}^- \cdot \left((\mathbf{u}_{\text{1I}}^h)_{n+1}^- - \mathbf{u}_{\text{2I}}^h\right) \mathrm{d}\Gamma = 0, \tag{7.4}$$

$$\begin{aligned}
&\int_{Q_n} (\mathbf{w}_{\text{1I}}^h)_{n+1}^- \cdot \rho \left(\frac{\partial \mathbf{u}^h}{\partial t} + \mathbf{u}^h \cdot \boldsymbol{\nabla} \mathbf{u}^h - \mathbf{f}^h\right) \mathrm{d}Q \\
&+ \int_{Q_n} \boldsymbol{\varepsilon}((\mathbf{w}_{\text{1I}}^h)_{n+1}^-) : \boldsymbol{\sigma}^h \mathrm{d}Q - \int_{(P_n)_{\text{h}}} (\mathbf{w}_{\text{1I}}^h)_{n+1}^- \cdot \mathbf{h}_{\text{1I}}^h \mathrm{d}P \\
&+ \sum_{e=1}^{(n_{\text{el}})_n} \int_{Q_n^e} \frac{\tau_{\text{SUPS}}}{\rho} \left(\rho \left(\frac{\partial (\mathbf{w}_{\text{1I}}^h)_{n+1}^-}{\partial t} + \mathbf{u}^h \cdot \boldsymbol{\nabla} (\mathbf{w}_{\text{1I}}^h)_{n+1}^-\right)\right) \cdot \mathbf{r}_{\text{M}}^h \mathrm{d}Q \\
&+ \sum_{e=1}^{(n_{\text{el}})_n} \int_{Q_n^e} \nu_{\text{LSIC}} \boldsymbol{\nabla} \cdot (\mathbf{w}_{\text{1I}}^h)_{n+1}^- \rho \boldsymbol{\nabla} \cdot \mathbf{u}^h \mathrm{d}Q \\
&- \sum_{e=1}^{(n_{\text{el}})_n} \int_{Q_n^e} \tau_{\text{SUPS}} (\mathbf{w}_{\text{1I}}^h)_{n+1}^- \cdot \left(\mathbf{r}_{\text{M}}^h \cdot \boldsymbol{\nabla} \mathbf{u}^h\right) \mathrm{d}Q \\
&- \sum_{e=1}^{(n_{\text{el}})_n} \int_{Q_n^e} \frac{\tau_{\text{SUPS}}^2}{\rho} \mathbf{r}_{\text{M}}^h \cdot \left(\boldsymbol{\nabla} (\mathbf{w}_{\text{1I}}^h)_{n+1}^-\right) \cdot \mathbf{r}_{\text{M}}^h \mathrm{d}Q = 0.
\end{aligned} \tag{7.5}$$

7.10 A Nonlinear-Elasticity Mesh Moving Method: Element-Based Mesh Relaxation (EBMR), 2013

The EBMR was introduced in 2013 [138] and named in [82, 139]. The EBMR improves the mesh quality. It can be used for improving the quality after the initial creation of a mesh, or occasionally during a mesh motion, or as part of the mesh motion. When used in connection with the mesh motion, it is less disruptive and less time-consuming than restoring the mesh quality by remeshing. In the EBMR, the new mesh has the same set of nodes and elements as before, but some of the nodes are moved slightly to improve the quality of some of the elements. The motion is determined from the nonlinear-elasticity equations of large-deformation mechanics and a locally-defined zero-stress state (ZSS) [140, 141]. Soon after its inception, the EBMR was used in a number of ST-SUPS parachute FSI computations, including those reported in [138, 142, 143]

7.11 First in 3D FSI Computation of Clusters of the Orion Spacecraft Parachute Final Design: ST-SUPS, 2013

The ST-SUPS, in combination with the HMGP-FG and SENCT-FC, was the first in 3D FSI computation of clusters of the Orion spacecraft parachute final design. The computations, reported in 2013 [138], were for the 2-parachute and 3-parachute clusters of the parachute described in Sect. 7.7. The SENCT-FC was used for dealing with the contact between the parachute canopies, handled as a near contact. The fluid mechanics volume mesh consists of four-node tetrahedral elements, and the membrane elements used in the parachute canopy structure are four-node quadrilaterals. The number of nodes and elements for the fluid and structure meshes are given in Table 7.4. The computations were with the linear-elasticity mesh moving method with MJBS and the EBMR. Partly because of the EBMR, the FSI computations were completed without remeshing. The quasi-direct coupling method was used in solving the coupled FSI equations. Figures 7.16 and 7.17 show, from the computations reported in [138], the clusters at different instants during the simulation.

Table 7.4 3D FSI computation of clusters of the Orion spacecraft parachute final design. Number of nodes (*nn*) and elements (*ne*) for the parachute clusters

		2-parachute	3-parachute
Structure			
	nn	58,157	87,230
Membrane	*ne*	48,480	72,720
Cable	*ne*	24,647	36,968
Truss	*ne*	26	26
Payload	*ne*	1	1
Interface	*nn*	55,104	82,656
Fluid			
Volume	*nn*	335,212	307,217
	ne	1,966,851	1,786,212
Interface	*nn*	4472	6708
	ne	8104	12,156

Fig. 7.16 3D FSI computation of clusters of the Orion spacecraft parachute final design. 2-parachute cluster at different instants during the FSI simulation

Fig. 7.17 3D FSI computation of clusters of the Orion spacecraft parachute final design. 3-parachute cluster at different instants during the FSI simulation

7.12 First in Flow Computation with Actual-Contact and High-Resolution Boundary-Layer Representations: Space–Time Topology Change (ST-TC) Method, 2013

The ST-TC, introduced in 2013 [52], was the first in flow computation with actual-contact and high-resolution boundary-layer representations. In flow computations where the moving solid surfaces come into contact, the nature of the problem dictates if the contact can be represented as a near contact or needs to be represented as an actual contact. If the nearness is close enough for obtaining a good flow solution, then representing the contact as a near contact would still leave a narrow mesh protection opening between the solid surfaces. With no topology change in the flow domain, mesh update in moving-mesh methods becomes less challenging. Many classes of problems can be solved that way, and we had many examples in earlier sections.

In some classes of problems, obtaining a meaningful flow solution requires the contact to be represented as an actual contact. That includes heart valve flow analysis, where the valve leaflets have to have an actual contact, wing-clapping aerodynamics, where the clapping wings have to have an actual contact, and tire aerodynamics with road contact, where the tire and the road have to have an actual contact. That is computationally challenging because we want to both represent the contact without leaving a mesh protection opening and represent the boundary layers with an accuracy we get from moving-mesh methods. The ST-TC has overcome this challenge.

Figure 7.18 illustrates the ST-TC mechanism. When two solid surfaces come into contact, the spatial mesh between them contracts and eventually the elements collapse over the contact area. As an element collapses, the ST element it belongs to remains active. For example, in Fig. 7.18, the spatial element with Points 1 and 2 has zero volume at the first time level. However, it has nonzero volume at the second time level, and therefore the corresponding ST element has nonzero volume and is active. As another example from Fig. 7.18, the spatial element with Points 3 and 4 has nonzero volume at the third time level and zero volume at the fourth time level, and thus the corresponding ST element has nonzero volume.

The ST-TC relies on a master–servant point system that maintains the connectivity of the "parent" mesh when there is a contact. Consequently, the distribution model in the parallel-computing environment does not change during the computations with moving meshes. That maintains the computational efficiency at a practical level. When a free point reaches a solid surface, it becomes a servant of a master point. When a servant point detaches from a solid surface, it becomes a free point.

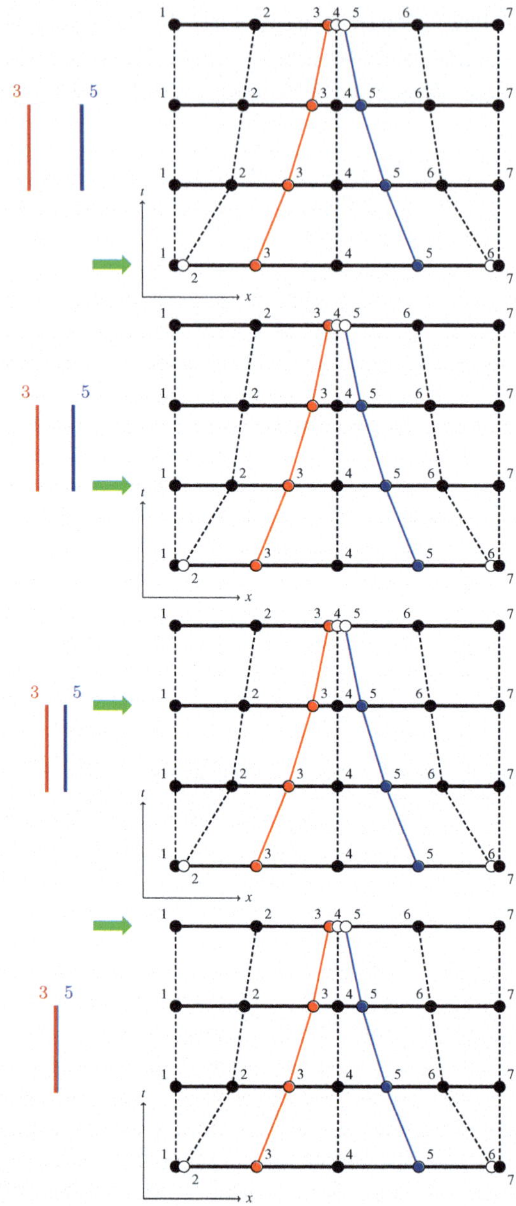

Fig. 7.18 ST-TC mechanism. As two internal solid surfaces (*red* and *blue*) come into contact, the inner spatial mesh between them, consisting of two elements, contracts and the outer meshes expand. While the two elements eventually collapse, the ST elements they belong to remain active. As the servant points (*white*) of the ST-TC master–servant point system detach from the boundaries, they become free points, and the free points reaching a solid surface become servants. The numbers are in reference to the parent mesh

7.13 First 2D ST-TC Flow Computations with Actual-Contact and High-Resolution Boundary-Layer Representations, 2013

The first 2D ST-TC flow computations with actual-contact and high-resolution boundary-layer representations were reported in [52], in the same article where the ST-TC was introduced. The computations were for a pair of symmetrically-flapping surfaces ("Flapping pair"). The pair of surfaces, with zero thickness, are undergoing a prescribed sinusoidal flapping, as shown in Fig. 7.19, with a period of $T = 18.0$ s. The projected width of the flapping surfaces along the horizontal axis is 1.0 m. The inflow velocity is 0.1 m/s. The Reynolds number based on these length and velocity scales is 6×10^3. The boundary conditions were no-slip on the flapping surfaces, uniform horizontal velocity at the inflow, zero-stress at the outflow, and slip at the upper and lower boundaries.

The computations were with the ST-SUPS and finite element discretization. Three different meshes were tested, each with a different time-step size, to see the influence of increased refinement in space and time. The meshes have a structured

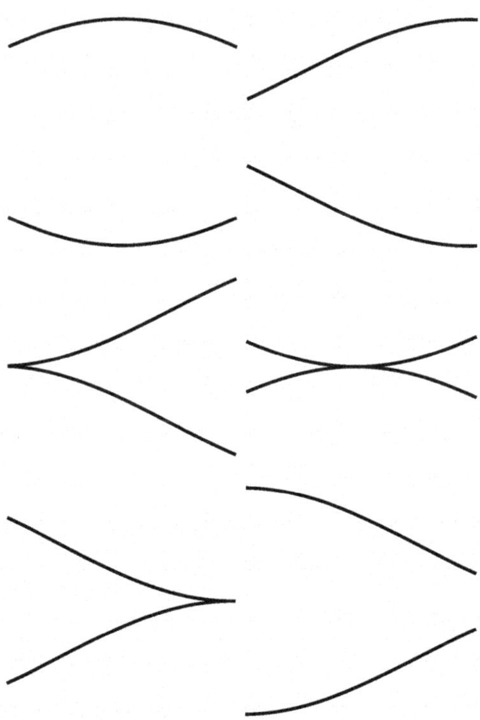

Fig. 7.19 2D ST-TC flow computation with actual-contact and high-resolution boundary-layer representations. Flapping pair. Surfaces at $t = 0$ (also 18), 3, 6, 9, 12, and 15 s

inner zone made of four-node quadrilateral elements and an unstructured outer zone made of three-node triangular elements. Figure 7.20 shows the inner-zone part around the flapping surfaces. During the flapping, only the inner mesh moves, with a special-purpose, algebraic mesh moving method. For clarity in illustrating how the elements collapse and are reborn, we show the results only for Mesh 1. Figure 7.21 shows the mesh around the flapping surfaces at the same instants as in Fig. 7.19. Figure 7.22 shows the velocity magnitude at the same instants as in Fig. 7.19.

7.14 First in 3D Heart Valve Flow Computations with Actual-Contact and High-Resolution Boundary-Layer Representations: ST-TC, 2014

The ST-TC was the first in 3D heart valve flow computations with actual-contact and high-resolution boundary-layer representations. The computations, reported in 2014 [53], were for an aortic valve model with coronary arteries and a mechanical aortic valve. In all occurrences of contact involving moving solid surfaces, the contact was represented as actual contact. We summarize the computation for the aortic valve model. The model, shown in Figs. 7.23 and 7.24, has three leaflets, two small outlets, corresponding to the coronary arteries, and one main outlet, corresponding to the aorta.

At each time step, which has four nonlinear iterations, the ST-SUPS is used in the first two iterations and the ST-VMS in the last two. The mesh is built from a structured mesh around the leaflets and an unstructured mesh elsewhere. The structured mesh is made of four-node tetrahedral, five-node pyramid-shaped, and six-node prismatic elements, and the unstructured mesh is made of four-node tetrahedral elements. The number of elements is 4,184,614, with 79 layers of nodes extruded from both the upper and lower surfaces of the leaflets. The computation was with a special-purpose, algebraic mesh moving method for the structured mesh and the linear-elasticity mesh moving method with MJBS for the unstructured mesh. Figure 7.25 shows the velocity magnitude. Figures 7.26 and 7.27 show the wall shear stress (WSS) on the leaflet surfaces.

7.15 First in 3D FSI Computations of Reefed Stages and Disreefing of the Orion Spacecraft Parachutes: ST-SUPS, 2014

The first 3D FSI computations of reefed stages and disreefing of the Orion spacecraft parachutes were reported in 2014 [142], for both the main parachute final design and the drogue parachute. Both the main and drogue parachutes have reefed stages "Stage 1" and "Stage 2," and a final stage "Stage 3." Reefed stages and disreefing are

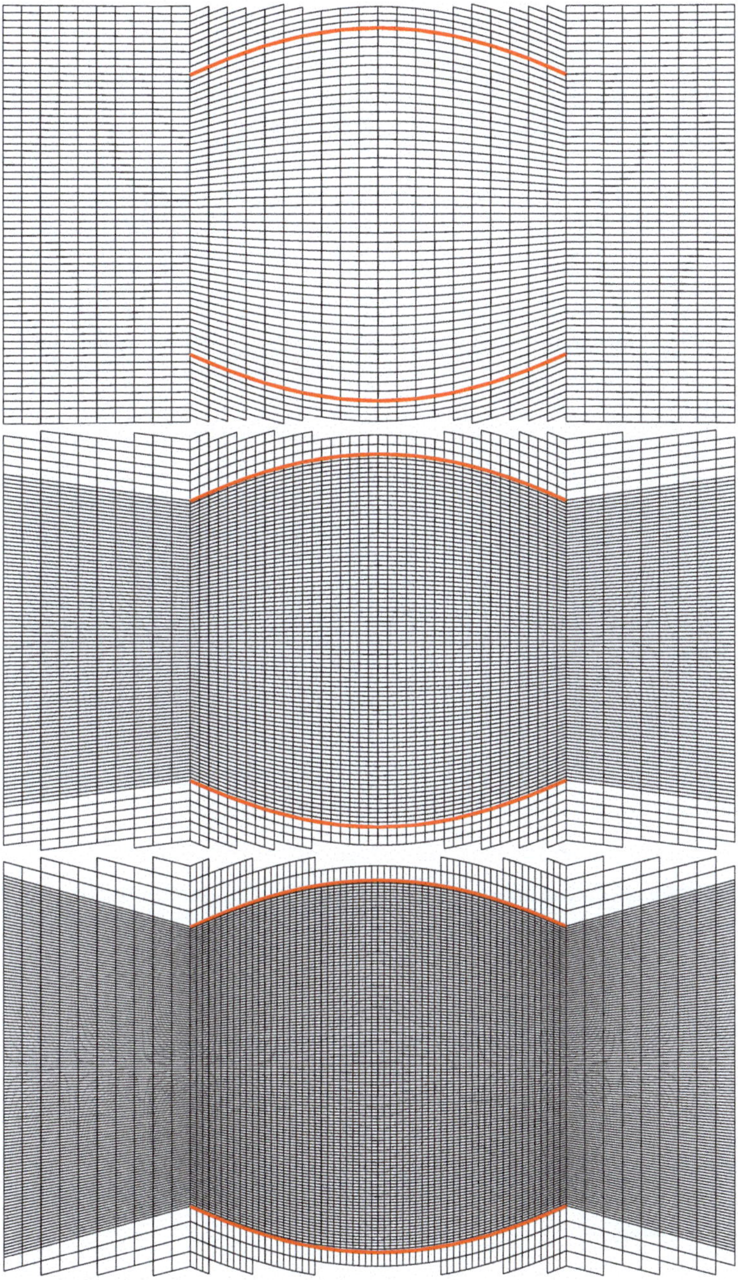

Fig. 7.20 2D ST-TC flow computation with actual-contact and high-resolution boundary-layer representations. Flapping pair. Mesh 1, Mesh 2, and Mesh 3. Inner-zone part around the flapping surfaces

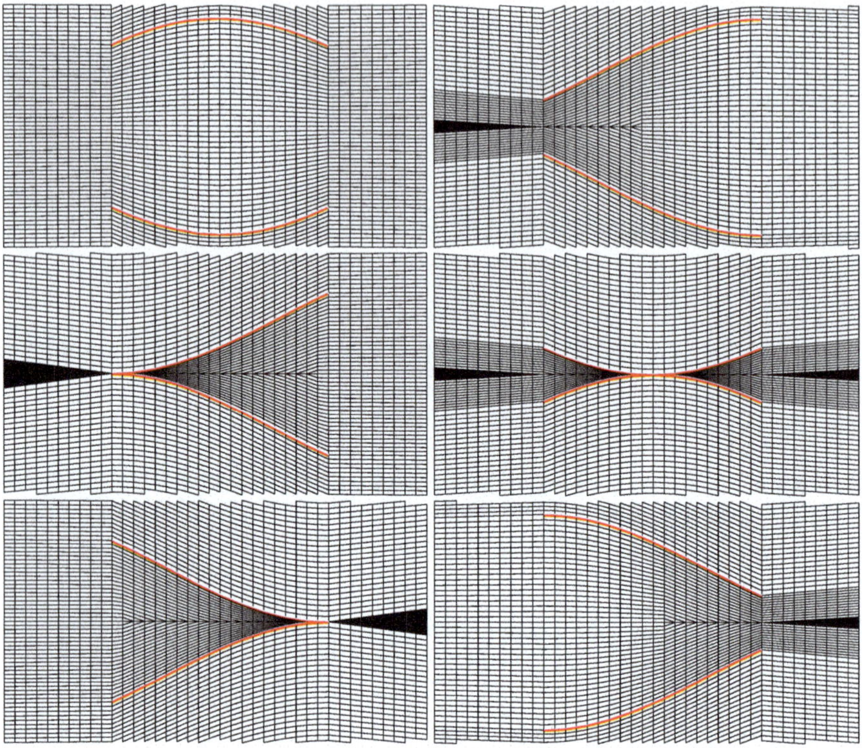

Fig. 7.21 2D ST-TC flow computation with actual-contact and high-resolution boundary-layer representations. Flapping pair. Mesh 1 at the same instants as in Fig. 7.19

computationally more challenging because of the increased geometric complexity and sudden, large changes in the parachute shape. We list the computed-cases reported in [142].

- Orion spacecraft main parachute, Stage 1 and Stage 2.
- Orion spacecraft main parachute, disreefing from Stage 1 to Stage 2 and from Stage 2 to Stage 3.
- 2-parachute cluster of the Orion spacecraft main parachutes, disreefing from Stage 2 to Stage 3.
- Orion spacecraft drogue parachute, Stage 1 and Stage 2.
- Orion spacecraft drogue parachute, disreefing from Stage 1 to Stage 2 and from Stage 2 to Stage 3.

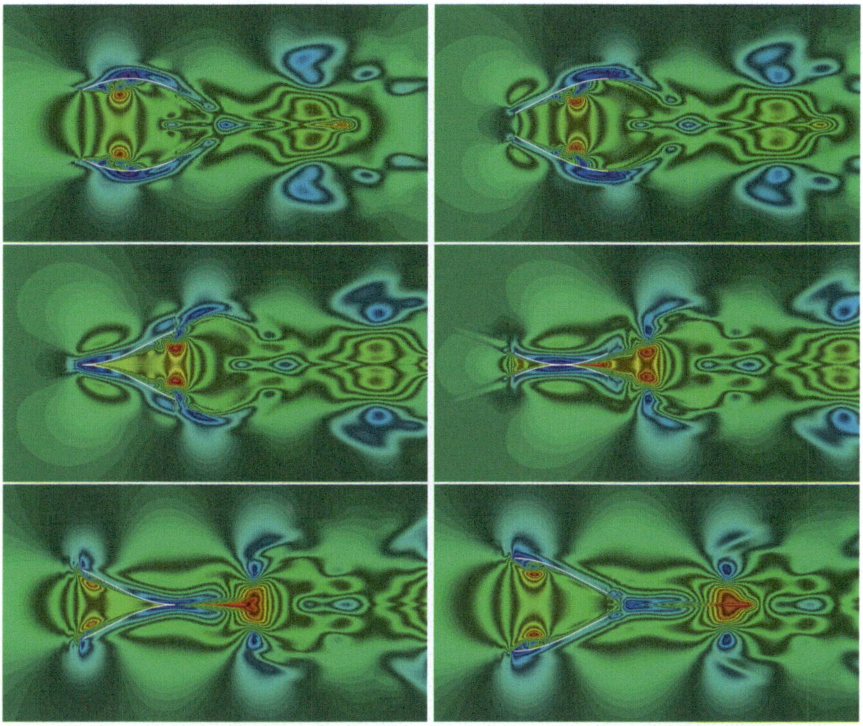

Fig. 7.22 2D ST-TC flow computation with actual-contact and high-resolution boundary-layer representations. Flapping pair. Velocity magnitude for Mesh 1 at the same instants as in Fig. 7.19. Color range from *blue* to *red* corresponds to the velocity range from low to high

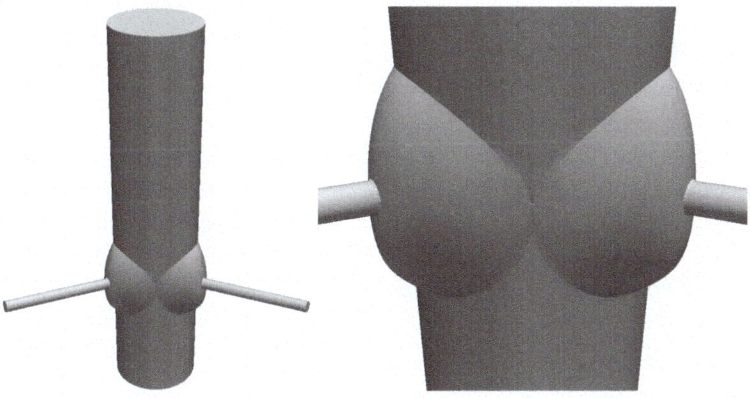

Fig. 7.23 3D heart valve flow computations with actual-contact and high-resolution boundary-layer representations. Aortic valve with coronary arteries. Model geometry. Aorta, leaflets, sinuses, and coronary arteries

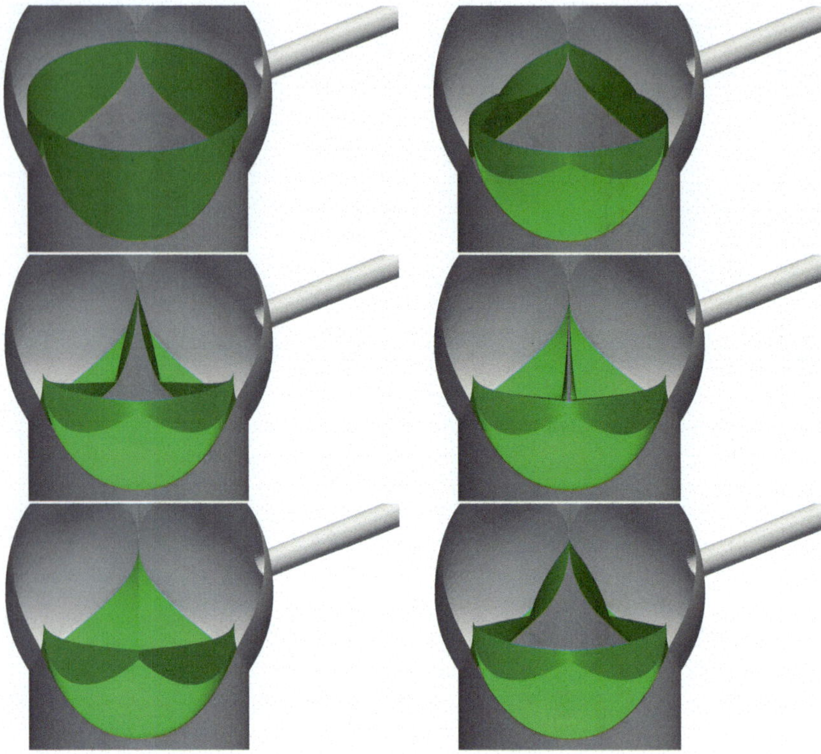

Fig. 7.24 3D heart valve flow computations with actual-contact and high-resolution boundary-layer representations. Aortic valve with coronary arteries. Leaflets at different instant during the cardiac cycle

7.16 First 3D Compressible-Flow ST-SUPG Computation of a Spacecraft Parachute, 2014

The first 3D compressible-flow ST-SUPG computation of a spacecraft parachute was reported in 2014 [143], for the Orion spacecraft drogue parachute. The computation was at Mach 0.5, with standard-day air properties at altitude 30,000 ft. The canopy representation includes gaps where the missing ribbons are located. Flow is resolved through the vent and the missing ribbons, but not through the slits between ribbons. No porosity model was used in this preliminary computation. The mesh is made of 6,042,373 four-node tetrahedral elements. Figure 7.28 shows the flow field.

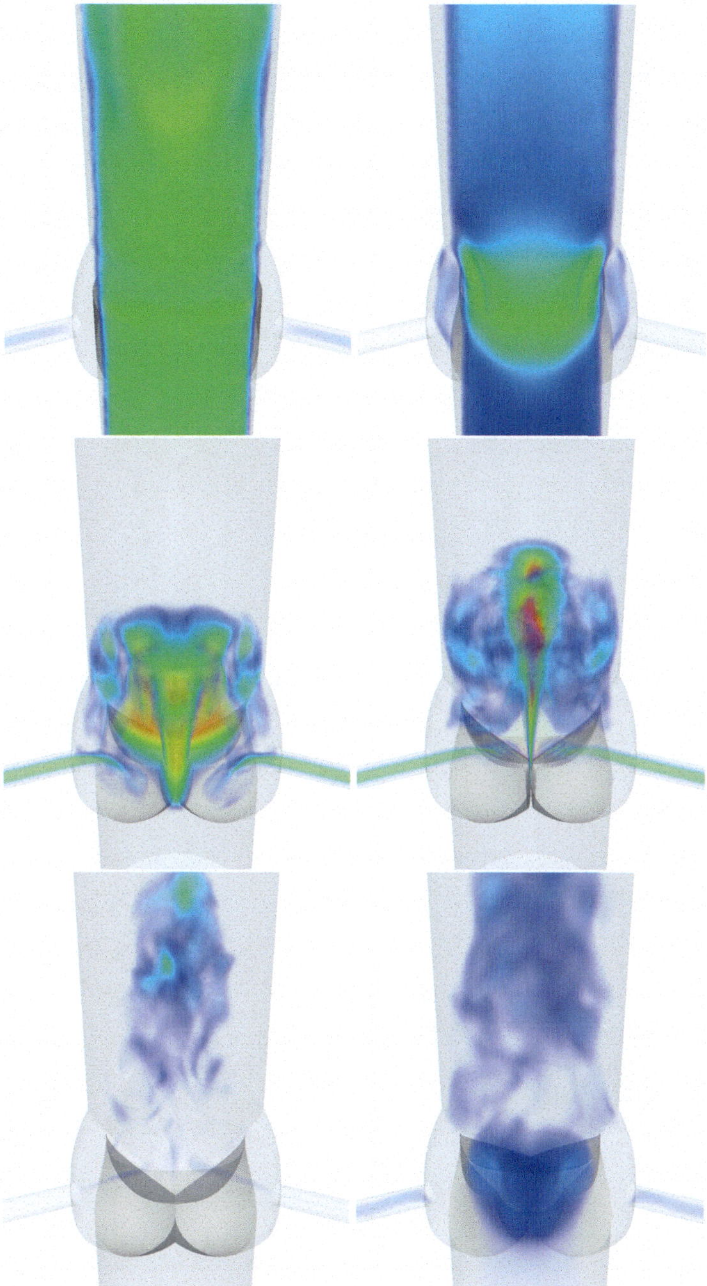

Fig. 7.25 3D heart valve flow computations with actual-contact and high-resolution boundary-layer representations. Aortic valve with coronary arteries. Velocity magnitude at the same instants as in Fig. 7.24. Color range from *blue* to *red* corresponds to the velocity range from low to high

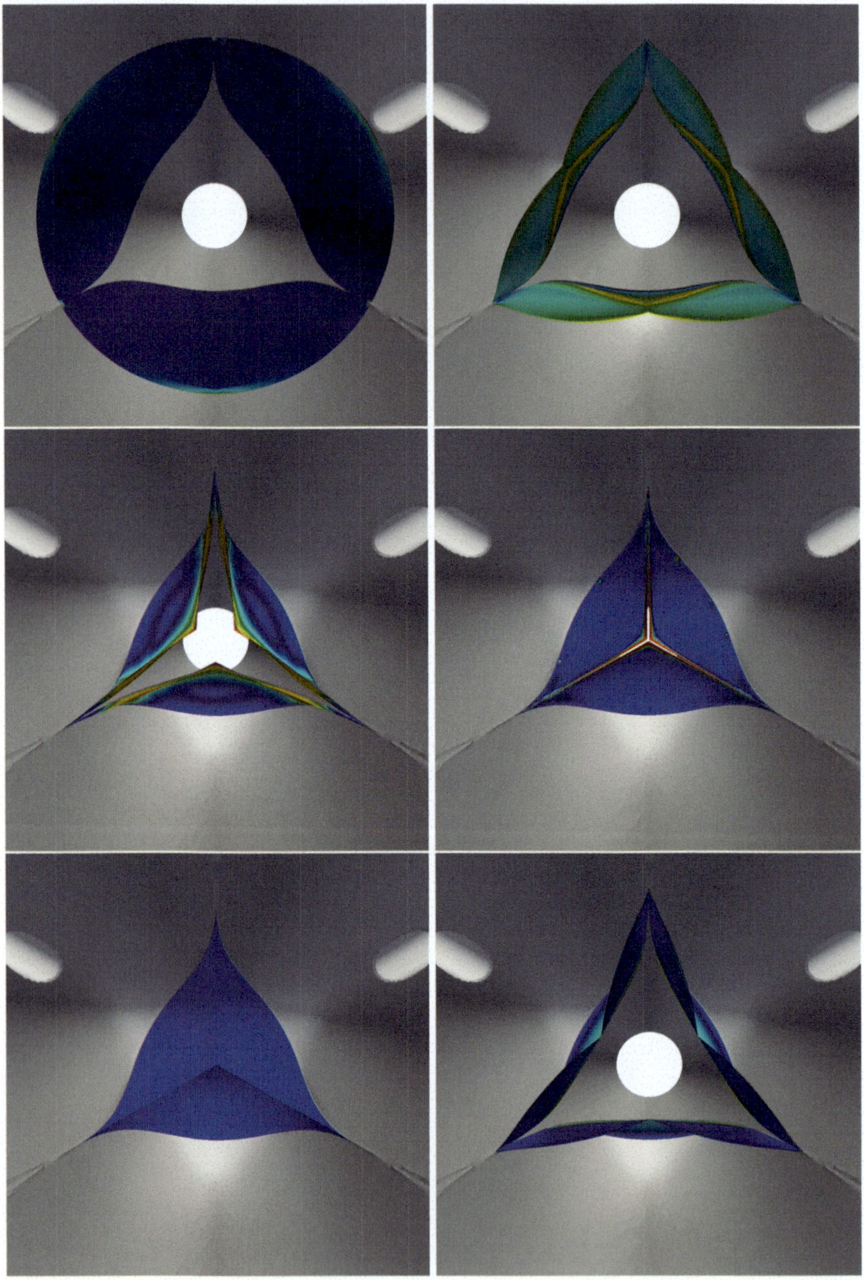

Fig. 7.26 3D heart valve flow computations with actual-contact and high-resolution boundary-layer representations. Aortic valve with coronary arteries. WSS on the lower surface of the leaflets at the same instants as in Fig. 7.24. Color range from *blue* to *red* corresponds to the WSS range from low to high

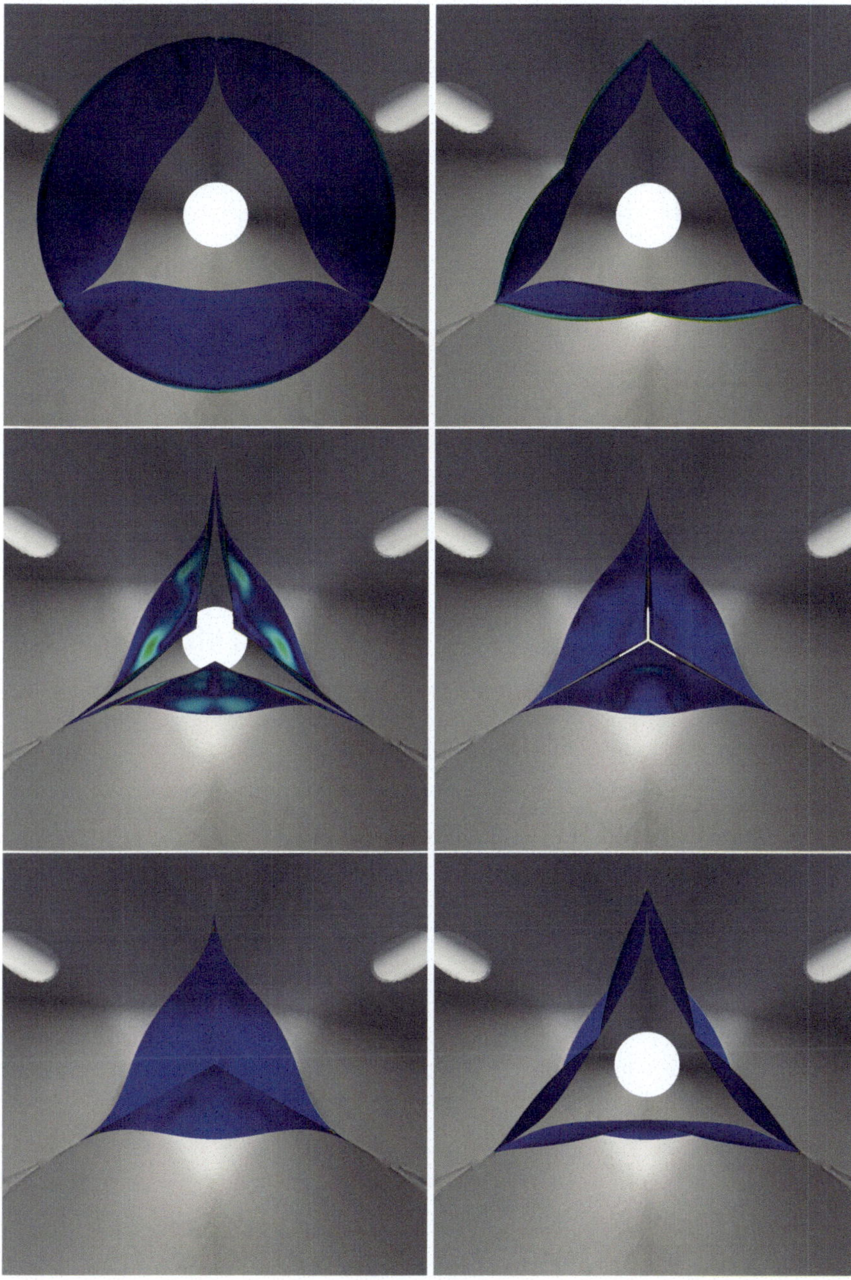

Fig. 7.27 3D heart valve flow computations with actual-contact and high-resolution boundary-layer representations. Aortic valve with coronary arteries. WSS on the upper surface of the leaflets at the same instants as in Figure 7.24. Color range from *blue* to *red* corresponds to the WSS range from low to high

Fig. 7.28 3D compressible-flow ST-SUPG computation of a spacecraft parachute. Orion spacecraft drogue parachute. Velocity vectors colored by magnitude. Color range from *blue* to *red* corresponds to the velocity range from low to high

7.17 Thermo-Fluid ST-VMS Method, 2015

The thermo-fluid ST-VMS method was introduced in 2015 [16]. This ST-VMS extension is for the Navier–Stokes equations of incompressible flows with thermal coupling and Boussinesq approximation and the thermal-transport (energy) equation. With the thermal coupling, the body force in the Navier–Stokes equations becomes $\rho\mathbf{f} = \rho\left(1 - \beta_\theta\left(\theta - \theta_{\text{ref}}\right)\right)\mathbf{a}_{\text{GRAV}}$, where β_θ is the thermal-expansion coefficient, θ is the temperature, θ_{ref} is the reference temperature, and $\mathbf{a}_{\text{GRAV}}$ is the gravitational acceleration. The thermo-fluid ST-VMS is written as

$$
\begin{aligned}
&\int_{Q_n} \mathbf{w}^h \cdot \rho \left(\frac{\partial \mathbf{u}^h}{\partial t} + \mathbf{u}^h \cdot \boldsymbol{\nabla}\mathbf{u}^h - \mathbf{f}^h \right) \mathrm{d}Q \\
&+ \int_{Q_n} \boldsymbol{\varepsilon}(\mathbf{w}^h) : \boldsymbol{\sigma}^h \mathrm{d}Q - \int_{(P_n)_h} \mathbf{w}^h \cdot \mathbf{h}^h \mathrm{d}P \\
&+ \int_{Q_n} q^h \boldsymbol{\nabla} \cdot \mathbf{u}^h \mathrm{d}Q + \int_{\Omega_n} (\mathbf{w}^h)_n^+ \cdot \rho \left((\mathbf{u}^h)_n^+ - (\mathbf{u}^h)_n^- \right) \mathrm{d}\Omega
\end{aligned}
$$

$$
\begin{aligned}
&+\sum_{e=1}^{(n_{\mathrm{el}})_n} \int_{Q_n^e} \frac{\tau_{\mathrm{SUPS}}}{\rho} \left(\rho \left(\frac{\partial \mathbf{w}^h}{\partial t} + \mathbf{u}^h \cdot \boldsymbol{\nabla} \mathbf{w}^h \right) + \boldsymbol{\nabla} q^h \right) \cdot \mathbf{r}_{\mathrm{M}}^h \mathrm{d}Q \\
&+\sum_{e=1}^{(n_{\mathrm{el}})_n} \int_{Q_n^e} \nu_{\mathrm{LSIC}} \boldsymbol{\nabla} \cdot \mathbf{w}^h \rho \boldsymbol{\nabla} \cdot \mathbf{u}^h \mathrm{d}Q \\
&-\sum_{e=1}^{(n_{\mathrm{el}})_n} \int_{Q_n^e} \tau_{\mathrm{SUPS}} \mathbf{w}^h \cdot \left(\mathbf{r}_{\mathrm{M}}^h \cdot \boldsymbol{\nabla} \mathbf{u}^h \right) \mathrm{d}Q \\
&-\sum_{e=1}^{(n_{\mathrm{el}})_n} \int_{Q_n^e} \frac{\tau_{\mathrm{SUPS}}^2}{\rho} \mathbf{r}_{\mathrm{M}}^h \cdot \left(\boldsymbol{\nabla} \mathbf{w}^h \right) \cdot \mathbf{r}_{\mathrm{M}}^h \mathrm{d}Q \\
&+\int_{Q_n} w_\theta^h \rho C_{\mathrm{p}} \left(\frac{\partial \theta^h}{\partial t} + \mathbf{u}^h \cdot \boldsymbol{\nabla} \theta^h \right) \mathrm{d}Q \\
&+\int_{Q_n} \boldsymbol{\nabla} w_\theta^h \cdot \kappa \boldsymbol{\nabla} \theta^h \mathrm{d}Q - \int_{(P_n)_{h_\theta}} w_\theta^h h_\theta^h \mathrm{d}P \\
&+\int_{\Omega_n} (w_\theta^h)_n^+ \rho C_{\mathrm{p}} \left((\theta^h)_n^+ - (\theta^h)_n^- \right) \mathrm{d}\Omega \\
&+\sum_{e=1}^{(n_{\mathrm{el}})_n} \int_{Q_n^e} (\tau_{\mathrm{SUPG}})_\theta \left(\frac{\partial w_\theta^h}{\partial t} + \mathbf{u}^h \cdot \boldsymbol{\nabla} w_\theta^h \right) r_{\mathrm{E}}^h \mathrm{d}Q \\
&-\sum_{e=1}^{(n_{\mathrm{el}})_n} \int_{Q_n^e} \frac{(\tau_{\mathrm{SUPG}})_\theta}{C_{\mathrm{p}}} \mathbf{w}^h \cdot \beta_\theta(\theta^h) r_{\mathrm{E}}^h \mathbf{a}_{\mathrm{GRAV}} \mathrm{d}Q \\
&-\sum_{e=1}^{(n_{\mathrm{el}})_n} \int_{Q_n^e} \tau_{\mathrm{SUPS}} C_{\mathrm{p}} w_\theta^h \left(\mathbf{r}_{\mathrm{M}}^h \cdot \boldsymbol{\nabla} \theta^h \right) \mathrm{d}Q \\
&-\sum_{e=1}^{(n_{\mathrm{el}})_n} \int_{Q_n^e} \frac{\tau_{\mathrm{SUPS}} (\tau_{\mathrm{SUPG}})_\theta}{\rho} \mathbf{r}_{\mathrm{M}}^h \cdot \boldsymbol{\nabla} w_\theta^h r_{\mathrm{E}}^h \mathrm{d}Q = 0.
\end{aligned}
\tag{7.6}
$$

Here, w_θ is the test function associated with θ, C_{p} is the constant-pressure specific heat, κ is the thermal conductivity, r_{E}^h is the residual of the energy equation, given as

$$
r_{\mathrm{E}}^h = \rho C_{\mathrm{p}} \left(\frac{\partial \theta^h}{\partial t} + \mathbf{u}^h \cdot \boldsymbol{\nabla} \theta^h \right) - \boldsymbol{\nabla} \cdot (\kappa \boldsymbol{\nabla} \theta^h), \tag{7.7}
$$

and the stabilization parameter $(\tau_{\mathrm{SUPG}})_\theta$ can be found in [16].

Part II
Methods and Solutions in Isogeometric Analysis

Chapter 8
Methods and Solutions in 2010–2012

8.1 Space–Time Isogeometric Analysis (ST-IGA), 2010

The ST-IGA was introduced in 2010 [14]. It is IGA discretization in the ST framework, with IGA basis functions in space or time or both space and time. Discretization with IGA basis functions in time is uniquely offered by the ST framework. Figure 8.1 shows, as an example in 2D space, a physical ST element and its parent element. The basis functions in time, which are part of the ST basis functions given in Eq. (3.1), can be B-splines or Bernstein polynomials, as shown in Fig. 8.2 (from [14]). They can also be as simple as linear functions in time, as shown in Fig. 8.3 (from [14]).

Superior accuracy of discretization with IGA basis functions in space [23, 144–146] is now textbook knowledge [24, 147]; it enables more accurate representation of the geometry and increased accuracy in the flow solution. Because that superior accuracy is reached with fewer control points, the effective element sizes will be larger. Consequently, we can take larger time steps in the computation without exceeding the Courant number we target for good accuracy.

Discretization with IGA basis functions in time leads to a number of advantages. Deriving full benefit from using IGA basis functions in space is among those advantages, and this was shown with the stability and accuracy analysis presented in 2010 [14]. With that specific advantage, the ST framework will increase the Courant number we should target for good accuracy. We will get that increase even if we use linear basis functions in time. We will get even more increase if we use higher-order B-splines in time. We mention two other advantages in using IGA basis functions in time. (i) As stated in 2010 [14] and can be seen in the ST-IGA computations reported in 2011 [50], using IGA basis functions in time brings higher accuracy in representing the motion of a solid surface, a mesh motion consistent

T. E. Tezduyar, K. Takizawa, *Space–Time Computational Flow Analysis*,
Modeling and Simulation in Science, Engineering and Technology,
https://doi.org/10.1007/978-3-031-88727-7_8

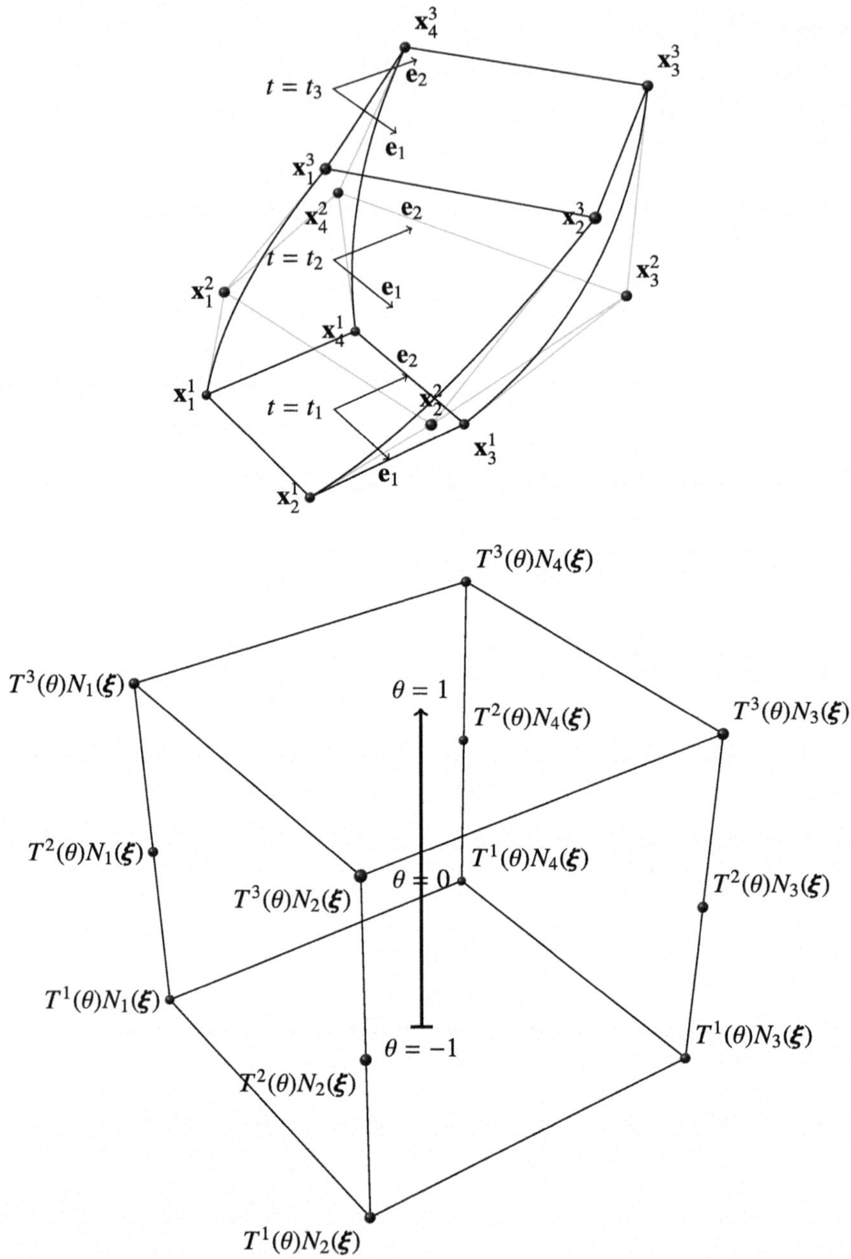

Fig. 8.1 A physical ST element and its parent element. The example is for an element that is quadrilateral in space and quadratic NURBS in time ($n_{\text{ens}} = 4$, $n_{\text{ent}} = 3$). While typically the element would be quadratic NURBS also in space, in this example, not to crowd figure, we used an element that is bilinear in space. See also Fig. 3.2

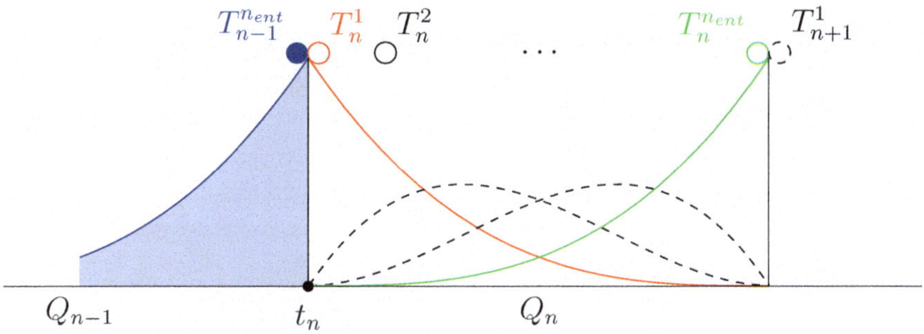

Fig. 8.2 ST-IGA basis functions in time. Cubic Bernstein polynomials

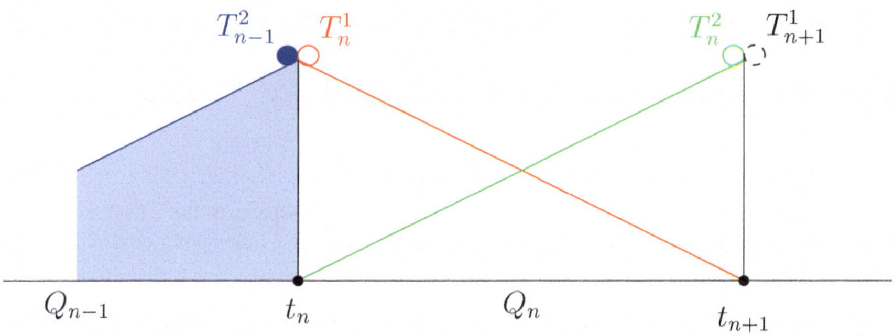

Fig. 8.3 ST-IGA basis functions in time. Linear functions

with that surface motion, and better efficiency in representing the mesh motion and in remeshing. As shown in 2011 [50], representation of a circular path will be exact with quadratic NURBS basis functions, achievable with as little as two patches. Furthermore, as explained briefly in 2010 [14] and in detail in 2011 [50], to have speeds that are invariant along the circular path, a constant angular speed can be prescribed with the aid of a secondary mapping. (ii) The "ST-C," introduced in 2013 [148], where the letter "C" means "continuous," is a method for extracting time-continuous data from the computed data. It can work as a data compression method in dealing with large volumes of data (see, for example, [149]). We will now illustrate some of these advantages.

Figure 8.4 illustrates, using the locust-wing flapping reported in 2011 [50] as a context, two key aspects of the ST-IGA: geometry and motion representations with IGA basis functions in space and time.

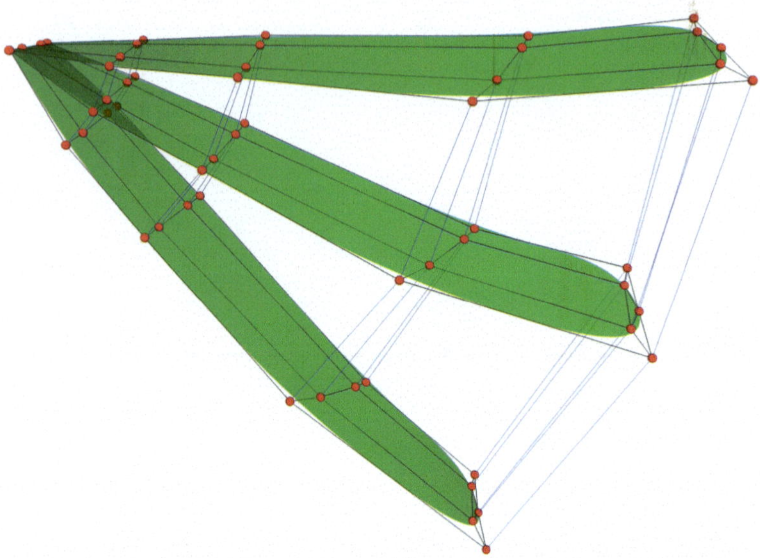

Fig. 8.4 Locust-wing flapping as a context to illustrate two key aspects of the ST-IGA: geometry and motion representations with IGA basis functions in space and time. Wing control mesh and corresponding wing geometry at three control points in time

Solution representation with IGA basis functions in space and in time are two other key aspects of the ST-IGA. A Fourier stability and accuracy analysis for spatial discretization with B-splines was given in [19]. A pure-advection Fourier stability and accuracy analysis for the ST-SUPG was performed in 2010 [14] for quadratic B-splines in space and various-order B-splines in time in combination with different Courant numbers. We provide a partial overview of that here. The Courant number is defined as $\mathrm{CFL} = \frac{c\Delta t}{h}$, where h is the element length, Δt is the time-step size, and $c = u - v$ is the velocity relative to the mesh. The different B-splines time discretizations are "SV" [13], linear B-splines, and quadratic B-splines. The SV, which uses linear B-splines, is equivalent to ALE with Crank–Nicolson time integration. Figure 8.5 shows, for different CFL values, the phase error $|1 - \tilde{\omega}/\omega|$ as a function of the dimensionless wave number kh. The figure shows that the ST-IGA, by using IGA basis functions also in time, enables getting full benefit from using IGA basis functions in space.

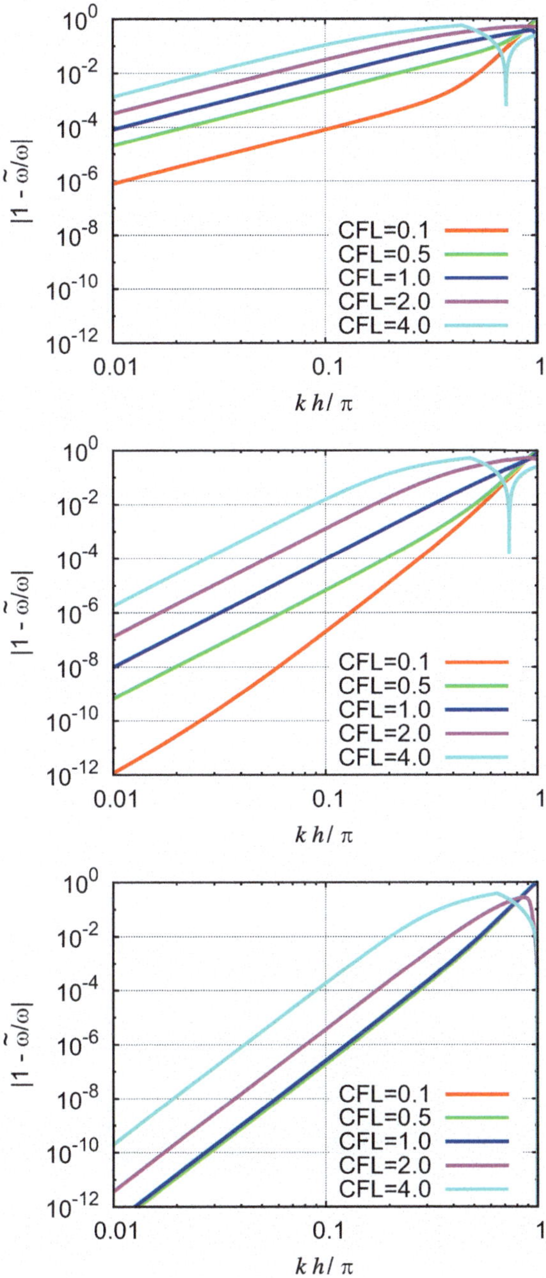

Fig. 8.5 Pure-advection Fourier analysis for the ST-SUPG. Phase error. Basis functions in time: SV, linear B-splines, and quadratic B-splines

8.2 First 2D Computations with the ST-IGA, 2010

The first 2D computations with the ST-IGA were reported in 2010 [14]. We highlight the computations from [14].

8.2.1 Flow Past a NACA 64-618 Airfoil

The Reynolds number is 6×10^6. Both the ST-SUPS and ST-VMS were tested, with a rather coarse mesh for such a high Reynolds number. The IGA discretization is with a mesh made of 1400 quadratic B-splines elements and with linear elements in time. Figure 8.6 shows the mesh around the airfoil. Figure 8.7 shows the velocity magnitude from the computations with the ST-SUPS and ST-VMS.

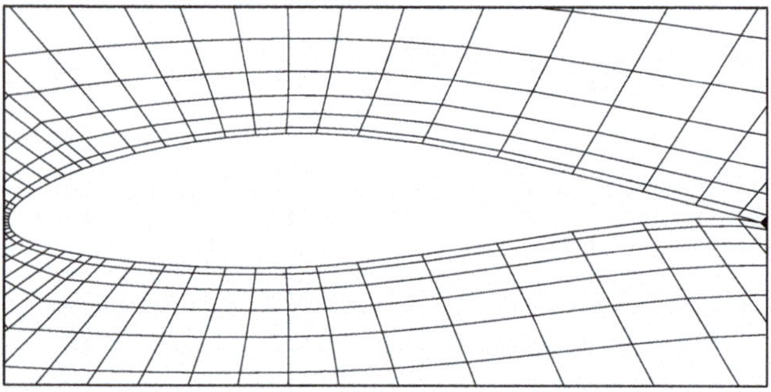

Fig. 8.6 Flow past a NACA 64-618 airfoil. Computed with the ST-IGA. Mesh near the airfoil

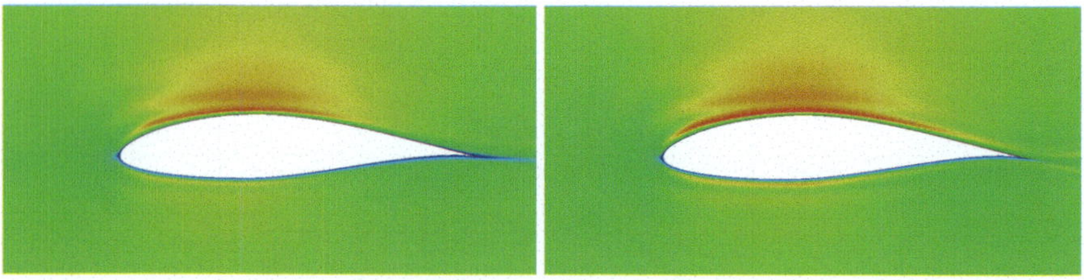

Fig. 8.7 Flow past a NACA 64-618 airfoil. Computed with the ST-IGA, and with the ST-SUPS and ST-VMS. Velocity magnitude

8.2.2 Advection Problems with a Rotational Flow Field

Two advection problems with a rotational flow field were computed: rotating cosine puff and Zalesak's solid-body rotation test [150]. Both were computed with the ST-SUPG, with no DC term. The cosine puff was computed over a square domain, and Zalesak's test over both square and donut-shaped domains. In the cosine puff, the IGA discretization is with a single-patch mesh made of 16,384 cubic B-splines elements and with linear, quadratic, and cubic B-splines in time. In Zalesak's test, for the square domain, the IGA discretization in space and time is the same as in the cosine puff, and for the donut-shaped domain, the discretization is with a four-patch mesh made of 4096 quadratic NURBS elements and with linear, quadratic, and cubic B-splines in time. We report the results from the Zalesak's test computed over the square domain. Figure 8.8 shows the initial condition and the solution after a full rotation, for linear, quadratic, and cubic B-splines in time, with 160 time steps per rotation.

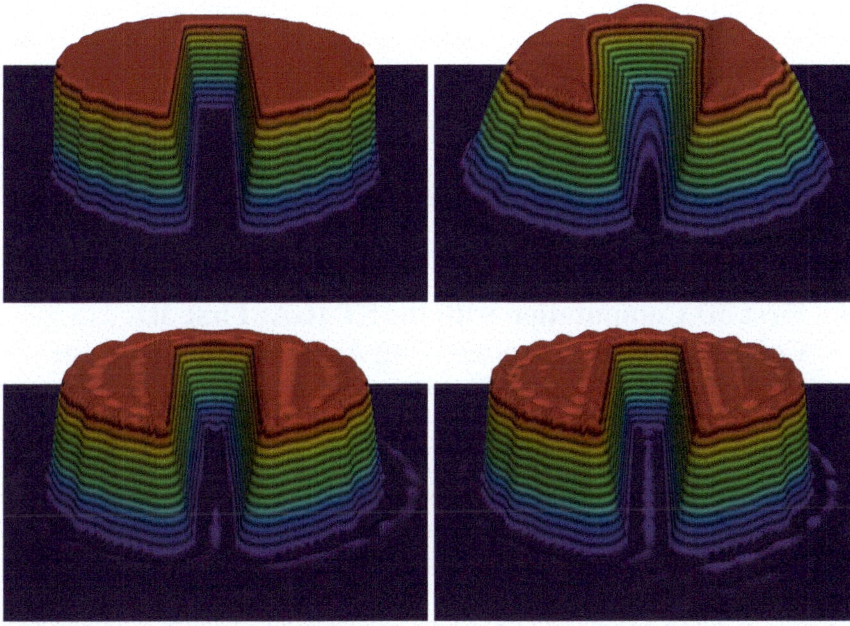

Fig. 8.8 Zalesak's solid-body rotation test. Computed over the square domain. Initial condition and the exact solution after one full rotation (*top-left*). Solution after a full rotation for linear (*top-right*), quadratic (*bottom-left*), and cubic B-splines (*bottom-right*) in time

8.3 Space–Time/NURBS Mesh Update Method (STNMUM), 2011

The STNMUM was introduced in 2011 [50] and named in [151]. It is a general-purpose mesh update method, where the motion of solid surfaces and mesh motion consistent with that are represented with IGA basis functions in time. The temporal-control-point meshes come from the meshes generated or meshes obtained by mesh moving methods such as the linear-elasticity mesh moving with MJBS. The mesh at any other point in time can be obtained from the representation with the IGA basis functions, where the order of the basis functions is determined from the required level of continuity. Figures 8.9 and 8.10 illustrate the STNMUM concept, using the locust-wing flapping reported in 2011 [50] as a context. Using IGA basis functions in time brings higher accuracy in representing the motion of a solid surfaces and mesh motion. With the cubic B-splines used in Figs. 8.9 and 8.10, we achieve continuity in the acceleration, which in turn provides continuity in the aerodynamic forces calculated.

If the STNMUM is used in computations involving a rotating solid surface, such as a wind turbine rotor, representation of the circular path of any point on that surface will be exact with quadratic NURBS basis functions, achievable with as little as two patches. Representation of the circular path of any point in a mesh rotating with that solid surface will also be exact. Furthermore, to have speeds that are invariant along those circular paths, a constant angular speed can be prescribed with the aid of a secondary mapping (see Sect. 8.1).

8.4 First 3D Computation with the ST-IGA: First 3D Computation of Flapping-Wing Aerodynamics with the Wing Motion Coming From an Actual Locust, 2011

The first 3D computation with the ST-IGA was reported in 2011 [50], which was also the first 3D computation of flapping-wing aerodynamics with the wing motion coming from an actual locust. The motion and deformation data for the wings was extracted from the high-speed, multi-camera video recordings of a locust in a wind tunnel at Baylor College of Medicine, Houston, Texas. The video was recorded by using a set of tracking points marked on the forewings and hindwings of the locust. How the wing motion and deformation data was extracted from the video data and represented using NURBS basis functions in space and time was described in detail in [50]. The wing geometries and motions are represented with IGA basis functions in time, as shown in Fig. 8.4 (Sect. 8.1), using quadratic NURBS. The mesh motion

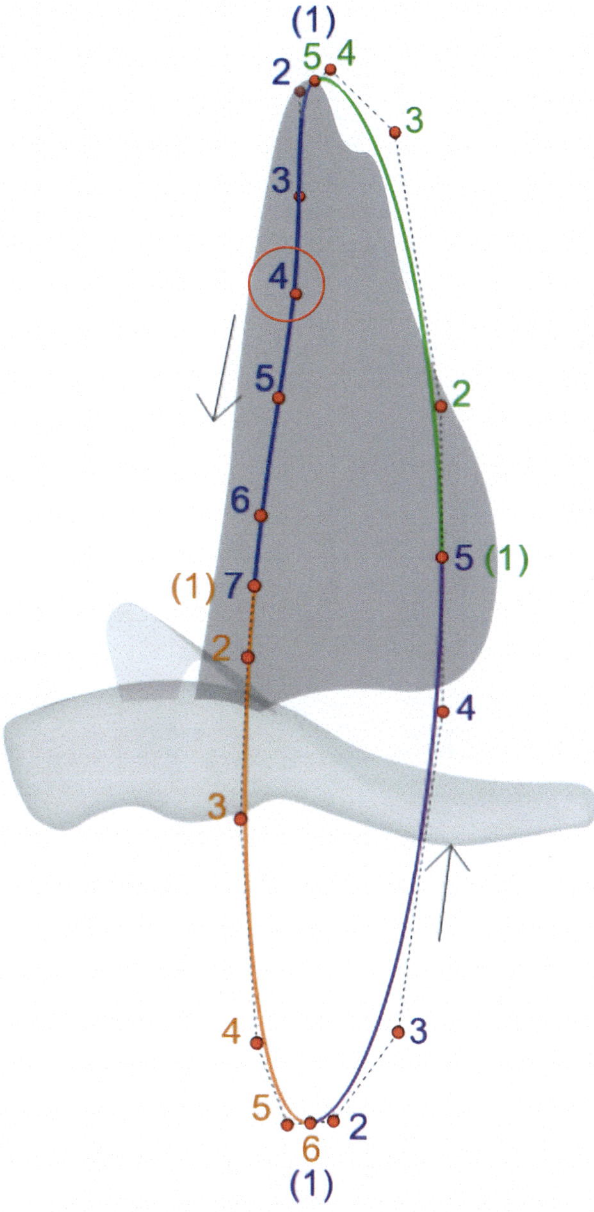

Fig. 8.9 Locust-wing flapping as a context to illustrate the STNMUM concept. The wing motion is represented with four temporal patches, using cubic B-splines, which provide continuity in the acceleration

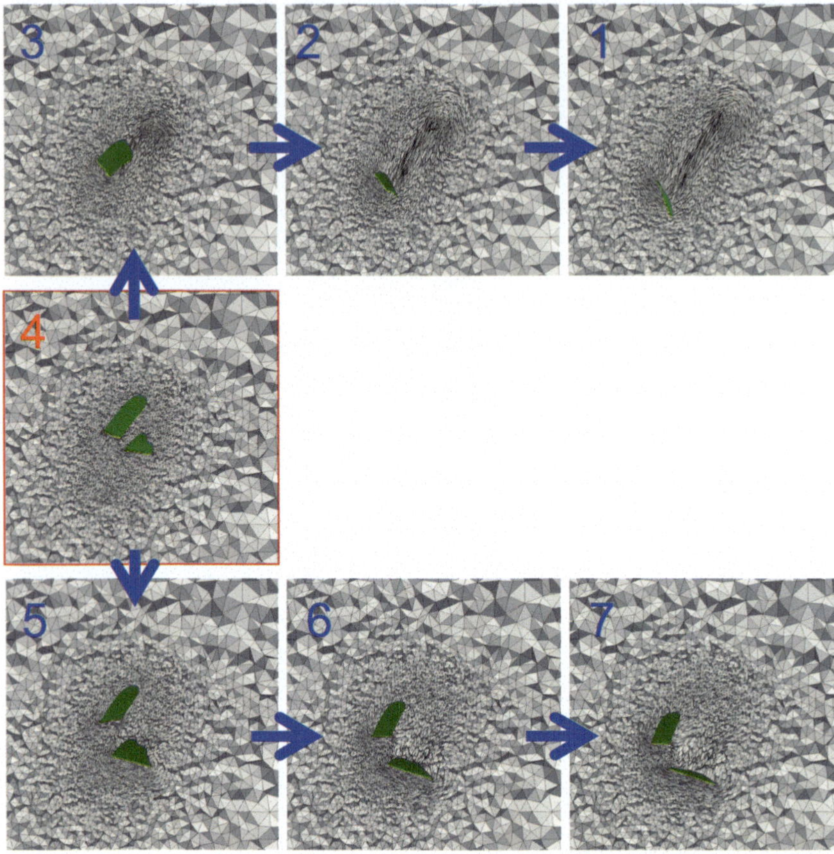

Fig. 8.10 Locust-wing flapping as a context to illustrate the STNMUM concept. Meshes associated with the 7 control points of the *blue* patch in Fig. 8.9. Mesh generation takes place at one of the control points in the middle of the patch, which is control point 4 in this case, and the meshes at the other control points are obtained by using a mesh moving method, which is the linear-elasticity mesh moving with MJBS in this case. In total for the simulation, mesh generation takes place only 4 times, and mesh moving takes place only 19 times

and remeshing are done with the STNMUM, conceptually as explained in Figs. 8.9 and 8.10 (Sect. 8.3). The volume mesh is made of linear tetrahedral elements, with the element count averaging 2.6 M. At each time step, which has four nonlinear iterations, the ST-SUPS is used in the first two iterations and the ST-VMS in the last two. Figure 8.11 shows, from [50], the flow patterns.

Fig. 8.11 3D computation of flapping-wing aerodynamics with the wing motion coming from an actual locust. Vorticity

8.5 First 3D Patient-Specific ST-VMS/ST-IGA Computation of Flow in a Cerebral Artery Segment with Stent-Treated Aneurysm and Flow-Resolved Stent, 2012

The first patient-specific ST-VMS/ST-IGA computation of flow in a cerebral artery segment with stent-treated aneurysm and flow-resolved stent was reported in 2012 [152]. The target in this kind of medical treatment is to reduce the blood circulation in the aneurysm, inducing thrombosis. The purpose in this class of simulations is to predict how much the circulation is reduced with a single stent and double staggered stents. The stent model used in [152] was generated using the geometry of a Cordis Precise Pro Rx nitinol self-expanding stent (PC0630RXC) with a wire diameter of about 0.1 mm. Influence of using stent models with zero and finite thickness in the normal direction was also studied in [152]. The surface over which the stent mesh is built is represented by NURBS. Figure 8.12 shows examples of surfaces for single and double full stents. Figures 8.13 and 8.14 show the lumen geometry for the artery segment computed and the mesh for the single-stent case. The volume meshes are made of linear tetrahedral elements, with an element count of 3,168,305, 3,300,182, and 3,736,603 for the no-stent, single-stent, and double-stent cases. Figures 8.15 and 8.16 show the vorticity in the aneurysm for all three cases.

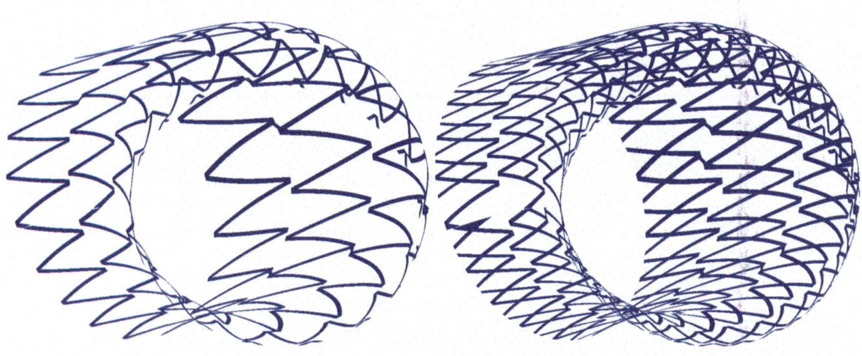

Fig. 8.12 Patient-specific ST-VMS/ST-IGA computation of flow in a cerebral artery segment with stent-treated aneurysm and flow-resolved stent. Examples of surfaces for single and double full stents

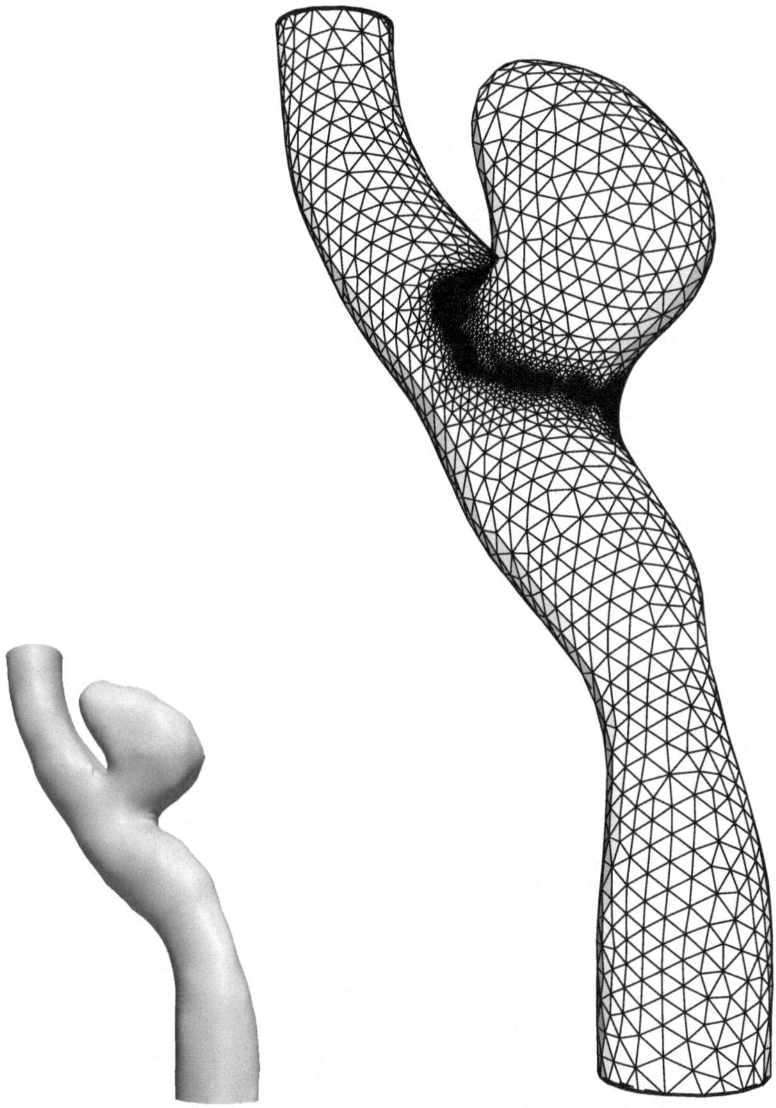

Fig. 8.13 Patient-specific ST-VMS/ST-IGA computation of flow in a cerebral artery segment with stent-treated aneurysm and flow-resolved stent. Arterial lumen geometry and the lumen surface mesh for the single-stent case

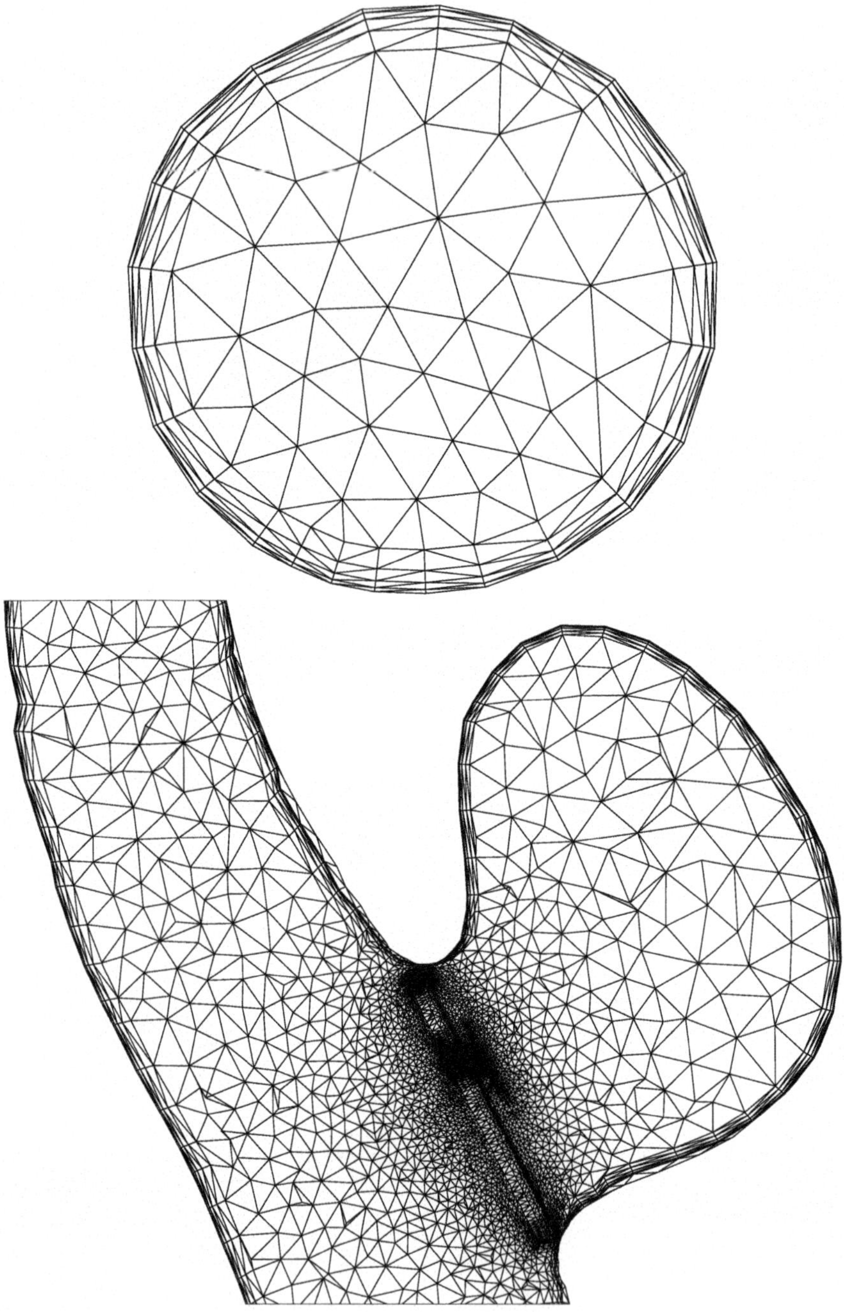

Fig. 8.14 Patient-specific ST-VMS/ST-IGA computation of flow in a cerebral artery segment with stent-treated aneurysm and flow-resolved stent. Mesh cross-sections for the single-stent case

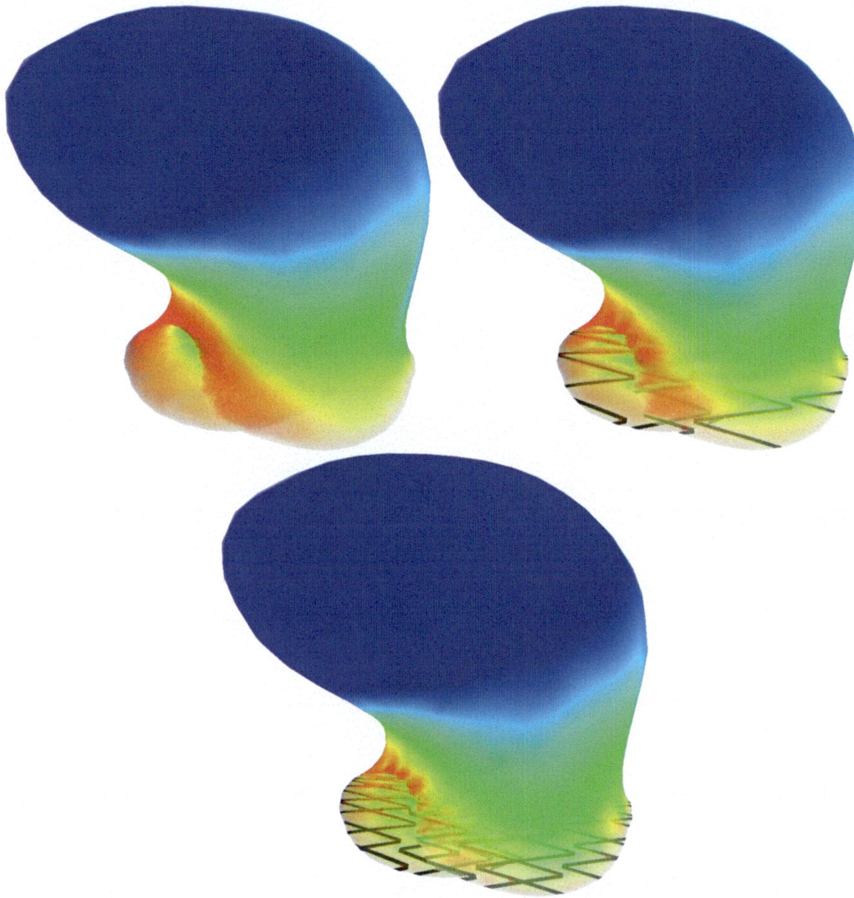

Fig. 8.15 Patient-specific ST-VMS/ST-IGA computation of flow in a cerebral artery segment with stent-treated aneurysm and flow-resolved stent. No stent, single stent, and double stent. Vorticity magnitude at peak flow into the aneurysm. Color range from *blue* to *red* corresponds to the vorticity range from low to high

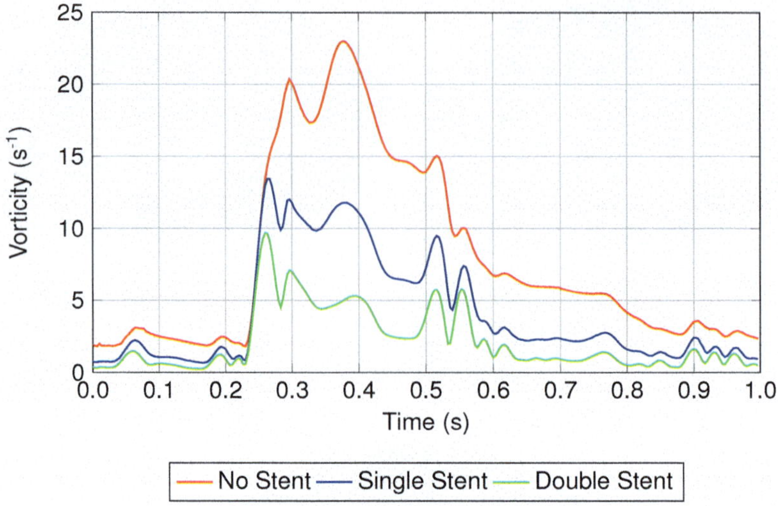

Fig. 8.16 Patient-specific ST-VMS/ST-IGA computation of flow in a cerebral artery segment with stent-treated aneurysm and flow-resolved stent. Spatially averaged vorticity magnitude in the aneurysm

Chapter 9
Methods and Solutions in 2013–2015

9.1 First 3D ST-VMS/ST-IGA Computation of Wind Turbine Rotor and Tower Aerodynamics, 2013

The first 3D ST-VMS/ST-IGA computation of wind turbine rotor and tower aerodynamics was reported in 2013 [151], for the NREL 5MW offshore baseline wind turbine. Figure 9.1 shows the model. The ST-IGA with IGA basis functions in time is used for the temporal representation of the rotor motion, enabling us to represent the circular paths associated with that motion exactly and specify a constant angular velocity corresponding to the invariant speeds along those paths. The STNMUM is used in representation of the motion and deformation of the volume meshes computed and also in remeshing. The rotor motion is defined in three identical 120° segments. Each 120° segment is divided into six patches to keep the mesh distortion under control. Each patch is a 20° arc, with three temporal-control points. Figure 9.2 shows how a blade-tip path is represented in time. Two different meshes, both made of linear tetrahedral elements, were used in [151] for the rotor and tower aerodynamics: one with boundary-layer mesh only for the blades and one for both the blades and the tower. Figure 9.3 shows the boundary-layer mesh at $\frac{3}{4}R$. The element count range for the mesh with boundary-layer mesh only for the blades is 2,493,524–2,553,100. Figure 9.4 shows the flow patterns.

9.2 ST-IGA with Continuous Data Representation in Time: ST-C, 2013

The ST-C was introduced in 2013 [148], where the letter "C" means "continuous." This is a method for extracting time-continuous data from the computed data, using IGA basis functions in time. This gives us a temporally smooth, IGA-based solution,

T. E. Tezduyar, K. Takizawa, *Space–Time Computational Flow Analysis*,
Modeling and Simulation in Science, Engineering and Technology,
https://doi.org/10.1007/978-3-031-88727-7_9

Fig. 9.1 3D ST-VMS/ST-IGA computation of wind turbine rotor and tower aerodynamics. Model

which is desirable in some cases, and a more efficient way of dealing with the computed data. It can work as a data compression method in dealing with large volumes of data (see, for example, [149]). The ST-C was used for that purpose in a diverse set of 3D computations, including multiscale thermo-fluid analysis of ground vehicles and tires in 2015 [16], flow-driven string dynamics in a pump 2018 [153], wind turbine long-wake flows in 2011 [154–156], and high-resolution multiscale car and tire aerodynamics in 2022 [149].

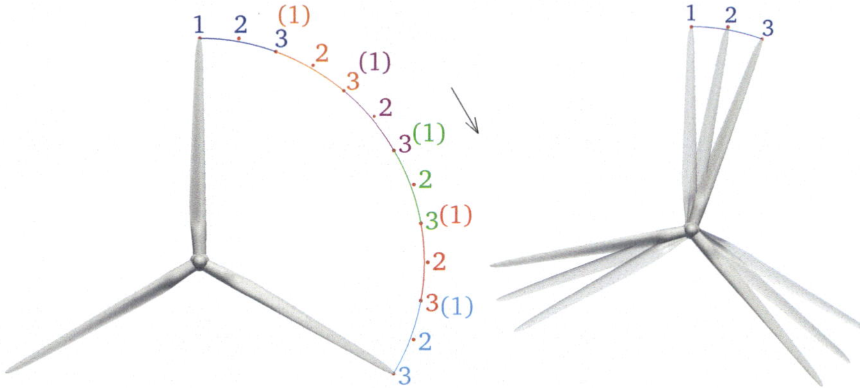

Fig. 9.2 3D ST-VMS/ST-IGA computation of wind turbine rotor and tower aerodynamics. Blade-tip path with temporal patches and control point numbering local to each patch. A control point at the start of a patch and colocated with a control point at the end of the previous patch is in parentheses. Mesh generation takes place at control point 2, and the meshes at control points 1 and 3 are obtained by using linear-elasticity mesh moving with MJBS

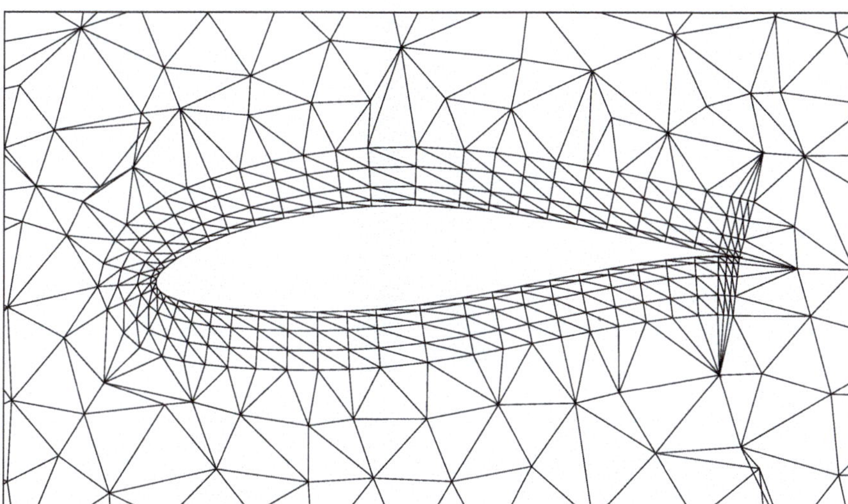

Fig. 9.3 3D ST-VMS/ST-IGA computation of wind turbine rotor and tower aerodynamics. Boundary-layer mesh at $\frac{3}{4}R$

Fig. 9.4 3D ST-VMS/ST-IGA computation of wind turbine rotor and tower aerodynamics. Vorticity magnitude. Color range from *blue* to *red* corresponds to the vorticity range from low to high

9.3 First 2D ST-TC/ST-IGA Flow Computation with Actual-Contact and High-Resolution Boundary-Layer Representations, 2013

The first 2D ST-TC/ST-IGA flow computation with actual-contact and high-resolution boundary-layer representations was reported in [52], in the same article where the ST-TC was introduced. The computation was for two pairs of symmetrically-flapping surfaces with coordinated opening/closing action ("Two opening/closing pairs"), intended to represent the mitral and aortic valves. The surfaces have zero thickness. The ST-IGA is used with quadratic NURBS basis functions in space to represent the external domain boundaries and cubic B-spline basis functions in time to represent the motion of those boundaries.

The computation was with the ST-VMS and finite element discretization. The mesh is built from structured meshes around the valves and an unstructured mesh elsewhere. The structured meshes are made of four-node quadrilateral elements and the unstructured mesh is made of three-node triangular elements. Figure 9.5 shows the mesh. The computation was with special-purpose, algebraic mesh moving methods for the structured meshes and the linear-elasticity mesh moving method with MJBS for the unstructured mesh. Figure 9.6 shows the velocity vectors and pressure around the closed mitral pair.

9.4 First 3D Computation with the Thermo-Fluid ST-VMS and ST-IGA: Thermo-Fluid Analysis of a Freight Truck and Its Rear Set of Tires, 2015

The first 3D computation with the thermo-fluid ST-VMS and ST-IGA was in 2015 [16], for thermo-fluid analysis of a freight truck and its rear set of tires, with emphasis on the tires. The analysis includes the heat from the engine and exhaust system, separated into the heat losses from the exhaust pipe surface and exhaust pipe end. The analysis also includes the heave motion of the truck body, prescribed as a periodic motion with a given semi-amplitude and frequency. A spatially multiscale version of the MDM is used in the analysis. The global domain containing the entire truck serves as the primary subdomain. The local domain containing the rear set of tires serves as the secondary subdomain, which is fully inside the primary subdomain. First the computation is carried out over the global domain, with a reasonable mesh refinement. The large amount of time-history data from the global computation is stored using the ST-C, with cubic B-splines in time. This is followed by a higher-resolution computation over the local domain, which gives us increased accuracy in the thermo-fluid analysis. At each time step of the computation, the boundary conditions are extracted from the stored global data. Figure 9.7 shows the truck model and the local domain containing the left half of the rear tires.

The computation was with the ST-SUPS for the global domain and ST-VMS for the local domain. The meshes for both the global and local domains are made of linear tetrahedral elements, with an element count of 4,857,973 and 32,986,249 for the global and local domains. Figure 9.8 shows the mesh near the tire. The tires have axial symmetry in the computation, and therefore there is no need for a moving mesh to represent the rotation. However, the truck-body heave motion is modeled, and a moving mesh is needed for that. The STNMUM is used for that. In antiperiodic representations of the heave motion and the corresponding mesh motion, in the ST-IGA, the IGA basis functions in time are cubic B-splines. There are five control

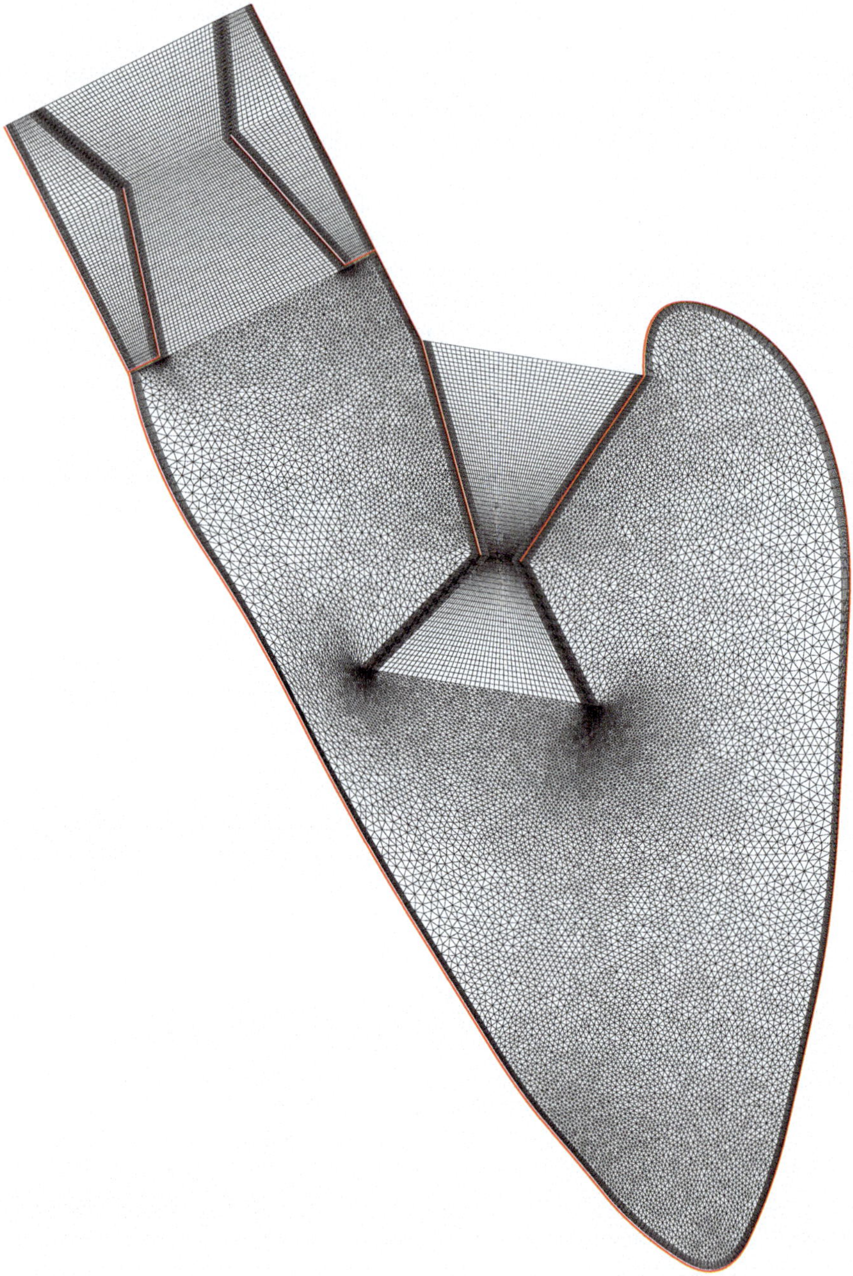

Fig. 9.5 2D ST-TC/ST-IGA flow computation with actual-contact and high-resolution boundary-layer representations. Two opening/closing pairs. Mesh

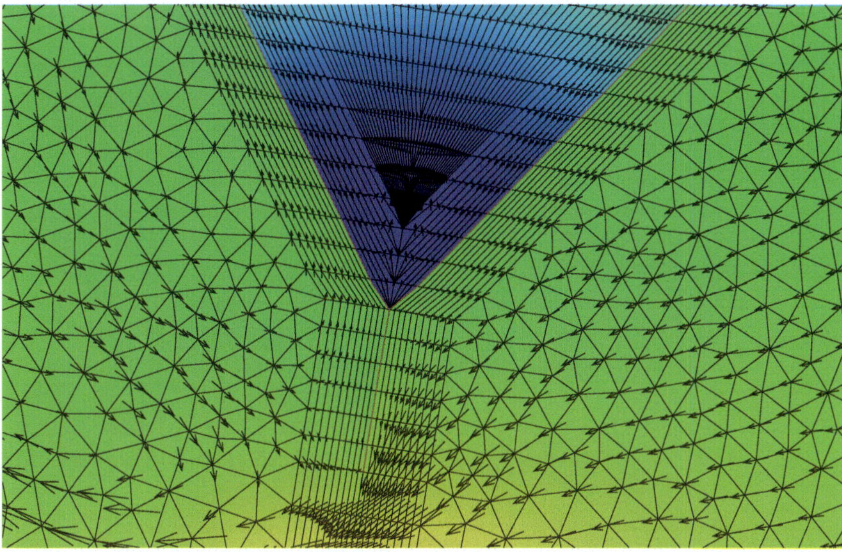

Fig. 9.6 2D ST-TC/ST-IGA flow computation with actual-contact and high-resolution boundary-layer representations. Two opening/closing pairs. Velocity vectors and pressure around the closed mitral pair. The *red* lines are the zero-thickness surfaces

Fig. 9.7 3D thermo-fluid analysis of a freight truck and its rear set of tires. Truck model and the local domain containing the left half of the rear tires

Fig. 9.8 3D thermo-fluid analysis of a freight truck and its rear set of tires. Mesh resolution in the tangential and normal directions for the global and local domains

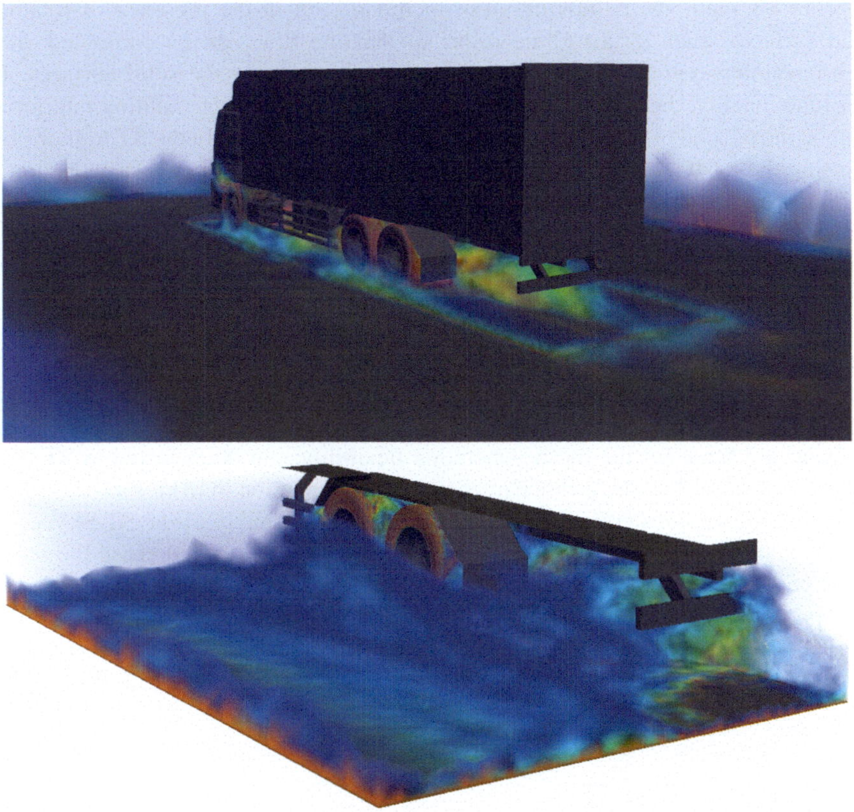

Fig. 9.9 3D thermo-fluid analysis of a freight truck and its rear set of tires. Temperature from the global and local computations. Color range from *blue* to *red* corresponds to the temperature range from low to high

points for the cycle, only three of which are independent, corresponding to the up, down, and middle points of the heave motion. The control meshes corresponding to the up and down points are calculated by solving the nonlinear-elasticity equations with the neo-Hookean model and MJBS. The mesh corresponding to the middle point serves as the reference configuration. Figure 9.9 shows the temperature from the global and local computations.

9.5 Space–Time Slip Interface (ST-SI) Method, 2015

The ST-SI was introduced in 2015 [54]. It is a multifunction method, accurately connecting the parts of the solution obtained over two mesh zones with nonmatching meshes at the SI between the zones. Its original function was to enable the use of the core ST methods as moving-mesh methods even in the presence of rotating solid surfaces, such as a turbine rotor, so that the flow can be computed with high-resolution representation of the boundary layers near the solid surfaces. In that functionality, the ST-SI is essentially the ST version of the "sliding interface" method introduced in [146] in connection with the ALE-VMS. In the ST framework, by using the ST-IGA with IGA basis functions in time, more specifically quadratic NURBS basis functions in time, we achieve exact representation of the solid-surface and mesh rotations. Furthermore, the synthesis of the ST-SI and ST-IGA with IGA basis functions in space meets the requirement of accurately connecting the parts of the solution obtained over two mesh zones with nonmatching meshes not because of rotating solid surfaces requiring slip between the two mesh zones but because we just want to be free from the matching requirement. That gives us IGA mesh generation flexibility and helps us address the IGA mesh generation challenge for complex geometries. Multiple SIs in a computation, some for the rotating solid surfaces and some for the mesh generation flexibility, can be expected in ST flow computations, such as the turbocharger turbine computation in [51].

The ST-SI versions going beyond the original functionality of the ST-SI version with the "fluid–fluid SI" work in comparable ways and further increase the scope of the core ST methods. The ST-SI version with "fluid–solid SI," also introduced in [54], is basically the ST counterpart of the weak-Dirichlet-condition method [157]. The ST-SI version with "porosity SI" [54] is for a thin porous structure with fluid on its two sides. We provide here an overview of all three versions.

9.5.1 Fluid–Fluid SI

In describing the fluid–fluid SI, labels "Side A" and "Side B" will represent the two sides of the SI. The ST-SI associated with the ST-VMS given by Eq. (7.1) includes integrals over the SI added to the left-hand side of the equation. The integrals for Side A and Side B are first added separately, using the test functions $\mathbf{w}_{\mathrm{A}}^h$, q_{A}^h and $\mathbf{w}_{\mathrm{B}}^h$, q_{B}^h. Then, putting together the integrals added for each side, the complete set of integrals added becomes

$$
\begin{aligned}
&-\int_{(P_n)_{\mathrm{SI}}} \left(q_{\mathrm{B}}^h \mathbf{n}_{\mathrm{B}} - q_{\mathrm{A}}^h \mathbf{n}_{\mathrm{A}}\right) \cdot \frac{1}{2}\left(\mathbf{u}_{\mathrm{B}}^h - \mathbf{u}_{\mathrm{A}}^h\right) \mathrm{d}P \\
&-\int_{(P_n)_{\mathrm{SI}}} \rho \mathbf{w}_{\mathrm{B}}^h \cdot \frac{1}{2}\left(\mathcal{F}_{\mathrm{B}}^h - \left|\mathcal{F}_{\mathrm{B}}^h\right|\right)\left(\mathbf{u}_{\mathrm{B}}^h - \mathbf{u}_{\mathrm{A}}^h\right) \mathrm{d}P \\
&-\int_{(P_n)_{\mathrm{SI}}} \rho \mathbf{w}_{\mathrm{A}}^h \cdot \frac{1}{2}\left(\mathcal{F}_{\mathrm{A}}^h - \left|\mathcal{F}_{\mathrm{A}}^h\right|\right)\left(\mathbf{u}_{\mathrm{A}}^h - \mathbf{u}_{\mathrm{B}}^h\right) \mathrm{d}P \\
&+\int_{(P_n)_{\mathrm{SI}}} \left(\mathbf{n}_{\mathrm{B}} \cdot \mathbf{w}_{\mathrm{B}}^h + \mathbf{n}_{\mathrm{A}} \cdot \mathbf{w}_{\mathrm{A}}^h\right) \frac{1}{2}\left(p_{\mathrm{B}}^h + p_{\mathrm{A}}^h\right) \mathrm{d}P
\end{aligned}
$$

$$
\begin{aligned}
&- \int_{(P_n)_{\mathrm{SI}}} \left(\mathbf{w}_{\mathrm{B}}^h - \mathbf{w}_{\mathrm{A}}^h\right) \cdot \left(\hat{\mathbf{n}}_{\mathrm{B}} \cdot \mu \left(\boldsymbol{\varepsilon}(\mathbf{u}_{\mathrm{B}}^h) + \boldsymbol{\varepsilon}(\mathbf{u}_{\mathrm{A}}^h)\right)\right) \mathrm{d}P \\
&- \gamma \int_{(P_n)_{\mathrm{SI}}} \hat{\mathbf{n}}_{\mathrm{B}} \cdot \mu \left(\boldsymbol{\varepsilon}\left(\mathbf{w}_{\mathrm{B}}^h\right) + \boldsymbol{\varepsilon}\left(\mathbf{w}_{\mathrm{A}}^h\right)\right) \cdot \left(\mathbf{u}_{\mathrm{B}}^h - \mathbf{u}_{\mathrm{A}}^h\right) \mathrm{d}P \\
&+ \int_{(P_n)_{\mathrm{SI}}} \frac{\mu C}{h} \left(\mathbf{w}_{\mathrm{B}}^h - \mathbf{w}_{\mathrm{A}}^h\right) \cdot \left(\mathbf{u}_{\mathrm{B}}^h - \mathbf{u}_{\mathrm{A}}^h\right) \mathrm{d}P,
\end{aligned}
\tag{9.1}
$$

where

$$
\mathcal{F}_{\mathrm{B}}^h = \mathbf{n}_{\mathrm{B}} \cdot \left(\mathbf{u}_{\mathrm{B}}^h - \mathbf{v}_{\mathrm{B}}^h\right), \tag{9.2}
$$

$$
\mathcal{F}_{\mathrm{A}}^h = \mathbf{n}_{\mathrm{A}} \cdot \left(\mathbf{u}_{\mathrm{A}}^h - \mathbf{v}_{\mathrm{A}}^h\right), \tag{9.3}
$$

$$
\hat{\mathbf{n}}_{\mathrm{B}} = \frac{\mathbf{n}_{\mathrm{B}} - \mathbf{n}_{\mathrm{A}}}{\|\mathbf{n}_{\mathrm{B}} - \mathbf{n}_{\mathrm{A}}\|}. \tag{9.4}
$$

Here, $(P_n)_{\mathrm{SI}}$ is the SI in the ST domain, $\mathbf{n}_{\mathrm{A}}$ and $\mathbf{n}_{\mathrm{B}}$ are the unit normal vectors for the two sides, $\mathbf{v}_{\mathrm{A}}^h$ and $\mathbf{v}_{\mathrm{B}}^h$ are the mesh velocities, $\gamma = 1$, and C is a nondimensional constant. The element lengths are

$$
h = \left(\frac{h_{\mathrm{B}}^{-2} + h_{\mathrm{A}}^{-2}}{2}\right)^{-\frac{1}{2}}, \tag{9.5}
$$

$$
h_{\mathrm{B}} = 2\left(\mathbf{n}_{\mathrm{B}}\mathbf{n}_{\mathrm{B}} : \mathbf{G}\right)^{-\frac{1}{2}} \qquad \text{(for Side B)}, \tag{9.6}
$$

$$
h_{\mathrm{A}} = 2\left(\mathbf{n}_{\mathrm{A}}\mathbf{n}_{\mathrm{A}} : \mathbf{G}\right)^{-\frac{1}{2}} \qquad \text{(for Side A)}, \tag{9.7}
$$

where $\mathbf{G}$ is the space-only element-metric tensor.

9.5.2 Fluid–Solid SI

The ST-SI with the fluid–solid SI is obtained (see [54]) by starting with the integrals added for Side B and replacing the Side A velocity with the velocity $\mathbf{g}^h$ coming from the solid surface. Then the SI integrals added to Eq. (7.1) to represent the weakly-enforced Dirichlet conditions become

$$
\begin{aligned}
&- \int_{(P_n)_{\mathrm{SI}}} q_{\mathrm{B}}^h \mathbf{n}_{\mathrm{B}} \cdot \left(\mathbf{u}_{\mathrm{B}}^h - \mathbf{g}^h\right) \mathrm{d}P \\
&- \int_{(P_n)_{\mathrm{SI}}} \rho \mathbf{w}_{\mathrm{B}}^h \cdot \frac{1}{2} \left(\mathcal{F}_{\mathrm{B}}^h - \left|\mathcal{F}_{\mathrm{B}}^h\right|\right) \left(\mathbf{u}_{\mathrm{B}}^h - \mathbf{g}^h\right) \mathrm{d}P
\end{aligned}
$$

$$
\begin{aligned}
&-\int_{(P_n)_{\mathrm{SI}}} \mathbf{w}_{\mathrm{B}}^h \cdot \left(\mathbf{n}_{\mathrm{B}} \cdot \boldsymbol{\sigma}_{\mathrm{B}}^h\right) \mathrm{d}P \\
&-\gamma \int_{(P_n)_{\mathrm{SI}}} \mathbf{n}_{\mathrm{B}} \cdot 2\mu\boldsymbol{\varepsilon}\left(\mathbf{w}_{\mathrm{B}}^h\right) \cdot \left(\mathbf{u}_{\mathrm{B}}^h - \mathbf{g}^h\right) \mathrm{d}P \\
&+\int_{(P_n)_{\mathrm{SI}}} \frac{\mu C}{h_{\mathrm{B}}} \mathbf{w}_{\mathrm{B}}^h \cdot \left(\mathbf{u}_{\mathrm{B}}^h - \mathbf{g}^h\right) \mathrm{d}P.
\end{aligned} \tag{9.8}
$$

9.5.3 Porosity SI

In the model here, the porosity velocity is expressed as a function of the pressure difference between the two sides of the SI:

$$
u_{\mathrm{PORO}}^h = -\mathcal{P}(\Delta p^h), \tag{9.9}
$$

where

$$
\Delta p^h \equiv p_{\mathrm{A}}^h - p_{\mathrm{B}}^h. \tag{9.10}
$$

The velocity in $\mathbf{n}_{\mathrm{B}}$ direction is defined as the positive porosity velocity. The expression $\mathcal{P}(\Delta p^h)$ can take different forms depending the nature of the porosity. The normal component of the velocity is assumed to be continuous, and the tangential component is set to the tangential component of $\mathbf{u}_{\mathrm{S}}^h$, where $\mathbf{u}_{\mathrm{S}}^h$ is the structure velocity, and the volume flux is imposed with the porosity velocity taken into account. Only the integrals added for Side B were given in [54], leaving it to the reader to do the same for Side A and put together the integrals added for each side. Here we give the complete set of integrals added:

$$
\begin{aligned}
&-\int_{(P_n)_{\mathrm{SI}}} q_{\mathrm{B}}^h \mathbf{n}_{\mathrm{B}} \cdot \left(\mathbf{u}_{\mathrm{B}}^h - u_{\mathrm{PORO}}^h \hat{\mathbf{n}}_{\mathrm{B}} - \mathbf{u}_{\mathrm{S}}^h\right) \mathrm{d}P \\
&-\int_{(P_n)_{\mathrm{SI}}} q_{\mathrm{A}}^h \mathbf{n}_{\mathrm{A}} \cdot \left(\mathbf{u}_{\mathrm{A}}^h - u_{\mathrm{PORO}}^h \hat{\mathbf{n}}_{\mathrm{B}} - \mathbf{u}_{\mathrm{S}}^h\right) \mathrm{d}P \\
&-\int_{(P_n)_{\mathrm{SI}}} \rho \mathbf{w}_{\mathrm{B}}^h \cdot \frac{1}{2}\left(\mathcal{F}_{\mathrm{B}}^h - \left|\mathcal{F}_{\mathrm{B}}^h\right|\right) \left(\mathbf{u}_{\mathrm{B}}^h - \hat{\mathbf{N}}_{\mathrm{B}} \cdot \mathbf{u}_{\mathrm{A}}^h - \mathbf{T}_{\mathrm{B}} \cdot \mathbf{u}_{\mathrm{S}}^h\right) \mathrm{d}P \\
&-\int_{(P_n)_{\mathrm{SI}}} \rho \mathbf{w}_{\mathrm{A}}^h \cdot \frac{1}{2}\left(\mathcal{F}_{\mathrm{A}}^h - \left|\mathcal{F}_{\mathrm{A}}^h\right|\right) \left(\mathbf{u}_{\mathrm{A}}^h - \hat{\mathbf{N}}_{\mathrm{B}} \cdot \mathbf{u}_{\mathrm{B}}^h - \mathbf{T}_{\mathrm{A}} \cdot \mathbf{u}_{\mathrm{S}}^h\right) \mathrm{d}P \\
&+\int_{(P_n)_{\mathrm{SI}}} \mathbf{w}_{\mathrm{B}}^h \cdot \mathbf{n}_{\mathrm{B}} p_{\mathrm{B}}^h \mathrm{d}P + \int_{(P_n)_{\mathrm{SI}}} \mathbf{w}_{\mathrm{A}}^h \cdot \mathbf{n}_{\mathrm{A}} p_{\mathrm{B}}^h \mathrm{d}P
\end{aligned}
$$

$$
\begin{aligned}
&- \int_{(P_n)_{\text{SI}}} \left(\hat{\mathbf{N}}_{\text{B}} \cdot \left(\mathbf{w}_{\text{B}}^h - \mathbf{w}_{\text{A}}^h\right)\right) \cdot \left(\hat{\mathbf{n}}_{\text{B}} \cdot \mu \left(\boldsymbol{\varepsilon}(\mathbf{u}_{\text{B}}^h) + \boldsymbol{\varepsilon}(\mathbf{u}_{\text{A}}^h)\right)\right) \mathrm{d}P \\
&- \int_{(P_n)_{\text{SI}}} \left(\mathbf{T}_{\text{B}} \cdot \mathbf{w}_{\text{B}}^h\right) \cdot \left(\mathbf{n}_{\text{B}} \cdot 2\mu \boldsymbol{\varepsilon}(\mathbf{u}_{\text{B}}^h)\right) \mathrm{d}P \\
&- \int_{(P_n)_{\text{SI}}} \left(\mathbf{T}_{\text{A}} \cdot \mathbf{w}_{\text{A}}\right) \cdot \left(\mathbf{n}_{\text{A}} \cdot 2\mu \boldsymbol{\varepsilon}(\mathbf{u}_{\text{A}}^h)\right) \mathrm{d}P \\
&- \gamma \int_{(P_n)_{\text{SI}}} \left(\hat{\mathbf{n}}_{\text{B}} \cdot \mu \left(\boldsymbol{\varepsilon}\left(\mathbf{w}_{\text{B}}^h\right) + \boldsymbol{\varepsilon}\left(\mathbf{w}_{\text{A}}^h\right)\right)\right) \cdot \left(\hat{\mathbf{N}}_{\text{B}} \cdot \left(\mathbf{u}_{\text{B}}^h - \mathbf{u}_{\text{A}}^h\right)\right) \mathrm{d}P \\
&- \gamma \int_{(P_n)_{\text{SI}}} \left(\mathbf{n}_{\text{B}} \cdot 2\mu \boldsymbol{\varepsilon}\left(\mathbf{w}_{\text{B}}^h\right)\right) \cdot \left(\mathbf{T}_{\text{B}} \cdot \left(\mathbf{u}_{\text{B}}^h - \mathbf{u}_{\text{S}}^h\right)\right) \mathrm{d}P \\
&- \gamma \int_{(P_n)_{\text{SI}}} \left(\mathbf{n}_{\text{A}} \cdot 2\mu \boldsymbol{\varepsilon}\left(\mathbf{w}_{\text{A}}^h\right)\right) \cdot \left(\mathbf{T}_{\text{A}} \cdot \left(\mathbf{u}_{\text{A}}^h - \mathbf{u}_{\text{S}}^h\right)\right) \mathrm{d}P \\
&+ \int_{(P_n)_{\text{SI}}} \frac{\mu C}{h} \left(\hat{\mathbf{N}}_{\text{B}} \cdot \left(\mathbf{w}_{\text{B}}^h - \mathbf{w}_{\text{B}}^h\right)\right) \cdot \left(\mathbf{u}_{\text{B}}^h - \mathbf{u}_{\text{A}}^h\right) \mathrm{d}P \\
&+ \int_{(P_n)_{\text{SI}}} \frac{\mu C}{h_{\text{B}}} \left(\mathbf{T}_{\text{B}} \cdot \mathbf{w}_{\text{B}}^h\right) \cdot \left(\mathbf{u}_{\text{B}}^h - \mathbf{u}_{\text{S}}^h\right) \mathrm{d}P \\
&+ \int_{(P_n)_{\text{SI}}} \frac{\mu C}{h_{\text{A}}} \left(\mathbf{T}_{\text{A}} \cdot \mathbf{w}_{\text{A}}^h\right) \cdot \left(\mathbf{u}_{\text{A}}^h - \mathbf{u}_{\text{S}}^h\right) \mathrm{d}P, \qquad (9.11)
\end{aligned}
$$

where

$$\hat{\mathbf{N}}_{\text{B}} = \hat{\mathbf{n}}_{\text{B}} \hat{\mathbf{n}}_{\text{B}}, \qquad (9.12)$$

$$\mathbf{T}_{\text{B}} = \mathbf{I} - \mathbf{n}_{\text{B}} \mathbf{n}_{\text{B}}, \qquad (9.13)$$

$$\mathbf{T}_{\text{A}} = \mathbf{I} - \mathbf{n}_{\text{A}} \mathbf{n}_{\text{A}}. \qquad (9.14)$$

9.6 First 3D ST-SI/ST-IGA Computation: Aerodynamics of a Tsunami-Shelter Vertical-Axis Wind Turbine, 2015

The first 3D ST-SI/ST-IGA computation was reported in 2015 [54], for the aerodynamics of a tsunami-shelter vertical-axis wind turbine (VAWT). Figure 9.10 shows the VAWT. The model is based on the wind turbine in [158]. The computation was at constant wind speed and a prescribed rotor motion at constant angular velocity. Both the fluid–fluid SI and fluid–solid SI versions of the ST-SI were used in the computation, with the ST-VMS. The ST-IGA with IGA basis functions in time, more specifically quadratic NURBS basis functions in time, are used in representing the position of the mesh in the zone that rotates with the turbine rotor. There are three patches, 120° each, and the secondary mapping mentioned in Sect. 8.1

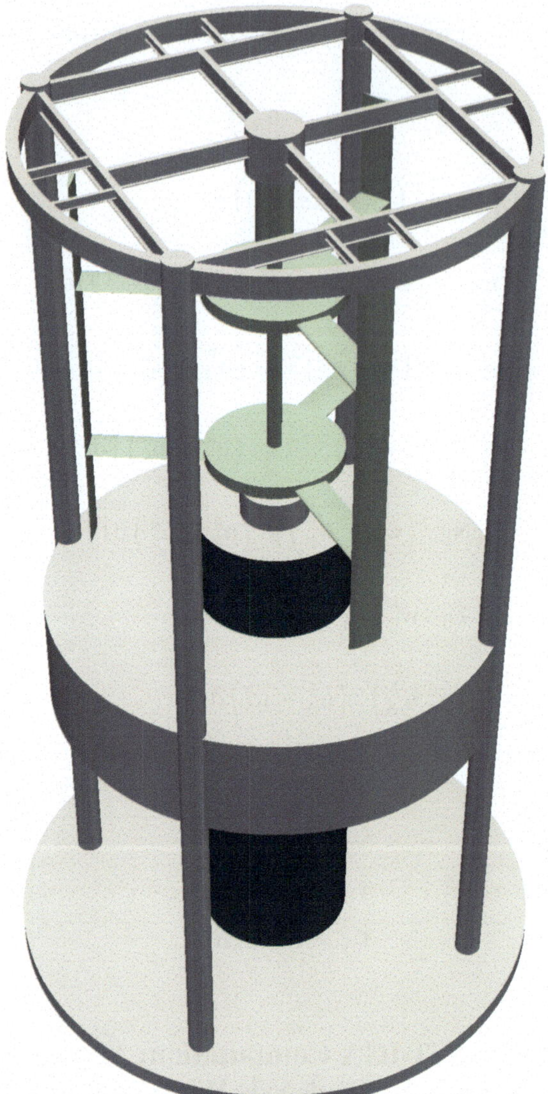

Fig. 9.10 3D ST-SI/ST-IGA computation of a tsunami-shelter VAWT. Model

is used to achieve the constant angular velocity. The mesh is made of four-node tetrahedral, five-node pyramid-shaped, and eight-node hexahedral elements, with an element count of 6,875,420. Figure 9.11 shows the mesh. The fluid velocity at the rotor surface is prescribed, with the fluid–solid SI, as a weakly-imposed Dirichlet boundary condition. Figure 9.12 shows the flow patterns.

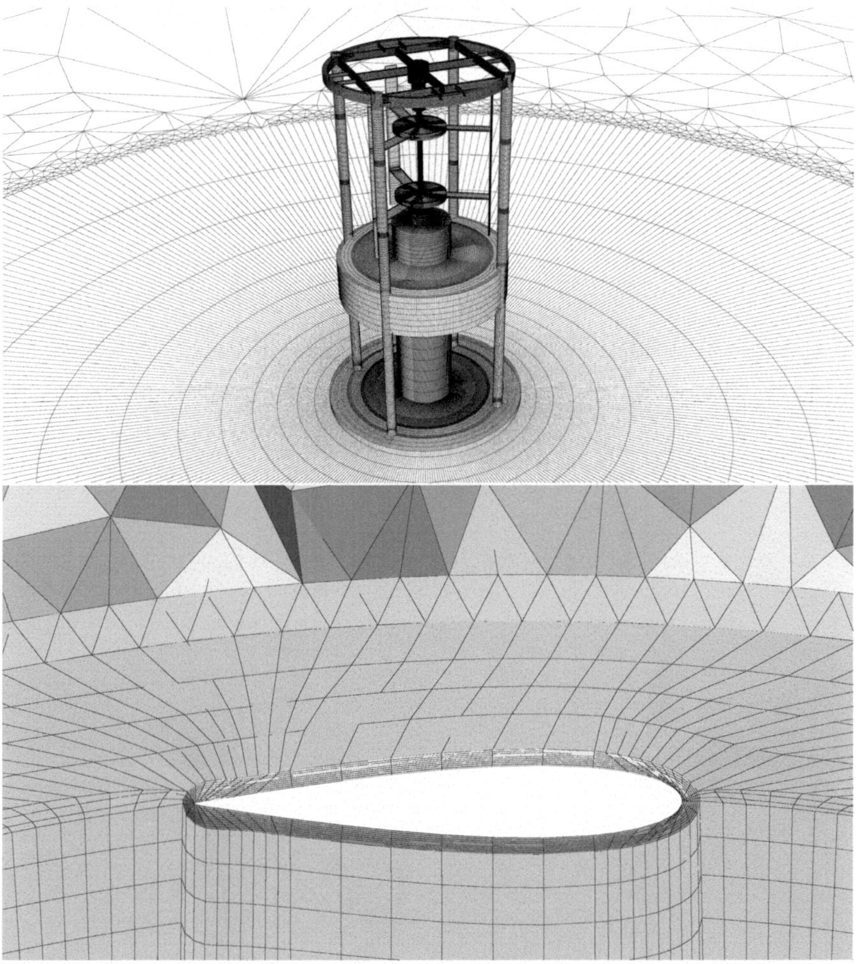

Fig. 9.11 3D ST-SI/ST-IGA computation of a tsunami-shelter VAWT. Mesh around the turbine and around a blade. The fluid–fluid SI is between the hexahedral and tetrahedral elements seen around the blade

9.7 Thermo-Fluid ST-SI Method, 2015

The thermo-fluid ST-SI was introduced in 2015 [55]. It is the ST-SI method associated with the thermo-fluid ST-VMS described in Sect. 7.17. It can also be seen as the thermo-fluid extension of the incompressible-flow ST-SI described in Sect. 9.5, and it is similarly multifunctional. When its function is to enable the use of the thermo-fluid ST-VMS as a moving-mesh method even in the presence of rotating solid surfaces, the thermo-fluid problem can be computed with high-resolution representation of the thermo-fluid boundary layers near the solid

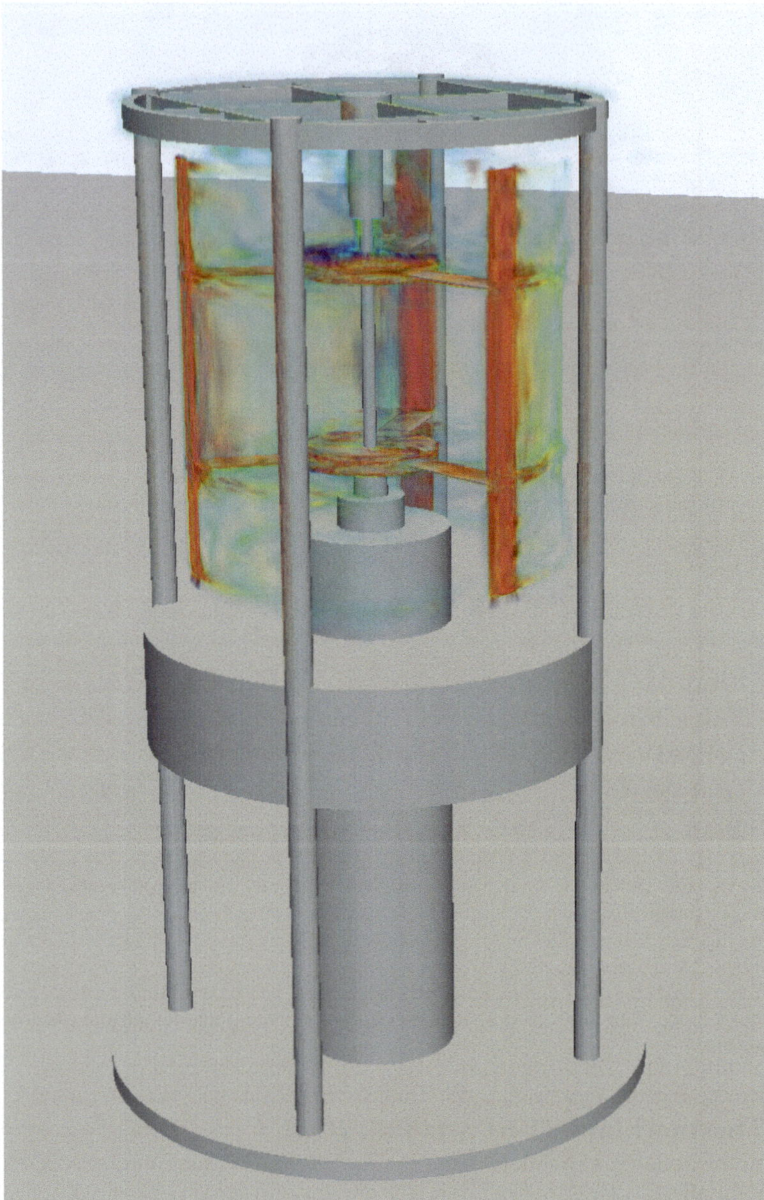

Fig. 9.12 3D ST-SI/ST-IGA computation of a tsunami-shelter VAWT. Vorticity magnitude. Color range from *blue* to *red* corresponds to the vorticity range from low to high. The flow is from *left* to *right*

surfaces. As in the incompressible-flow ST-SI, by using the ST-IGA with IGA basis functions in time, more specifically quadratic NURBS basis functions in time, we achieve exact representation of the solid-surface and mesh rotations. Also as in the incompressible-flow ST-SI, the synthesis of the ST-SI and ST-IGA with IGA basis functions in space meets the requirement of accurately connecting the parts of the solution obtained over two mesh zones with nonmatching meshes even when there are no rotating solid surfaces but we want mesh generation flexibility. In addition to the thermo-fluid ST-SI version with the fluid–fluid SI, the version with fluid–solid SI was introduced in [55], for weak enforcement of the thermo-fluid Dirichlet conditions. We provide here an overview of both versions.

9.7.1 Thermo-Fluid Fluid–Fluid SI

The integrals for Side A and Side B are first added separately, using the test functions $\mathbf{w}_{\mathrm{A}}^h$, q_{A}^h, $(w_\theta)_{\mathrm{A}}^h$ and $\mathbf{w}_{\mathrm{B}}^h$, q_{B}^h, $(w_\theta)_{\mathrm{B}}^h$. Then, putting together the integrals added for each side, the complete set of integrals added to the left-hand side of Eq. (7.6) becomes

$$
\begin{aligned}
&-\int_{(P_n)_{\mathrm{SI}}} \left(q_{\mathrm{B}}^h \mathbf{n}_{\mathrm{B}} - q_{\mathrm{A}}^h \mathbf{n}_{\mathrm{A}}\right) \cdot \frac{1}{2}\left(\mathbf{u}_{\mathrm{B}}^h - \mathbf{u}_{\mathrm{A}}^h\right) \mathrm{d}P \\
&-\int_{(P_n)_{\mathrm{SI}}} \frac{1}{2}\rho \left(\mathbf{w}_{\mathrm{B}}^h \left(\mathcal{F}_{\mathrm{B}}^h - \left|\mathcal{F}_{\mathrm{B}}^h\right|\right) - \mathbf{w}_{\mathrm{A}}^h \left(\mathcal{F}_{\mathrm{A}}^h - \left|\mathcal{F}_{\mathrm{A}}^h\right|\right)\right) \\
&\qquad\qquad \cdot \left(\mathbf{u}_{\mathrm{B}}^h - \mathbf{u}_{\mathrm{A}}^h\right) \mathrm{d}P \\
&-\int_{(P_n)_{\mathrm{SI}}} \frac{1}{2}\rho C_{\mathrm{p}} \left((w_\theta)_{\mathrm{B}}^h \left(\mathcal{F}_{\mathrm{B}}^h - \left|\mathcal{F}_{\mathrm{B}}^h\right|\right)\right. \\
&\qquad\qquad \left. - (w_\theta)_{\mathrm{A}}^h \left(\mathcal{F}_{\mathrm{A}}^h - \left|\mathcal{F}_{\mathrm{A}}^h\right|\right)\right) \left(\theta_{\mathrm{B}}^h - \theta_{\mathrm{A}}^h\right) \mathrm{d}P \\
&+\int_{(P_n)_{\mathrm{SI}}} \left(\mathbf{n}_{\mathrm{B}} \cdot \mathbf{w}_{\mathrm{B}}^h + \mathbf{n}_{\mathrm{A}} \cdot \mathbf{w}_{\mathrm{A}}^h\right) \frac{1}{2}\left(p_{\mathrm{B}}^h + p_{\mathrm{A}}^h\right) \mathrm{d}P \\
&-\int_{(P_n)_{\mathrm{SI}}} \left(\mathbf{w}_{\mathrm{B}}^h - \mathbf{w}_{\mathrm{A}}^h\right) \cdot \left(\hat{\mathbf{n}}_{\mathrm{B}} \cdot \mu \left(\boldsymbol{\varepsilon}(\mathbf{u}_{\mathrm{B}}^h) + \boldsymbol{\varepsilon}(\mathbf{u}_{\mathrm{A}}^h)\right)\right) \mathrm{d}P \\
&-\gamma \int_{(P_n)_{\mathrm{SI}}} \hat{\mathbf{n}}_{\mathrm{B}} \cdot \mu \left(\boldsymbol{\varepsilon}\left(\mathbf{w}_{\mathrm{B}}^h\right) + \boldsymbol{\varepsilon}\left(\mathbf{w}_{\mathrm{A}}^h\right)\right) \cdot \left(\mathbf{u}_{\mathrm{B}}^h - \mathbf{u}_{\mathrm{A}}^h\right) \mathrm{d}P \\
&+\int_{(P_n)_{\mathrm{SI}}} \frac{\mu C}{h} \left(\mathbf{w}_{\mathrm{B}}^h - \mathbf{w}_{\mathrm{A}}^h\right) \cdot \left(\mathbf{u}_{\mathrm{B}}^h - \mathbf{u}_{\mathrm{A}}^h\right) \mathrm{d}P \\
&-\int_{(P_n)_{\mathrm{SI}}} \left((w_\theta)_{\mathrm{B}}^h - (w_\theta)_{\mathrm{A}}^h\right) \left(\hat{\mathbf{n}}_{\mathrm{B}} \cdot \frac{\kappa}{2}\left(\boldsymbol{\nabla}\theta_{\mathrm{B}}^h + \boldsymbol{\nabla}\theta_{\mathrm{A}}^h\right)\right) \mathrm{d}P
\end{aligned}
$$

$$
\begin{aligned}
& - \gamma_\theta \int_{(P_n)_{\mathrm{SI}}} \hat{\mathbf{n}}_{\mathrm{B}} \cdot \frac{\kappa}{2} \left(\boldsymbol{\nabla} \left(w_\theta \right)_{\mathrm{B}}^h + \boldsymbol{\nabla} \left(w_\theta \right)_{\mathrm{A}}^h \right) \left(\theta_{\mathrm{B}}^h - \theta_{\mathrm{A}}^h \right) \mathrm{d}P \\
& + \int_{(P_n)_{\mathrm{SI}}} \frac{\kappa C_\theta}{h} \left(\left(w_\theta \right)_{\mathrm{B}}^h - \left(w_\theta \right)_{\mathrm{A}}^h \right) \left(\theta_{\mathrm{B}}^h - \theta_{\mathrm{A}}^h \right) \mathrm{d}P,
\end{aligned} \tag{9.15}
$$

where $\gamma_\theta = 1$ and C_θ is a nondimensional constant.

9.7.2 Thermo-Fluid Fluid–Solid SI

The thermo-fluid ST-SI with the fluid–solid SI is obtained (see [55]) by starting with the integrals added for Side B and replacing the Side A velocity and temperature with the velocity $\mathbf{g}^h$ and temperature g_θ^h coming from the solid surface. Then the SI integrals added to the left-hand side of Eq. (7.6) to represent the weakly-enforced Dirichlet conditions become

$$
\begin{aligned}
& - \int_{(P_n)_{\mathrm{SI}}} q_{\mathrm{B}}^h \mathbf{n}_{\mathrm{B}} \cdot \left(\mathbf{u}_{\mathrm{B}}^h - \mathbf{g}^h \right) \mathrm{d}P \\
& - \int_{(P_n)_{\mathrm{SI}}} \rho \mathbf{w}_{\mathrm{B}}^h \cdot \frac{1}{2} \left(\mathcal{F}_{\mathrm{B}}^h - \left| \mathcal{F}_{\mathrm{B}}^h \right| \right) \left(\mathbf{u}_{\mathrm{B}}^h - \mathbf{g}^h \right) \mathrm{d}P \\
& - \int_{(P_n)_{\mathrm{SI}}} \rho C_{\mathrm{p}} \left(w_\theta \right)_{\mathrm{B}}^h \frac{1}{2} \left(\mathcal{F}_{\mathrm{B}}^h - \left| \mathcal{F}_{\mathrm{B}}^h \right| \right) \left(\theta_{\mathrm{B}}^h - g_\theta^h \right) \mathrm{d}P \\
& - \int_{(P_n)_{\mathrm{SI}}} \mathbf{w}_{\mathrm{B}}^h \cdot \left(\mathbf{n}_{\mathrm{B}} \cdot \boldsymbol{\sigma}_{\mathrm{B}}^h \right) \mathrm{d}P \\
& - \gamma \int_{(P_n)_{\mathrm{SI}}} \mathbf{n}_{\mathrm{B}} \cdot 2\mu \boldsymbol{\varepsilon} \left(\mathbf{w}_{\mathrm{B}}^h \right) \cdot \left(\mathbf{u}_{\mathrm{B}}^h - \mathbf{g}^h \right) \mathrm{d}P \\
& + \int_{(P_n)_{\mathrm{SI}}} \frac{\mu C}{h_{\mathrm{B}}} \mathbf{w}_{\mathrm{B}}^h \cdot \left(\mathbf{u}_{\mathrm{B}}^h - \mathbf{g}^h \right) \mathrm{d}P \\
& - \int_{(P_n)_{\mathrm{SI}}} \left(w_\theta \right)_{\mathrm{B}}^h \left(\mathbf{n}_{\mathrm{B}} \cdot \kappa \boldsymbol{\nabla} \theta_{\mathrm{B}}^h \right) \mathrm{d}P \\
& - \gamma_\theta \int_{(P_n)_{\mathrm{SI}}} \mathbf{n}_{\mathrm{B}} \cdot \kappa \boldsymbol{\nabla} \left(w_\theta \right)_{\mathrm{B}}^h \left(\theta_{\mathrm{B}}^h - g_\theta^h \right) \mathrm{d}P \\
& + \int_{(P_n)_{\mathrm{SI}}} \frac{\kappa C_\theta}{h_{\mathrm{B}}} \left(w_\theta \right)_{\mathrm{B}}^h \left(\theta_{\mathrm{B}}^h - g_\theta^h \right) \mathrm{d}P.
\end{aligned} \tag{9.16}
$$

9.8 First 3D Thermo-Fluid ST-SI Computation: Thermo-Fluid Analysis of a Disk Brake, 2015

The first 3D thermo-fluid ST-SI computation was reported in 2015 [55], for the thermo-fluid analysis of a disk brake. Figure 9.13 shows the disk brake model. The rate of heat generation from the brake-pad friction is given as $\frac{\mathrm{d}}{\mathrm{d}t}(\frac{1}{2}I\omega^2)$, where I is the moment of inertia and ω is the rotation speed. The computation was with the thermo-fluid ST-VMS, at three rotation speeds in the ramp-down profile of the disk spin, from the start of the breaking until the spinning stops. The mesh is made of 26,138,697 four-node tetrahedral elements. Figures 9.14 and 9.15 show the flow temperature.

Fig. 9.13 3D thermo-fluid ST-SI computation of a disk brake. Model. The disk design includes fins. The *green* arrow indicates the disk rotation direction, and the *blue* arrow indicates the gravity and external-flow directions. The external-flow direction was based on an experimental setup

Fig. 9.14 3D thermo-fluid ST-SI computation of a disk brake. Mesh and flow temperature on a plane perpendicular to the disk axis and bisecting the fins, at an instant during the computation for the full rotation speed. The *red* line is the SI

9.9 First 3D ST-VMS/ST-SI/ST-IGA Computation of Fluid–Cable Interaction with One-Way-Dependence: Flow-Driven String Dynamics in Turbomachinery, 2015

The first 3D ST-VMS/ST-SI/ST-IGA computation of fluid–cable interaction with one-way-dependence was reported in 2015 [159], for flow-driven string dynamics in turbomachinery. To understand how strings get stuck on or around the blades in turbomachinery, a string is dropped into a fan placed horizontally. Figure 9.16 shows the fan model. By using the ST-IGA with IGA basis functions in time, more specifically quadratic NURBS basis functions in time, we achieve exact representation of the rotor and mesh motions. The one-way-dependence model was described in Sect. 6.9. In the simulation here based on that model, as the flow field is computed, the time-dependent data obtained is stored by using the ST-C, with cubic B-splines in time. Then, using that data, flow-driven dynamics of the string is computed. Contact between the string and solid surfaces is handled with the SENCT-FC. The fluid mechanics mesh is made of 10,726,390 four-node tetrahedral elements. The string mesh is made of 50 bending-stabilized cable elements [160], using the IGA with cubic NURBS basis functions. The IGA gives us not only a higher-order method and smoothness in the structure shape, but also smoothness in the fluid dynamics forces acting on the string. Figure 9.17 shows the flow velocity and string position at the beginning of the string dynamics computation. Figure 9.18 shows the string at three instants.

Fig. 9.15 3D thermo-fluid ST-SI computation of a disk brake. Flow temperature patterns at an instant during the computation for the full rotation speed

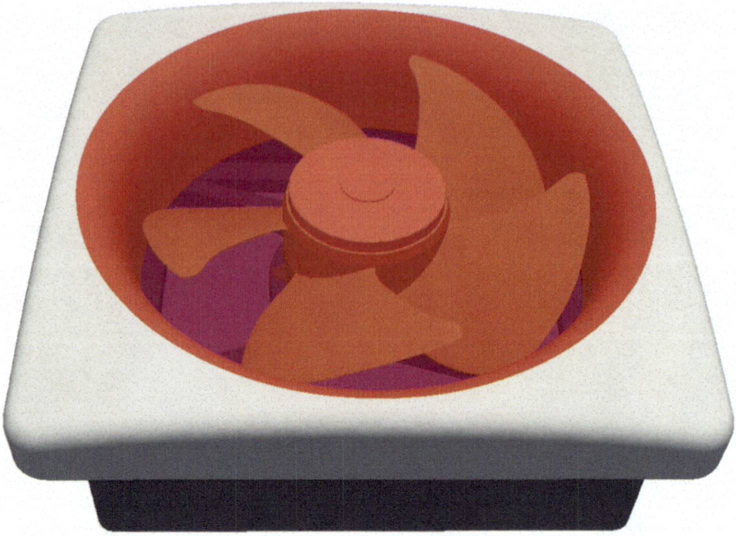

Fig. 9.16 3D ST-VMS/ST-SI/ST-IGA computation of fluid–cable interaction with one-way-dependence. Flow-driven string dynamics in a fan. Fan model. The *red* surfaces enclose the mesh spinning with the rotor

9.10 Synthesis of the ST-SI and ST-IGA with IGA Basis Functions in Space: ST-SI-IGA, 2015

The ST-SI-IGA, introduced in 2015 [51], is the synthesis of the ST-SI and ST-IGA with IGA basis functions in space. It is a multifunction method, accurately connecting the parts of the solution obtained over two IGA mesh zones with nonmatching meshes at the SI between the zones. In the original functionality of the ST-SI, the ST-SI-IGA enables the use of the core ST methods as moving-mesh methods even in the presence of rotating solid surfaces, such as a turbine rotor, so that the flow can be computed with high-resolution IGA representation of the boundary layers near the solid surfaces. Sometimes we have two IGA mesh zones with nonmatching meshes at the SI between the zones not because we have rotating solid surfaces requiring slip between the two IGA mesh zones but because we just want to be free from the matching requirement. This IGA mesh generation flexibility helps us address the IGA mesh generation challenge for complex geometries. When complex-geometry flow problems are computed with the ST-IGA with IGA basis functions in space, we are likely to have multiple SIs in the computation, some for the rotating surfaces and some for the mesh generation flexibility. That was the case, for example, in the turbocharger turbine computation reported in [51].

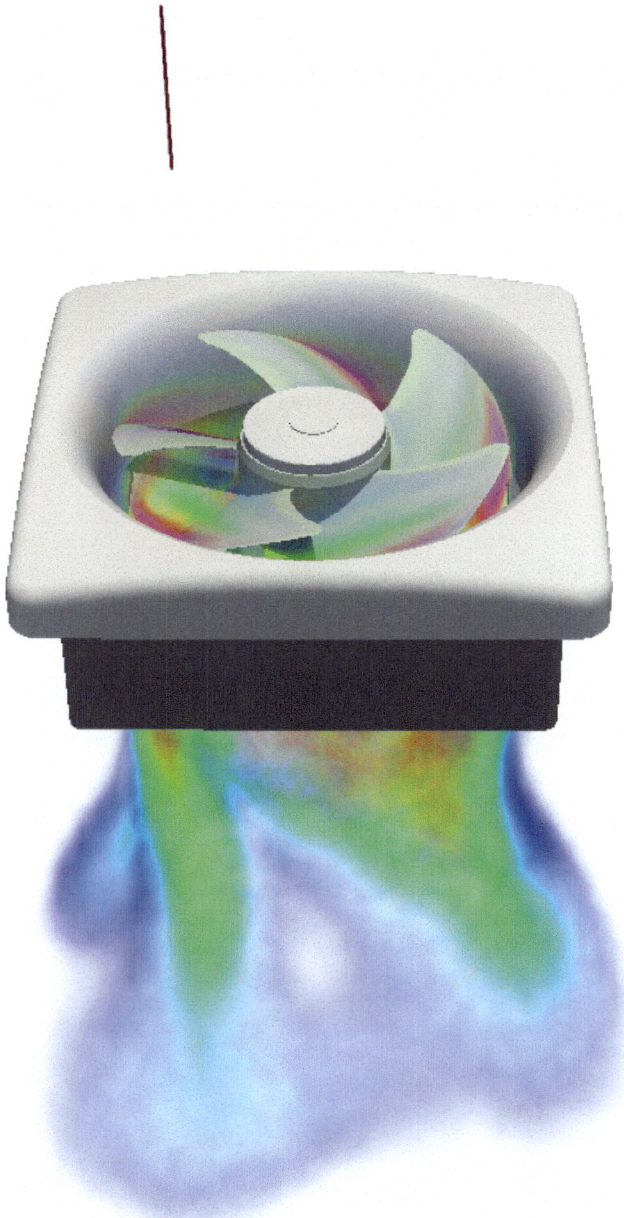

Fig. 9.17 3D ST-VMS/ST-SI/ST-IGA computation of fluid–cable interaction with one-way-dependence. Flow-driven string dynamics in a fan. Flow-velocity magnitude and string position at the beginning of the string dynamics computation. Color range from *blue* to *red* corresponds to the velocity range from low to high

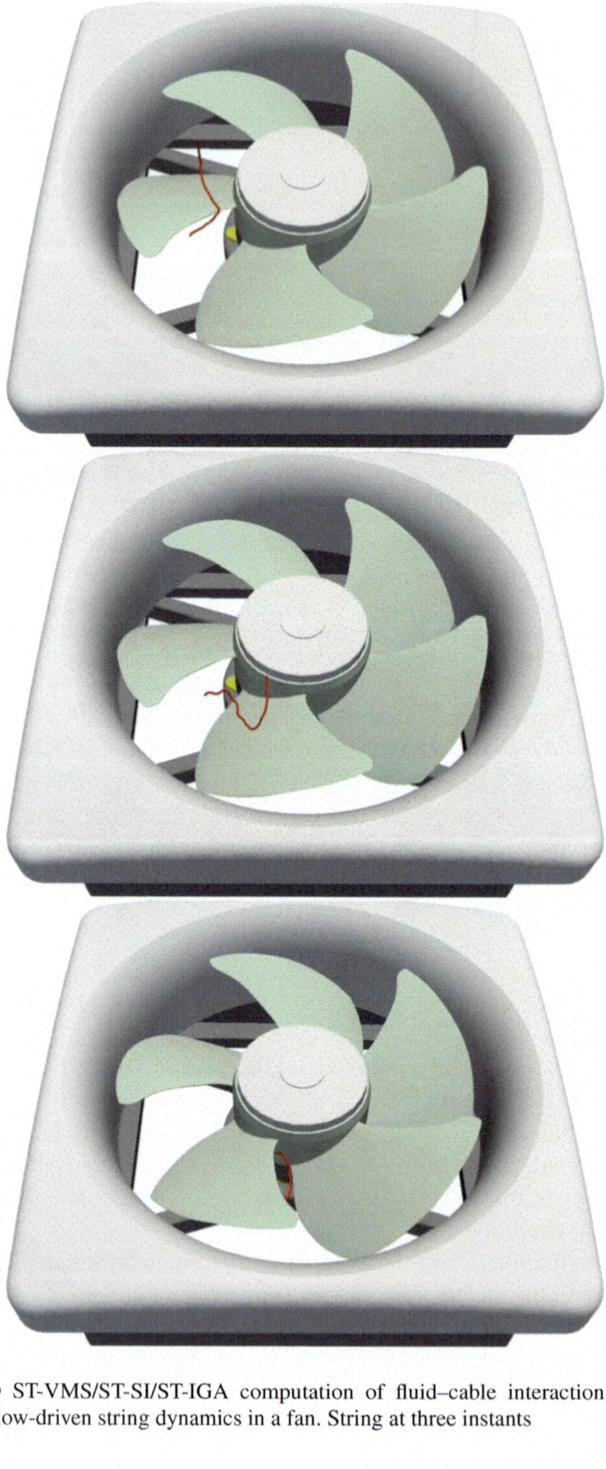

Fig. 9.18 3D ST-VMS/ST-SI/ST-IGA computation of fluid–cable interaction with one-way-dependence. Flow-driven string dynamics in a fan. String at three instants

9.11 First 3D Computation with the ST-SI-IGA: Flow Analysis of a Turbocharger Turbine, 2015

The first 3D computation with the ST-SI-IGA was reported in 2015 [51], in the same article where the ST-SI-IGA was introduced. The computation was for flow analysis of a turbocharger turbine. The mesh is made of quadratic NURBS elements. Figure 9.19 shows the turbine geometry and the NURBS control mesh. The computation was with the ST-VMS. With the ST-IGA with IGA basis functions in time, more specifically quadratic NURBS basis functions in time, we achieve exact representation of the rotor and mesh motions. Figure 9.20 shows the flow velocity.

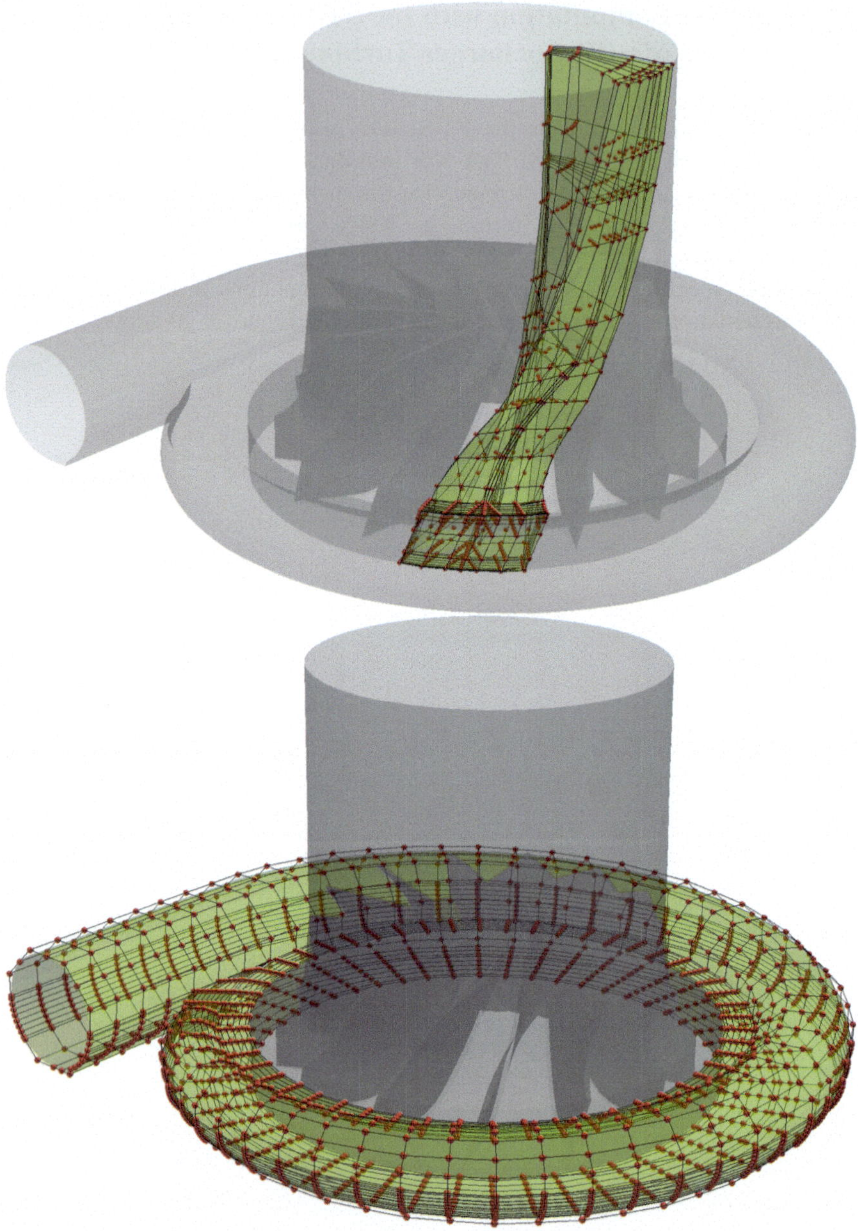

Fig. 9.19 3D ST-SI-IGA computation of flow in a turbocharger turbine. Turbine geometry and the NURBS control mesh for the rotor and volute

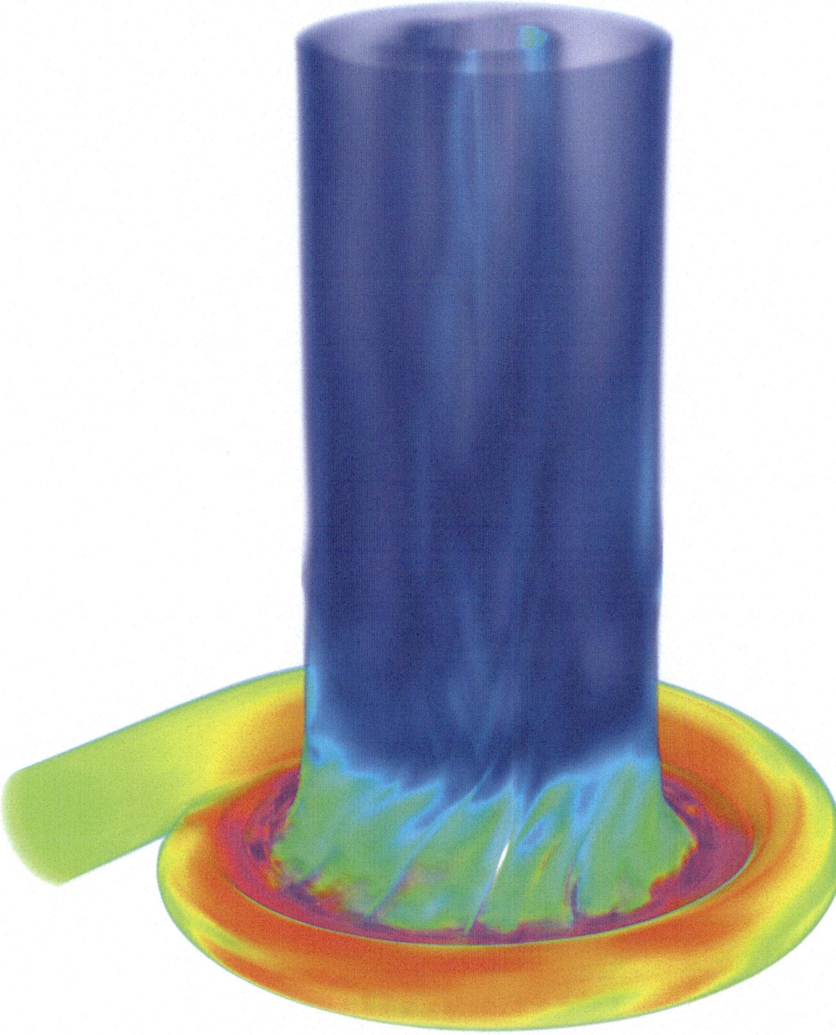

Fig. 9.20 3D ST-SI-IGA computation of flow in a turbocharger turbine. Velocity magnitude. Color range from *blue* to *red* corresponds to the velocity range from low to high

Chapter 10
Methods and Solutions in 2016–2018

10.1 First 3D Ram-Air Parachute Aerodynamics Computations with the ST-IGA with IGA Basis Functions in Space, 2016

The first 3D ram-air parachute aerodynamics computations with the ST-IGA with IGA basis functions in space were reported in 2016 [161] and included the first ST-SI-IGA computation with porosity SI, for a case with parachute fabric porosity. The aerodynamics computations reported in [161] were preceded by a shape-determination structural mechanics computation, also with IGA discretization in space, using membrane and cable elements. The parachute shape shown in Fig. 10.1 is from that computation. The aerodynamics computations were with the ST-VMS, for angle of attack values ranging from -2° to 12° in increments of 2°. The mesh, generated both inside and outside the parachute, is made of 233,378 quadratic NURBS elements. Using NURBS meshes addresses the challenge of mesh generation in the narrowing trailing-edge space and keeps the element density there at a reasonable level. Figures 10.2 and 10.3 show, from the computation with parachute fabric porosity and 0° angle of attack, the vorticity magnitude and the normal component of the velocity.

10.2 Synthesis of the ST-SI and ST-TC: ST-SI-TC, 2016

The ST-SI-TC, introduced in 2016 [162], is the synthesis of the ST-SI and ST-TC. Some classes of problems will need both the ST-SI and ST-TC for computation with actual-contact and high-resolution boundary-layer representations. In the ST-SI version with the fluid–fluid SI, we need elements on both sides of the SI. The SI might be between the solid surfaces coming into contact, or in a more general

T. E. Tezduyar, K. Takizawa, *Space–Time Computational Flow Analysis*,
Modeling and Simulation in Science, Engineering and Technology,
https://doi.org/10.1007/978-3-031-88727-7_10

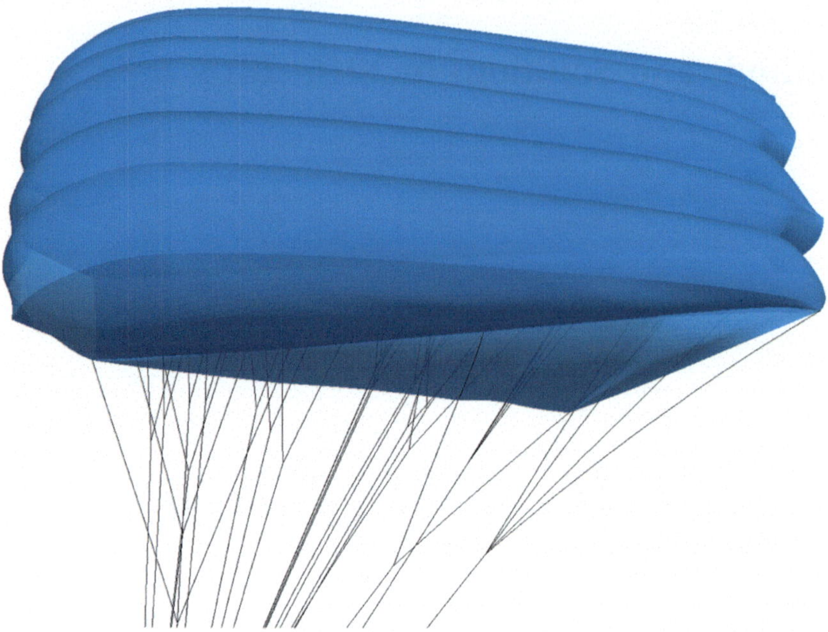

Fig. 10.1 3D ram-air parachute aerodynamics computations with the ST-IGA with IGA basis functions in space. Parachute shape from the structural mechanics computation

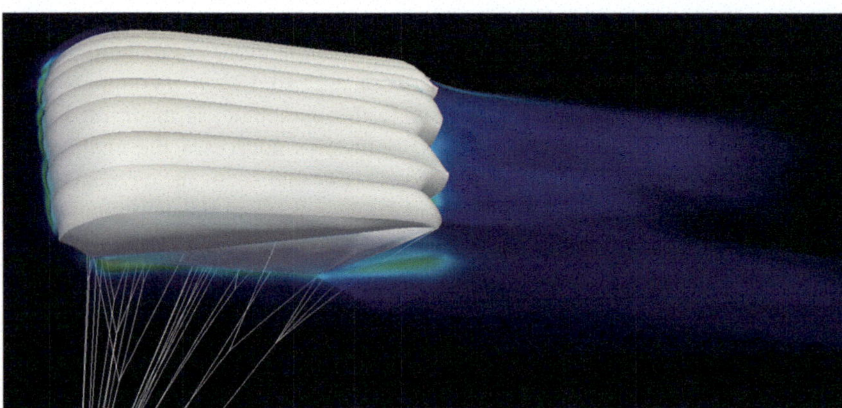

Fig. 10.2 3D ram-air parachute aerodynamics computations with the ST-IGA with IGA basis functions in space. Computation with parachute fabric porosity and 0° angle of attack. Vorticity magnitude. Color range from *blue* to *red* corresponds to the vorticity range from low to high

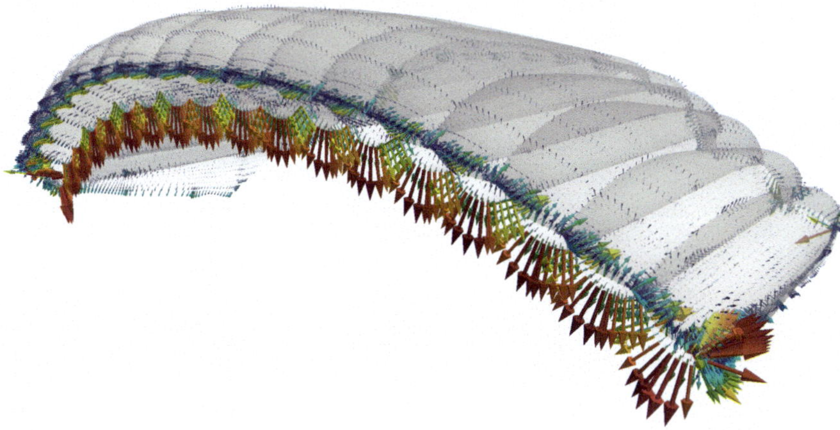

Fig. 10.3 3D ram-air parachute aerodynamics computations with the ST-IGA with IGA basis functions in space. Computation with parachute fabric porosity and 0° angle of attack. Normal component of the velocity. Color range from *blue* to *red* corresponds to the velocity range from low to high

context, might be merging with a fluid–solid interface. The elements between the solid surface and the coinciding SI segment collapse in the ST-TC process, and the SI segment switches from the fluid–fluid SI to the fluid–solid SI, creating an SI that is a mixture of the two SI types. With the ST-SI-TC, (i) the ST-TC element collapse and rebirth process gains independence from the solid-surface nodes, (ii) high-resolution boundary-layer representation near the fluid–solid interfaces is enabled even when an SI segment coincides with the solid surface, and (iii) effective management of contact location change and contact sliding is enabled (see Fig. 10.4).

10.3 First 3D Computation with the ST-SI-TC: Tire Aerodynamics with Transverse Grooves, Road Contact, Tire Deformation, and Actual-Contact and High-Resolution Boundary-Layer Representations, 2016

The first 3D computation with the ST-SI-TC was reported in 2016 [162], in the same article where the ST-SI-TC was introduced. The computation was for tire aerodynamics with transverse grooves, road contact, tire deformation, and actual-contact and high-resolution boundary-layer representations. The geometry of the model, which has 12 treads and an invariant cross-section, and the deformation pattern are shown in Fig. 10.5. Figure 10.6 shows the computational domain,

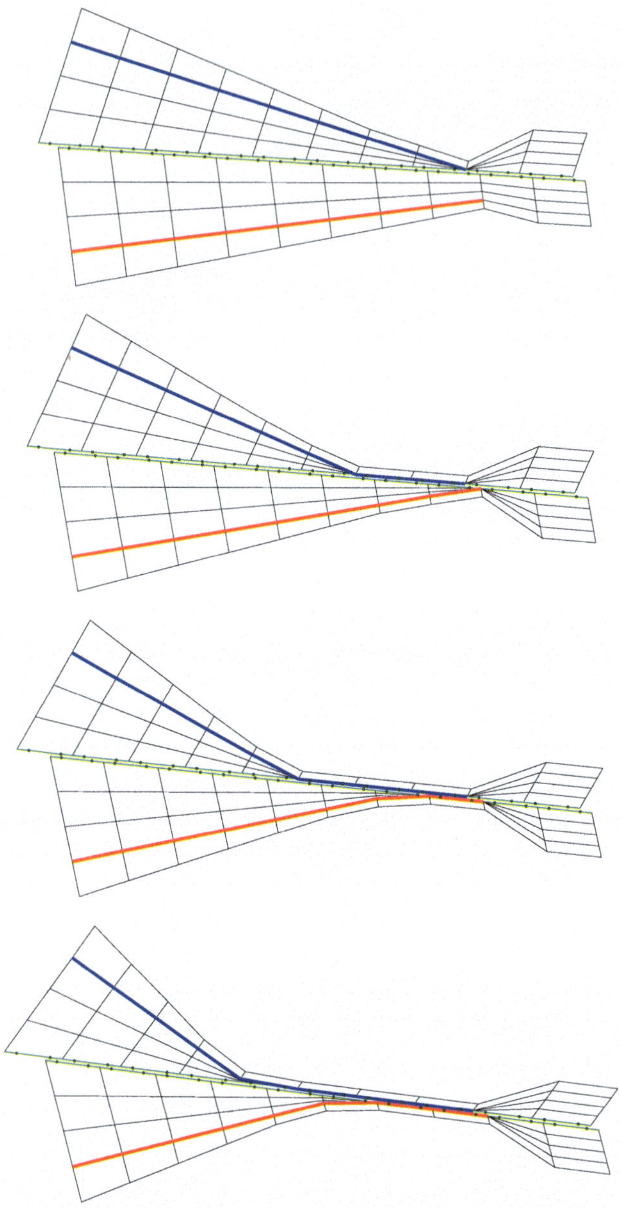

Fig. 10.4 ST-SI-TC contact and sliding. Illustration of how the method works in dealing with two bars (*blue* and *red*) contacting, deforming, and sliding on each other, in two mesh zones that start with a fluid–fluid SI between them. The elements near the bars are shown. The two *green* lines are the two sides of the SI. The circles are the surface-integration points. When a fluid–fluid SI merges with a fluid–solid interface, it switches to the fluid–solid SI, creating an SI that is a mixture of the two SI types

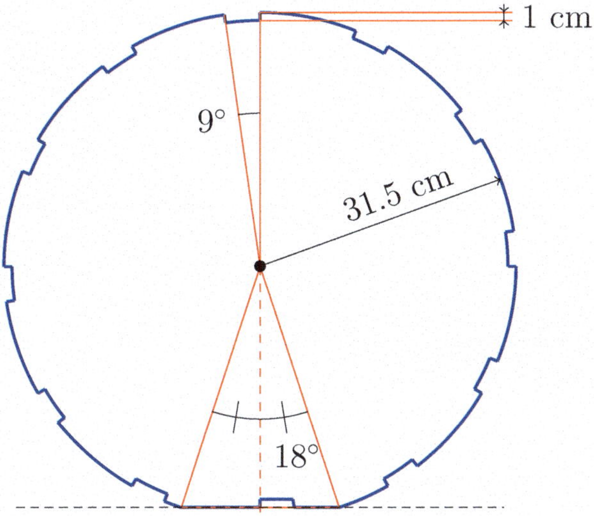

Fig. 10.5 3D ST-SI-TC computation of tire aerodynamics with transverse grooves, road contact, tire deformation, and actual-contact and high-resolution boundary-layer representations. Model cross-section, which is invariant. The model dimension in the axial direction is 27.5 cm

Fig. 10.6 3D ST-SI-TC computation of tire aerodynamics with transverse grooves, road contact, tire deformation, and actual-contact and high-resolution boundary-layer representations. Boundary conditions. The *left* (*red*) and *bottom* (*yellow*) boundaries represent the inflow and road, where the velocity is set to the linear speed at the tire periphery. The innermost (*blue*) circle is the tire surface, which is rotating. The larger (*pink*) circle is the SI. The bottom of the SI is coinciding with the road and the tire–road interface. The conditions at the *right* (*green*) and *top* (*cyan*) boundaries are zero-stress and slip, respectively. The condition at the two side boundaries is also slip

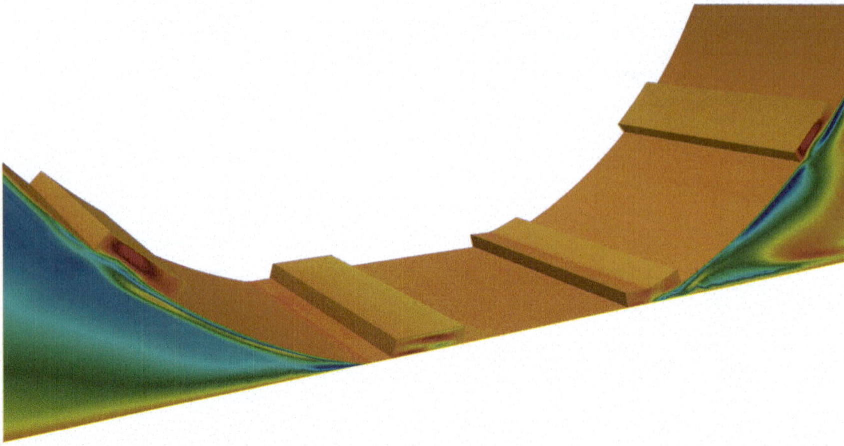

Fig. 10.7 3D ST-SI-TC computation of tire aerodynamics with transverse grooves, road contact, tire deformation, and actual-contact and high-resolution boundary-layer representations. Velocity magnitude. Color range from *blue* to *red* corresponds to the velocity range from low to high

boundary conditions, and the SI. The computation was with the ST-SUPS. The inner mesh around the tire is made of eight-node hexahedral elements, generated manually, rotated with the tire, and connected with an SI to the outer mesh. The outer mesh has a first layer of eight-node hexahedral elements, generated manually, next a layer of five-node pyramid elements, also generated manually, and after that four-node tetrahedral elements, generated automatically. The total element count is 5,415,467. Figure 10.7 shows the flow velocity.

10.4 Synthesis of the ST-SI-TC and ST-IGA: ST-SI-TC-IGA, 2016

The ST-SI-TC-IGA, introduced in 2016 [56], is the synthesis of the ST-SI-TC and ST-IGA. It increases the scope and accuracy in computing with actual-contact and high-resolution boundary-layer representations. (i) The classes of problems requiring computation with the ST-SI-TC quite often involve narrow and curved gaps between solid surfaces. The ST-SI-TC-IGA keeps element density required for good accuracy at a reasonable level. That in turn keeps the computational cost at a reasonable level. (ii) Computations with the ST-SI-TC also quite often involve elements with high aspect ratios. The ST-SI-TC-IGA brings more robustness to computations with such elements. (iii) With the special feature introduced in 2018 [163], the ST-SI-TC-IGA has a built-in Reynolds-equation limit. Computations involving fluid films between solid surfaces coming into contact can be carried out without the need to use separately a Reynolds-equation model. The ST-SI-TC-

IGA can emulate the Reynolds-equation model with a comparable solution quality and computational cost, working also in regions of the problem domain beyond the scope of the Reynolds-equation model.

10.5 First 3D Computations with the ST-SI-TC-IGA: Aortic Valve Flow Analysis with a Realistic Model Geometry, Leaflet Contact, and Actual-Contact and High-Resolution Boundary-Layer Representations, 2016

The first 3D computations with the ST-SI-TC-IGA were reported in 2016 [56], in the same article where the ST-SI-TC-IGA was introduced. To illustrate how the ST-SI-TC-IGA works, we show with a 2D example how we deal with two valve leaflets flapping and coming into contact in an asymmetric fashion. Figure 10.8 shows the mesh motion, and Fig. 10.9 shows the corresponding control mesh. The computations were for aortic valve flow analysis with a realistic model geometry, leaflet contact, and actual-contact and high-resolution boundary-layer representations. The model, shown in Fig. 10.10, has three leaflets and one main outlet, corresponding to the beginning of the aorta. The leaflet motion is prescribed in the computation. The computations reported in [56] were for two cases of leaflet motion: symmetric and asymmetric. In the symmetric case, the leaflet motions has 3-fold rotational periodicity, and in the asymmetric case, they do not. We summarize the asymmetric case here. The leaflets have the same closing motion but have a small time lag between them as they open. The computation was with the ST-SUPS and IGA discretization in space. The mesh is made of 54,000 quadratic NURBS elements, created with five SIs, which are also shown in Fig. 10.10. Three of the SIs connect the mesh sectors containing the leaflets in the valve region. The other two, the top and bottom circular planes in the figure, are for mesh generation flexibility. The mesh moving is with a special-purpose mesh moving method. Figure 10.11 shows the mesh at three instants. Figure 10.12 shows the flow patterns. Figure 10.13 shows the WSS on the lower leaflet surfaces.

10.6 Compressible-Flow ST-SI Method, 2017

The compressible-flow ST-SI was introduced in 2017 [164]. It is the ST-SI method associated with the compressible-flow ST-SUPG. As the incompressible-flow SI described in Sect. 9.5, it is multifunctional. When its function is to enable the use of the compressible-flow ST-SUPG as a moving-mesh method even in the presence of rotating solid surfaces, the compressible-flow problem can be computed with high-resolution representation of the boundary layers near the solid surfaces. As

Fig. 10.8 3D ST-SI-TC-IGA computation of aortic valve flow with a realistic model geometry, leaflet contact, and actual-contact and high-resolution boundary-layer representations. A 2D illustration of how the ST-SI-TC-IGA works. An example of contact and sliding (element representation). Element colors indicate the NURBS patches the elements belong to. The thick *red* curves are the leaflets. The two *green* vertical lines, with the narrow space between them added only for illustration purposes, represent the two sides of an SI

Fig. 10.9 3D ST-SI-TC-IGA computation of aortic valve flow with a realistic model geometry, leaflet contact, and actual-contact and high-resolution boundary-layer representations. A 2D illustration of how the ST-SI-TC-IGA works. An example of contact and sliding (control mesh). Shading colors indicate the NURBS patches. The *black* curves, which are actually collections of line segments, represent the control-mesh surfaces corresponding to the leaflets. The control points on the two sides of the narrow vertical gap correspond to the two sides of the SI. When control points coincide, they are in a master–servant relationship

Fig. 10.10 3D ST-SI-TC-IGA computation of aortic valve flow with a realistic model geometry, leaflet contact, and actual-contact and high-resolution boundary-layer representations. Aorta, leaflets, sinuses, and the five SIs

in the incompressible-flow ST-SI, by using the ST-IGA with IGA basis functions in time, more specifically quadratic NURBS basis functions in time, we achieve exact representation of the solid-surface and mesh rotations. Also as in the incompressible-flow ST-SI, the synthesis of the ST-SI and ST-IGA with IGA basis functions in space meets the requirement of accurately connecting the parts of the solution obtained over two mesh zones with nonmatching meshes even when there are no rotating solid surfaces but we want mesh generation flexibility. In addition to the compressible-flow ST-SI version with the fluid–fluid SI, the version with fluid–solid SI was introduced in [164], for weak enforcement of the Dirichlet conditions. That has two categories: weak enforcement for flow velocity and temperature conditions and weak enforcement for flow velocity and adiabatic conditions. The compressible-flow ST-SI version with "porosity SI," also introduced in [164], like its incompressible-flow predecessor, is for a thin porous structure with fluid on its two sides. It is with weakly-enforced flow velocity and adiabatic conditions.

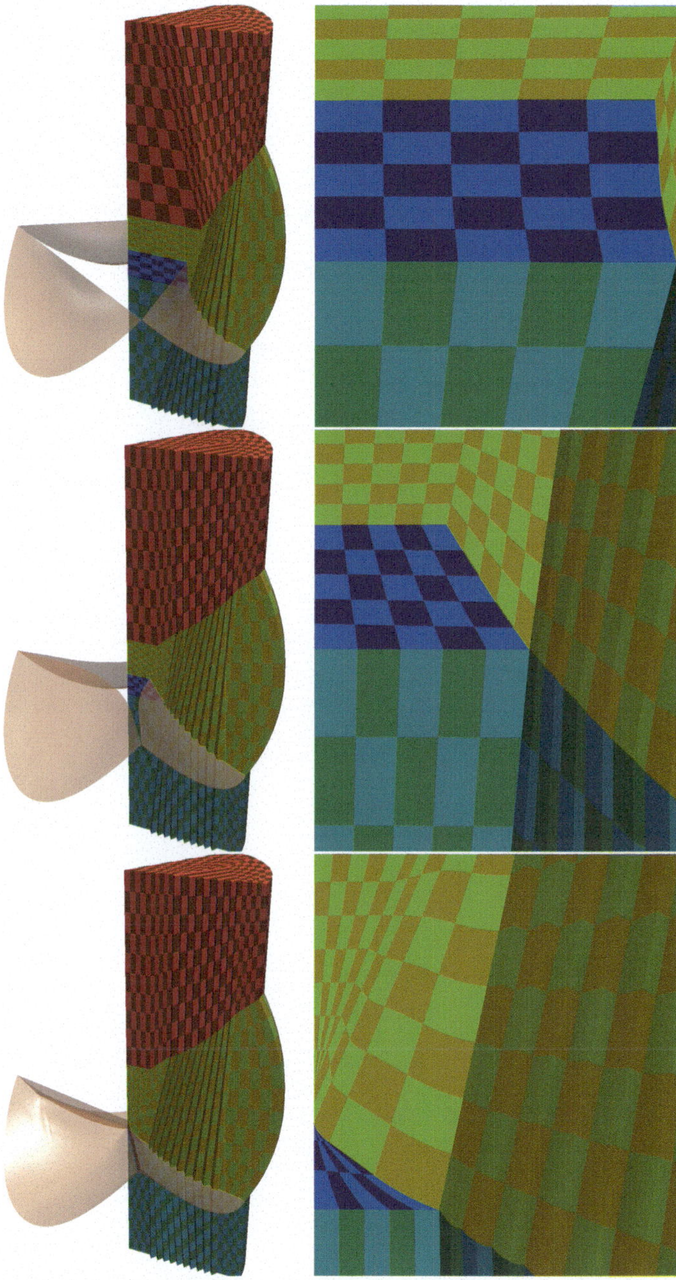

Fig. 10.11 3D ST-SI-TC-IGA computation of aortic valve flow with a realistic model geometry, leaflet contact, and actual-contact and high-resolution boundary-layer representations. The mesh at three instants, from when the valve is fully open (*top*) to when it is fully closed (*bottom*). The *right* pictures are the zoomed views around the leaflet

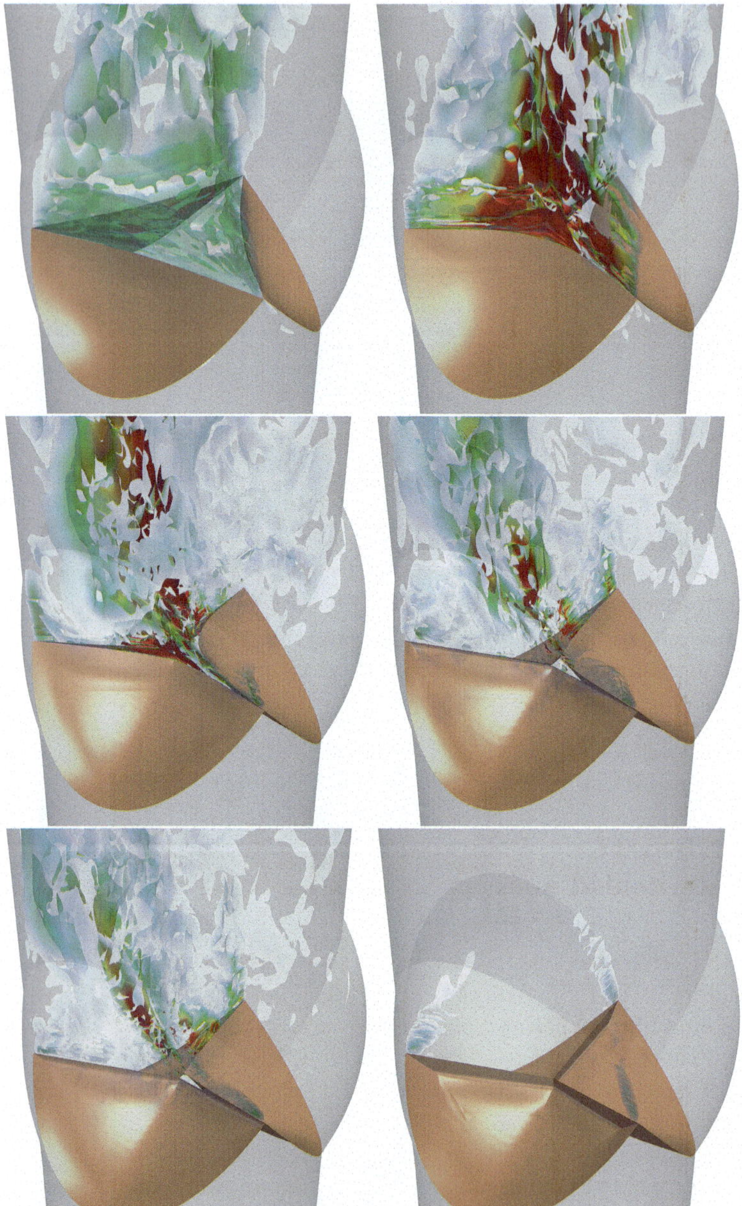

Fig. 10.12 3D ST-SI-TC-IGA computation of aortic valve flow with a realistic model geometry, leaflet contact, and actual-contact and high-resolution boundary-layer representations. Isosurfaces corresponding to a positive value of the second invariant of the velocity gradient tensor, colored by the velocity magnitude. Color range from *blue* to *red* corresponds to the velocity range from low to high

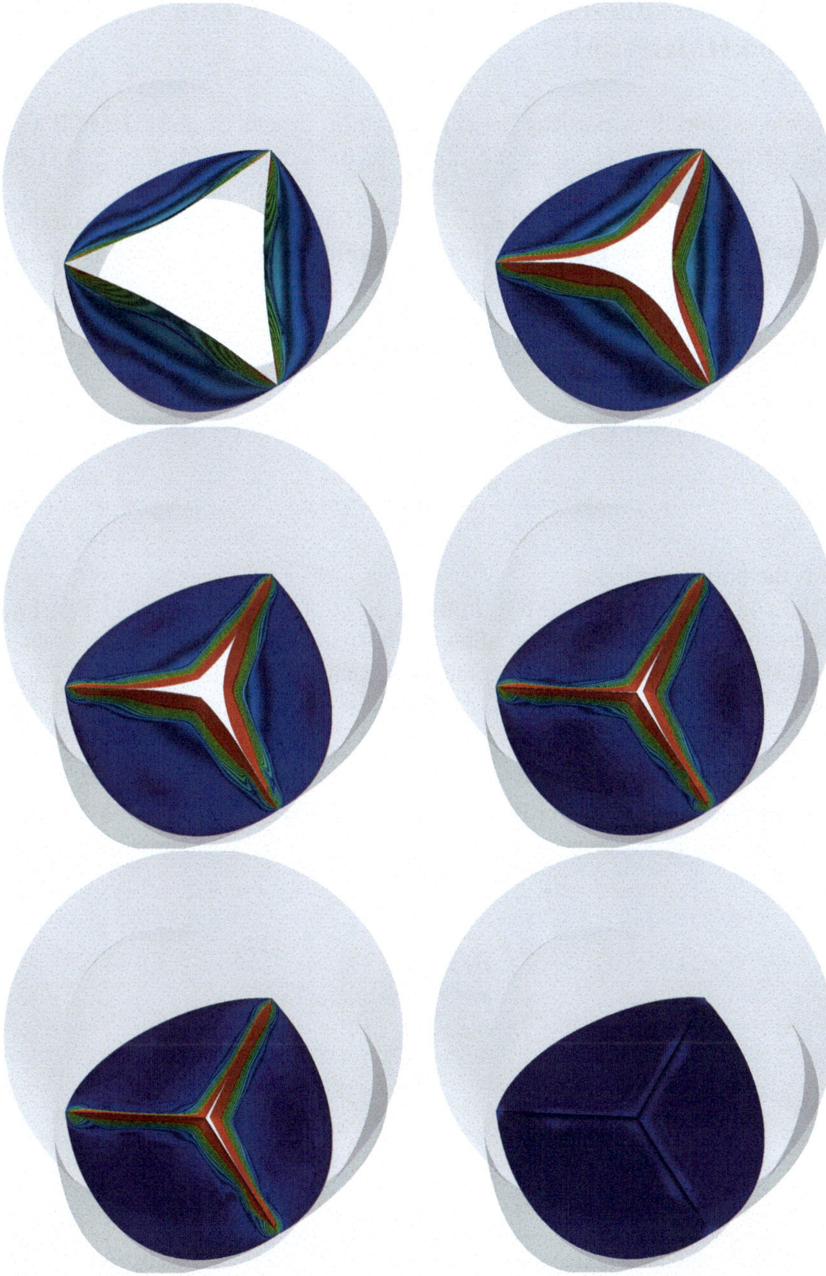

Fig. 10.13 3D ST-SI-TC-IGA computation of aortic valve flow with a realistic model geometry, leaflet contact, and actual-contact and high-resolution boundary-layer representations. WSS on the lower leaflet surfaces. The frames are for the same instants as in Fig. 10.12

10.7 Compressible-Flow Porosity Models for Thin Structures and HMGP, 2017

The compressible-flow porosity models for thin structures and HMGP were introduced in 2017 [164], in connection with the compressible-flow ST-SUPG. Figure 10.14 shows schematic representation a porous media. The derivation of the porosity model starts with the Darcy–Forchheimer model and, assuming a polytropic process in the porous medium of the thin structure, results in a quadratic equation for the mass flow rate across the structure:

$$\beta\,|\dot{m}|^2 + \frac{\mu}{D}\,|\dot{m}| - \mathcal{M}^2 = 0. \tag{10.1}$$

Here β and $\frac{\mu}{D}$ are the two porosity coefficients of the model,

$$\mathcal{M}^2 \equiv \frac{n}{1+n}\,|\rho_A p_A - \rho_B p_B|\,, \tag{10.2}$$

and n is the polytropic index.

In the compressible-flow HMGP in the context of parachutes, we first perform flow computations for an n-gore slice of the parachute canopy under a set of different flight conditions. That generates for each patch of the n-gore slice a set of patch-averaged $\mathcal{M}^2$ and $|\dot{m}|$ values, where $\dot{m}$ is the flow through the gaps of the patch. For each patch, the correlation between $\mathcal{M}^2$ and $|\dot{m}|$ is established by least-squares curve fitting based on Eq. (10.1), which yields $\left(\frac{\mu}{D}\right)_{\mathrm{G}}$ and β_{G} that represent the homogenized geometric porosity of the gaps. We set $\left(\frac{D}{\mu}\right)_{\mathrm{F}}$ equal to the fabric porosity and $\beta_{\mathrm{F}} = 0$.

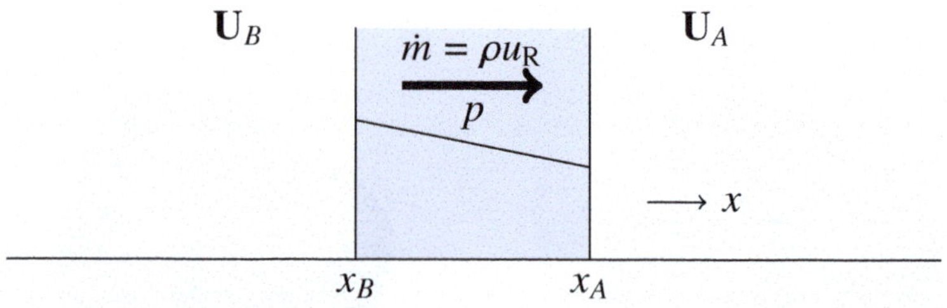

Fig. 10.14 Compressible-flow porosity models for thin structures and HMGP. Schematic representation a porous media. The coordinates x_{B} and x_{A} represent the "B" ("below") and "A" ("above") sides of the media. The flow direction from the B side to the A side is taken as the positive flow direction. Here, u_{R} is the velocity relative to the media, which is only in the normal direction. Across the media, the mass flow rate is invariant and the pressure is continuous

10.8 First 3D Computations with the Compressible-Flow ST-SI and Compressible-Flow HMPG: Aerodynamics of the Orion Spacecraft Drogue Parachute, 2017

The first 3D computations with the compressible-flow ST-SI and compressible-flow HMPG were in 2017 [164], in the same article where these two methods were introduced. The computations were for the aerodynamics of the Orion spacecraft drogue parachute, which has a ribbon construction and 24 gores. Additionally, there are three locations along the gore where ribbons are removed, creating wider gaps. The four sections of the gore separated by the wider gaps serve as the patches in the HMGP. The computations were carried out in two stages. The computations in the first stage were for one-gore slice of the parachute canopy, for nine sets of flight conditions, to calculate the porosity coefficients of the HMGP for the four patches of the canopy. The computation in the second stage was for the full canopy, for one of the nine sets of flight conditions, with modeled geometric porosity for the gaps between the ribbons and resolved geometric porosity for the three wider gaps between the patches. In the set of flight of conditions for the full-canopy computation, the altitude is 35,000 ft and the Mach number is 0.3. The set of flight conditions is identified with the label "AM31." The two-stage aerodynamics computations were preceded by structural mechanics shape-determination computations, using finite element meshes made of four-node quadrilateral membrane elements and two-node cable elements, with the uniform, dynamic pressure corresponding to each of the nine sets of flight conditions.

All flow computations were with the compressible-flow ST-SUPG. In the one-gore computations, the meshes are made of four-node tetrahedral elements, with the element count ranging from 5,127,466 to 5,583,272. Figure 10.15 shows the flow velocity from the one-gore computation for one of the nine sets of flight conditions. In the full-canopy computation for AM31, the mesh is made of 5,999,782 four-node tetrahedral elements. Figure 10.16 shows the surface mesh. Figures 10.17 and 10.18 show the temperature and density.

10.9 Complex-Geometry IGA Mesh Generation (CGIMG) Method, 2017

The CGIMG was introduced in 2017 [165]. The method name and abbreviation were coined in [149, 166]. Most real-world problems require dealing with complex geometries. In such cases, generating an IGA mesh is more challenging than generating a finite element mesh. The CGIMG was introduced to overcome that challenge and widens the application range of the ST-IGA.

The CGIMG consists of three steps. In the first step, a block-structured mesh is generated using existing techniques for such meshes. In the second step, that mesh is projected to a NURBS mesh made of patches corresponding to the blocks of the

Fig. 10.15 3D computations with the compressible-flow ST-SI and compressible-flow HMPG. Aerodynamics of the Orion spacecraft drogue parachute. Velocity magnitude from the one-gore computation for one of the nine sets of flight conditions, namely AM31. Color range from *blue* to *red* corresponds to the velocity range from low to high

block-structured mesh. In the third step, the original model surfaces are recovered, to the extent the nature of the recovered surfaces does not impede the robustness or accuracy of the flow computations. The CGIMG is expected to preserve the element quality and refinement distribution of the block-structured mesh. Mesh generation and mesh quality tests included in [165], with more presented in [167], show that the CGIMG is a practical IGA mesh generation method with good performance (see Fig. 10.19).

10.10 First 3D Computation with the ST-SI-IGA and CGIMG: Flow Analysis of a Turbocharger Turbine and Exhaust Manifold, 2017

The first 3D computation with the ST-SI-IGA and CGIMG was reported in 2017 [165], in the same article where the CGIMG was introduced. The computation

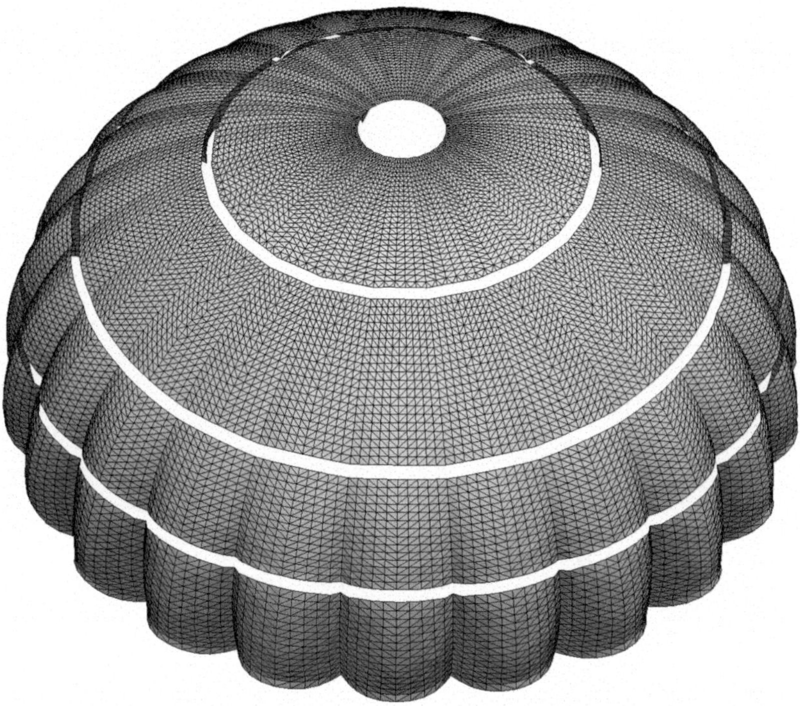

Fig. 10.16 3D computations with the compressible-flow ST-SI and compressible-flow HMPG. Aerodynamics of the Orion spacecraft drogue parachute. Surface mesh for the full-canopy computation for AM31

was for flow analysis of a turbocharger turbine and exhaust manifold. The mesh is made of 285,519 quadratic NURBS elements. Figure 10.20 shows the control mesh. Figure 10.21 shows the four SIs of the mesh; two have an actual slip and two are for mesh generation flexibility. With the ST-IGA with IGA basis functions in time, more specifically quadratic NURBS basis functions in time, we achieve exact representation of the rotor and mesh motions. At each time step, which has three nonlinear iterations, the ST-SUPS is used in the first two iterations and the ST-VMS in the last. Figure 10.22 shows the flow velocity.

10.11 Local Length Scales and Stabilization Parameters Targeting IGA and ST-IGA Discretizations, 2017

Local length scales and stabilization parameters targeting IGA discretization, including ST-IGA with IGA basis functions in space and time, were introduced in 2017 [78]. It is widely known that local length scales and stabilization param-

Fig. 10.17 3D computations with the compressible-flow ST-SI and compressible-flow HMPG. Aerodynamics of the Orion spacecraft drogue parachute. Temperature from the full-canopy computation for AM31. Color range from *blue* to *red* corresponds to the temperature range from low to high

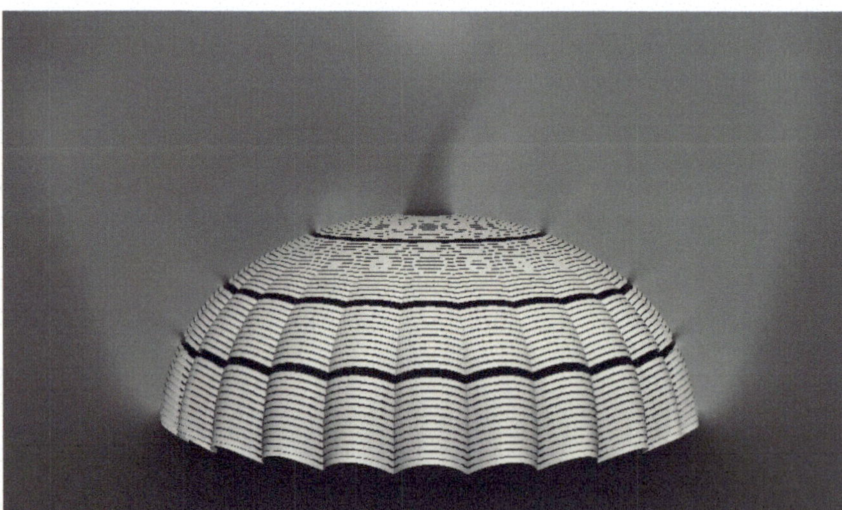

Fig. 10.18 3D computations with the compressible-flow ST-SI and compressible-flow HMPG. Aerodynamics of the Orion spacecraft drogue parachute. Density from the full-canopy computation for AM31. Gray range from *light* to *dark* corresponds to the density range from low to high

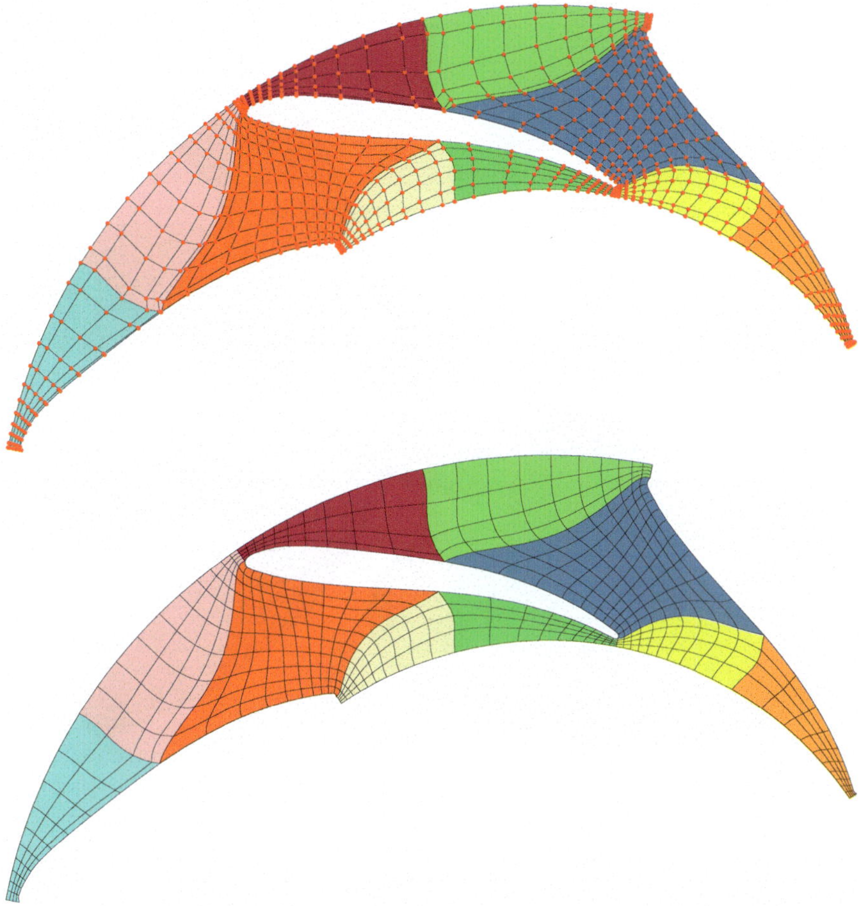

Fig. 10.19 Complex-Geometry IGA Mesh Generation (CGIMG) method. Cross-section of a NURBS control mesh around a stator blade of a turbocharger turbine (*top*). The *red* circles represent the control points, and the different-colored areas are the NURBS patches. The corresponding physical mesh (*bottom*). The lines represent the element boundaries

eters play an essential role in the SUPS, VMS, and other stabilization methods. In addition to having good stability and accuracy properties, they should have good invariance properties, for example, invariance with respect to element node reordering. There are typically two local length scales. They are the advection and diffusion length scales, used in the expressions for the advective and diffusive limits of the stabilization parameter. The advection length scales in use have always been in the flow direction. The diffusion length scales in use have mostly been just element-geometry-dependent. The reasons for also making that direction-dependent so that the spatial variation of the solution is taken into account were articulated in [168]. Those were the reasons behind the introduction of the diffusion length scale

Fig. 10.20 3D computation with the ST-SI-IGA and CGIMG. Flow analysis of a turbocharger turbine and exhaust manifold. Control mesh

in the solution-gradient direction, which was given in 2001 [169] without explicitly stating the reasons and identified as the diffusion length scale in [9]. Stabilization parameters targeting ST discretization were introduced in [9], to be used with the ST-SUPG and ST-SUPS. They have also been in use with the ST-VMS. We note that the integrals in the ST-SI also involve element length, and that is in the direction normal to the SI.

Until 2017, local length scales and stabilization parameters used in computations with IGA basis functions were those originally developed for computations with finite element basis functions. The local length scales and stabilization parameters targeting IGA discretization, despite being relatively new, have been used in computation of many classes of problems. The overview given in [149] on that topic includes how they started, how they evolved over the years, how the versions targeting IGA discretization were derived, and the classes of problems computed with these versions.

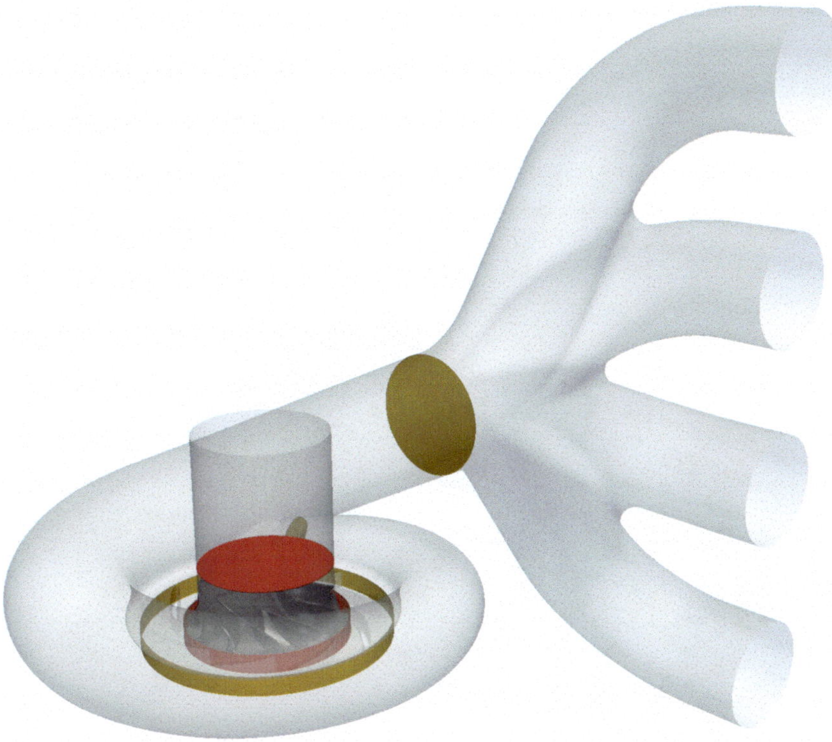

Fig. 10.21 3D computation with the ST-SI-IGA and CGIMG. Flow analysis of a turbocharger turbine and exhaust manifold. The four SIs of the mesh. The *red* SIs have an actual slip, and the *brown* SIs are for mesh generation flexibility

The τ_{SUPS} expression we now use is mostly from [78]:

$$\tau_{\mathrm{SUPS}} = \left(\tau_{\mathrm{SUGN12}}^{-2} + \tau_{\mathrm{SUGN3}}^{-2} + \tau_{\mathrm{SUGN4}}^{-2}\right)^{-\frac{1}{2}}. \tag{10.3}$$

The first two components are given as

$$\tau_{\mathrm{SUGN12}}^{-2} = \begin{bmatrix} 1 \\ \mathbf{u}^h \end{bmatrix} \begin{bmatrix} 1 \\ \mathbf{u}^h \end{bmatrix} : \mathbf{G}^{\mathrm{ST}} \tag{10.4}$$

and

$$\tau_{\mathrm{SUGN3}}^{-1} = \nu \frac{\boldsymbol{\varepsilon}^2(\mathbf{u}^h) : \mathbf{G}}{\boldsymbol{\varepsilon}(\mathbf{u}^h) : \boldsymbol{\varepsilon}(\mathbf{u}^h)}. \tag{10.5}$$

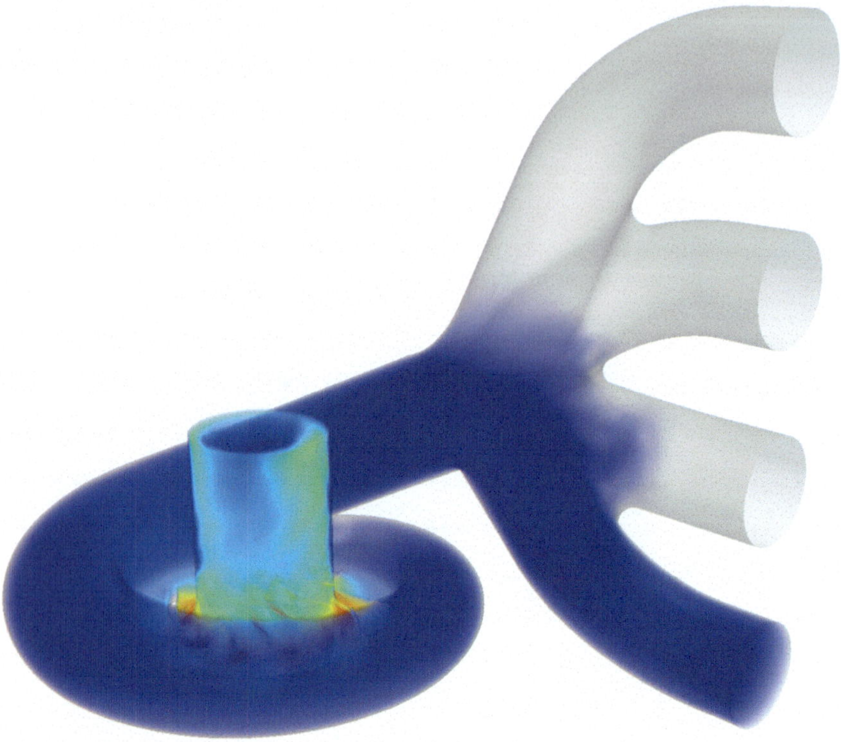

Fig. 10.22 3D computation with the ST-SI-IGA and CGIMG. Flow analysis of a turbocharger turbine and exhaust manifold. Velocity magnitude. Color range from *blue* to *red* corresponds to the velocity range from low to high

The second expression, which is our latest for $\tau_{\mathrm{SUGN3}}^{-1}$, is from [168]. Here the space-only and ST element metric tensors, $\mathbf{G}$ and $\mathbf{G}^{\mathrm{ST}}$, are defined as

$$\mathbf{G} = \hat{\mathbf{Q}}^{-\mathsf{T}} \cdot \hat{\mathbf{Q}}^{-1}, \tag{10.6}$$

$$\mathbf{G}^{\mathrm{ST}} = \left(\hat{\mathbf{Q}}^{\mathrm{ST}}\right)^{-\mathsf{T}} \cdot \left(\hat{\mathbf{Q}}^{\mathrm{ST}}\right)^{-1}, \tag{10.7}$$

where

$$\hat{\mathbf{Q}} = \mathbf{Q} \cdot \mathbf{D}^{-1}, \tag{10.8}$$

$$\hat{\mathbf{Q}}^{\mathrm{ST}} = \mathbf{Q}^{\mathrm{ST}} \cdot \left(\mathbf{D}^{\mathrm{ST}}\right)^{-1}. \tag{10.9}$$

The tensors $\mathbf{D}$ and $\mathbf{D}^{\mathrm{ST}}$ relate the integration and "preferred" parametric spaces (see [170]), and $\mathbf{D}^{\mathrm{ST}}$ is given as

$$\mathbf{D}^{\mathrm{ST}} = \begin{bmatrix} D_\theta & \mathbf{0}^{\mathrm{T}} \\ \mathbf{0} & \mathbf{D} \end{bmatrix} \tag{10.10}$$

once D_θ and $\mathbf{D}$ are selected. With that, we can rewrite Eqs. (10.6) and (10.7) as

$$\mathbf{G} = \mathbf{Q}^{-\mathrm{T}} \cdot \mathbf{D}^{\mathrm{T}} \cdot \mathbf{D} \cdot \mathbf{Q}^{-1}, \tag{10.11}$$

$$\mathbf{G}^{\mathrm{ST}} = \left(\mathbf{Q}^{\mathrm{ST}}\right)^{-\mathrm{T}} \cdot \begin{bmatrix} D_\theta^2 & \mathbf{0}^{\mathrm{T}} \\ \mathbf{0} & \mathbf{D}^{\mathrm{T}} \cdot \mathbf{D} \end{bmatrix} \cdot \left(\mathbf{Q}^{\mathrm{ST}}\right)^{-1}. \tag{10.12}$$

The selections for D_θ and $\mathbf{D}$ play an important role, especially in higher-order IGA discretization [78, 170] and simplex elements [171].

The third component, originating from [16], is defined as

$$\tau_{\mathrm{SUGN4}} = \left\| \boldsymbol{\nabla} \mathbf{u}^h \right\|_F^{-1}, \tag{10.13}$$

where $\| \cdot \|_F$ represents the Frobenius norm.

The stabilization parameter ν_{LSIC} is from [151]:

$$\nu_{\mathrm{LSIC}} = \frac{h_{\mathrm{LSIC}}^2}{\tau_{\mathrm{SUPS}}}, \tag{10.14}$$

where the element length h_{LSIC} is typically set equal to the minimum element length:

$$h_{\mathrm{MIN}} = 2 \left(\max_{\mathbf{r}} \left(\mathbf{r}\mathbf{r} : \mathbf{G} \right) \right)^{-\frac{1}{2}}. \tag{10.15}$$

10.12 First 3D Computations with the Compressible-Flow ST-SI-IGA and Compressible-Flow HMPG: Aerodynamics of the Orion Spacecraft Drogue Parachute, 2018

The first 3D computations with the compressible-flow ST-SI-IGA and compressible-flow HMPG were in 2018 [172], for aerodynamics of the Orion spacecraft drogue parachute. The computational framework is similar to the one described in Sect. 10.8. The main difference is that the spatial discretizations, in both structural mechanics and aerodynamics computations, are with IGA instead of finite elements. The parachute construction is the same as in Sect. 10.8, and again

Fig. 10.23 3D computations with the compressible-flow ST-SI-IGA and compressible-flow HMPG. Aerodynamics of the Orion spacecraft drogue parachute. Velocity magnitude from the one-gore computation for one of the three sets of flight conditions, namely AM31. Color range from *blue* to *red* corresponds to the velocity range from low to high

the four sections of the gore separated by the wider gaps serve as the patches in the HMGP. The computations in the first stage were for one-gore slice of the parachute canopy, for three of the nine sets of flight conditions in Sect. 10.8, to calculate the porosity coefficients of the HMGP for the four patches of the canopy. The computation in the second stage was for the full canopy, for the flight conditions AM31, with modeled geometric porosity for the gaps between the ribbons and resolved geometric porosity for the three wider gaps between the patches. The two-stage aerodynamics computations were preceded by structural mechanics shape-determination computations, using cubic NURBS meshes made of membrane and cable elements, with the uniform, dynamic pressure corresponding to each of the three sets of flight conditions.

All flow computations were with the compressible-flow ST-SUPG. In both the one-gore and full-canopy computations, the meshes are made of quadratic NURBS elements. In the one-gore computations, the element count ranges from 111,500 to 117,806. Figures 10.23 and 10.24 show the velocity and density from the one-gore computation for one of the three sets of flight conditions. In the full-canopy computation for AM31, the element count is 190,464. Figure 10.25 shows the surface mesh. Figures 10.26 and 10.27 show the velocity and density.

Fig. 10.24 3D computations with the compressible-flow ST-SI-IGA and compressible-flow HMPG. Aerodynamics of the Orion spacecraft drogue parachute. Density from the one-gore computation for one of the three sets of flight conditions, namely AM31. Color range from *blue* to *red* corresponds to the density range from low to high

10.13 First 3D Computation of Tire Aerodynamics with Near-Actual Tire Geometry, Road Contact, Tire Deformation, Actual-Contact and High-Resolution Boundary-Layer Representations, and IGA Discretization in Space, 2018

The first 3D computation of tire aerodynamics with near-actual tire geometry, road contact, tire deformation, actual-contact and high-resolution boundary-layer representations, and IGA discretization in space was reported in 2018 [77]. The tire model includes longitudinal and transverse grooves. The deformation pattern is realistic, provided by the tire company. The tire model, in its deformed configuration, is shown in Fig. 10.28. The tire deformation was provided at five uniformly spaced instants of a 5° rotation (see [77]). Cubic NURBS basis functions are used in the temporal representation of the deformation, extracted with the ST-C from the data at those five instants. The computation was with the ST-VMS and ST-SI-TC-IGA. The mesh is made of 2,921,552 quadratic NURBS elements. Figure 10.29 shows the mesh. The main complexity of the mesh generation was near the tire surface and that part of the mesh was generated manually. Although the rest of the mesh

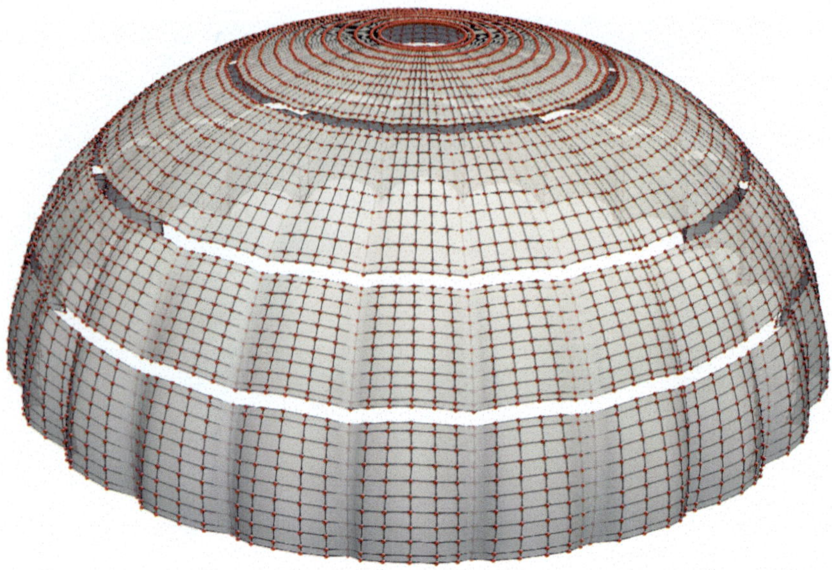

Fig. 10.25 3D computations with the compressible-flow ST-SI-IGA and compressible-flow HMPG. Aerodynamics of the Orion spacecraft drogue parachute. Surface and control mesh for the full-canopy computation for AM31

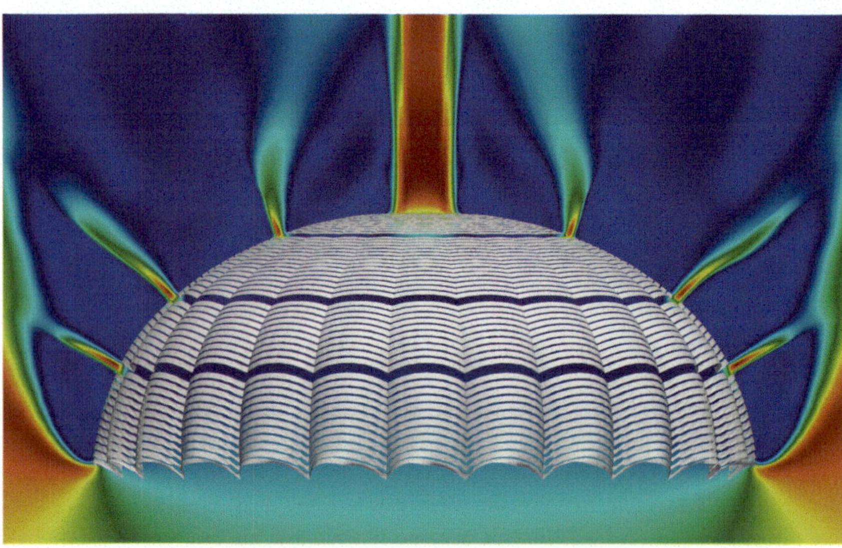

Fig. 10.26 3D computations with the compressible-flow ST-SI-IGA and compressible-flow HMPG. Aerodynamics of the Orion spacecraft drogue parachute. Velocity magnitude from the full-canopy computation for AM31. Color range from *blue* to *red* corresponds to the velocity range from low to high

Fig. 10.27 3D computations with the compressible-flow ST-SI-IGA and compressible-flow HMPG. Aerodynamics of the Orion spacecraft drogue parachute. Density from the full-canopy computation for AM31. Color range from *blue* to *red* corresponds to the density range from low to high

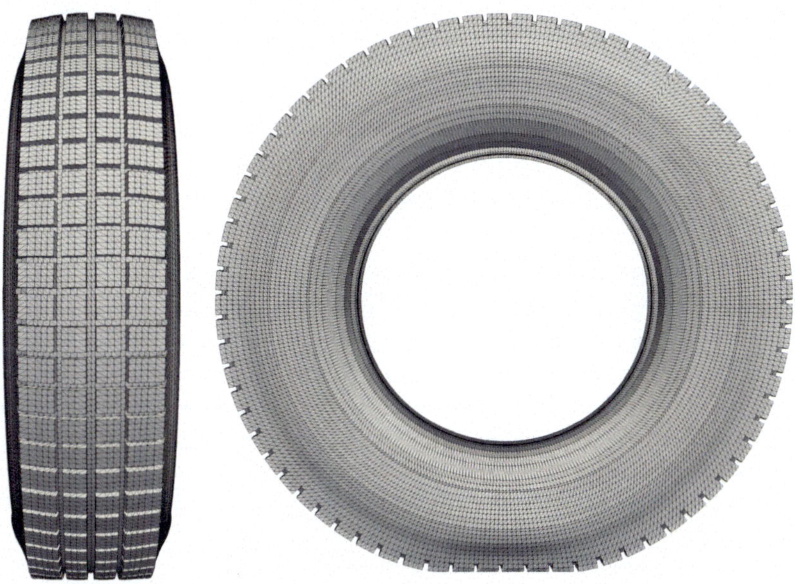

Fig. 10.28 3D computation of tire aerodynamics with near-actual tire geometry, road contact, tire deformation, actual-contact and high-resolution boundary-layer representations, and IGA discretization. Tire model in its deformed configuration

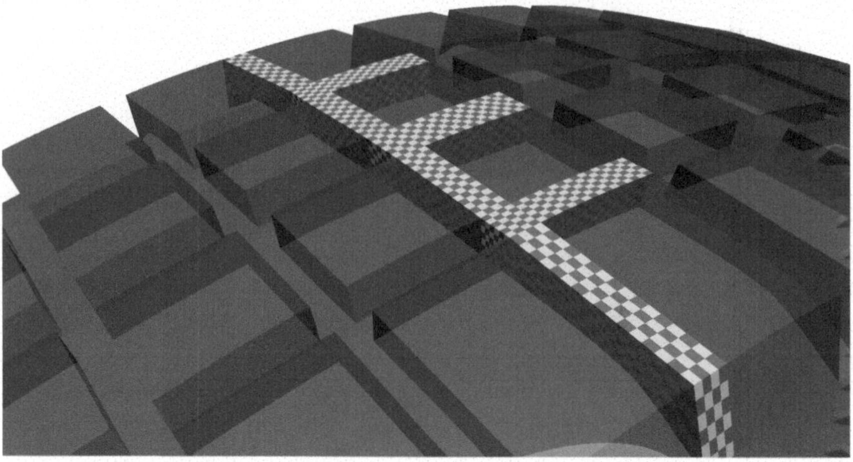

Fig. 10.29 3D computation of tire aerodynamics with near-actual tire geometry, road contact, tire deformation, actual-contact and high-resolution boundary-layer representations, and IGA discretization. Mesh. The checkerboard pattern is for differentiating between the NURBS elements

could have been generated with the CGIMG, because it was not difficult, that was also done manually. Figure 10.30 shows the flow velocity near the tire–road contact area. Figure 10.31 shows the flow patterns near the tire–road contact area.

10.14 First 3D ST-SI-IGA Computation of Fluid–Cable Interaction with One-Way-Dependence: Flow-Driven String Dynamics in Turbomachinery, 2018

The first 3D ST-SI-IGA computation of fluid–cable interaction with one-way-dependence was reported in 2018 [153], for flow-driven string dynamics in turbomachinery. The computational framework is similar to the one described in Sect. 9.9. The main difference is that spatial discretization in fluid mechanics computation is also with IGA instead of finite elements. The computation was with the ST-VMS. To understand how strings get stuck on or around the blades in turbomachinery, a string is dropped into a pump placed horizontally. By using the ST-IGA with IGA basis functions in time, more specifically quadratic NURBS basis functions in time, we achieve exact representation of the rotor and mesh motions. The one-way-dependence model was described in Sect. 6.9. As in Sect. 9.9, as the flow field is computed, the time-dependent data obtained is stored by using the ST-C, with cubic B-splines in time. Then, using that data, flow-driven dynamics of the string is computed. Contact between the string and solid surfaces is handled with the SENCT-FC. The fluid mechanics mesh is made of 544,466 quadratic NURBS

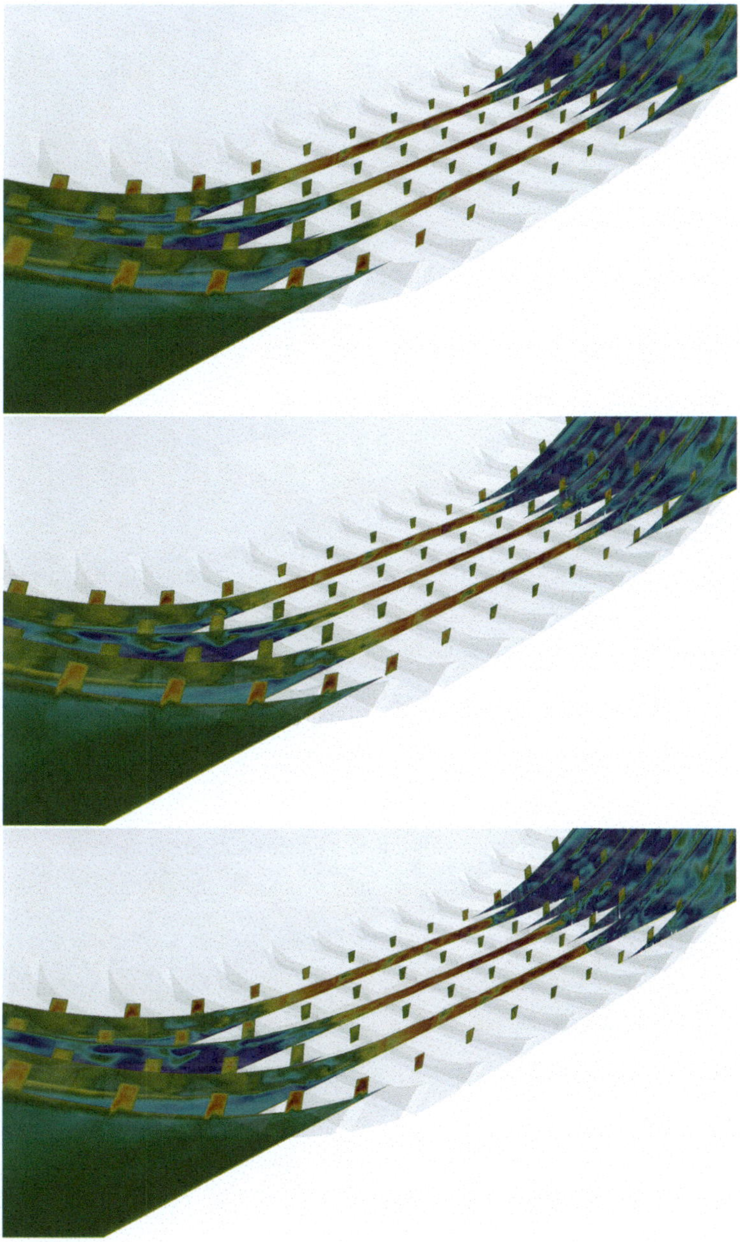

Fig. 10.30 3D computation of tire aerodynamics with near-actual tire geometry, road contact, tire deformation, actual-contact and high-resolution boundary-layer representations, and IGA discretization. Velocity magnitude at three uniformly spaced instants. Contact area. Data displayed on planes perpendicular to the tire axis. Color range from *blue* to *red* corresponds to the velocity range from low to high

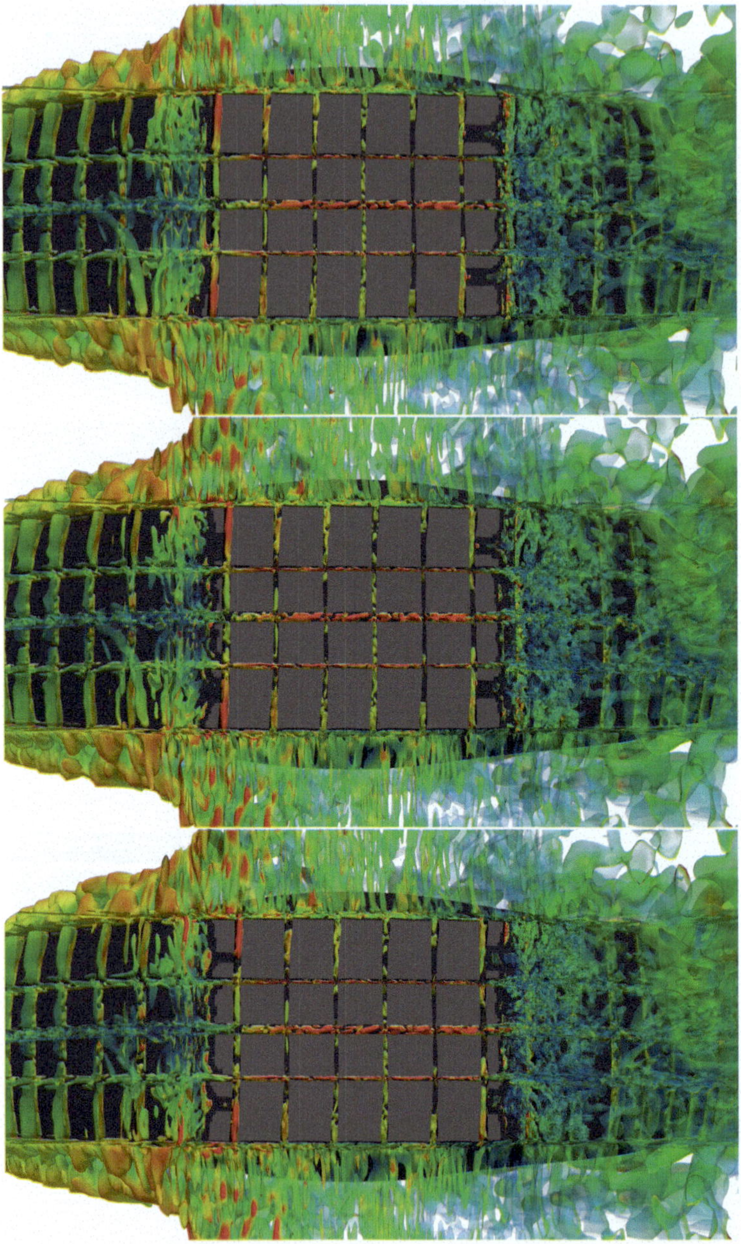

Fig. 10.31 3D computation of tire aerodynamics with near-actual tire geometry, road contact, tire deformation, actual-contact and high-resolution boundary-layer representations, and IGA discretization. Isosurfaces corresponding to a positive value of the second invariant of the velocity gradient tensor, colored by the velocity magnitude, viewed from the bottom, at three uniformly spaced instants. The *gray* zones are the contact areas. Color range from *blue* to *red* corresponds to the velocity range from low to high

elements. The string mesh is made of 16 bending-stabilized cable elements, using the IGA with cubic NURBS basis functions. The IGA gives us not only a higher-order method and smoothness in the structure shape, but also smoothness in the fluid dynamics forces acting on the string. Figure 10.32 shows the fluid mechanics mesh. Figure 10.33 shows the inlet, outlet, and SI, and Fig. 10.34 shows the flow patterns. The string-dynamics computations in [153] were for three different string lengths and 17 different initial string positions. Figure 10.35 shows the results from one of those 51 cases.

10.15 ST-SI-TC-IGA Special Feature: Built-in Reynolds-Equation Limit, 2018

The ST-SI-TC-IGA has a built-in Reynolds-equation limit. That special feature was introduced in 2018 [163]. With the built-in Reynolds-equation limit, computations involving fluid films in narrow spaces between solid surfaces forming a channel or coming into contact would not need a separate Reynolds-equation model. The ST-SI-TC-IGA can emulate the Reynolds-equation model with a comparable solution quality and computational cost and would also work in flow regions beyond the scope of the Reynolds-equation model. Reaching a solution quality comparable to that of the Reynolds-equation model can be achieved even with just one quadratic NURBS element across the gap of the lubrication fluid dynamics problem. An extensive set of test computations were reported in [163] to show how the method performs compared to the Reynolds-equation model, and also compared to finite element discretization. The tests demonstrate the effectiveness of the method.

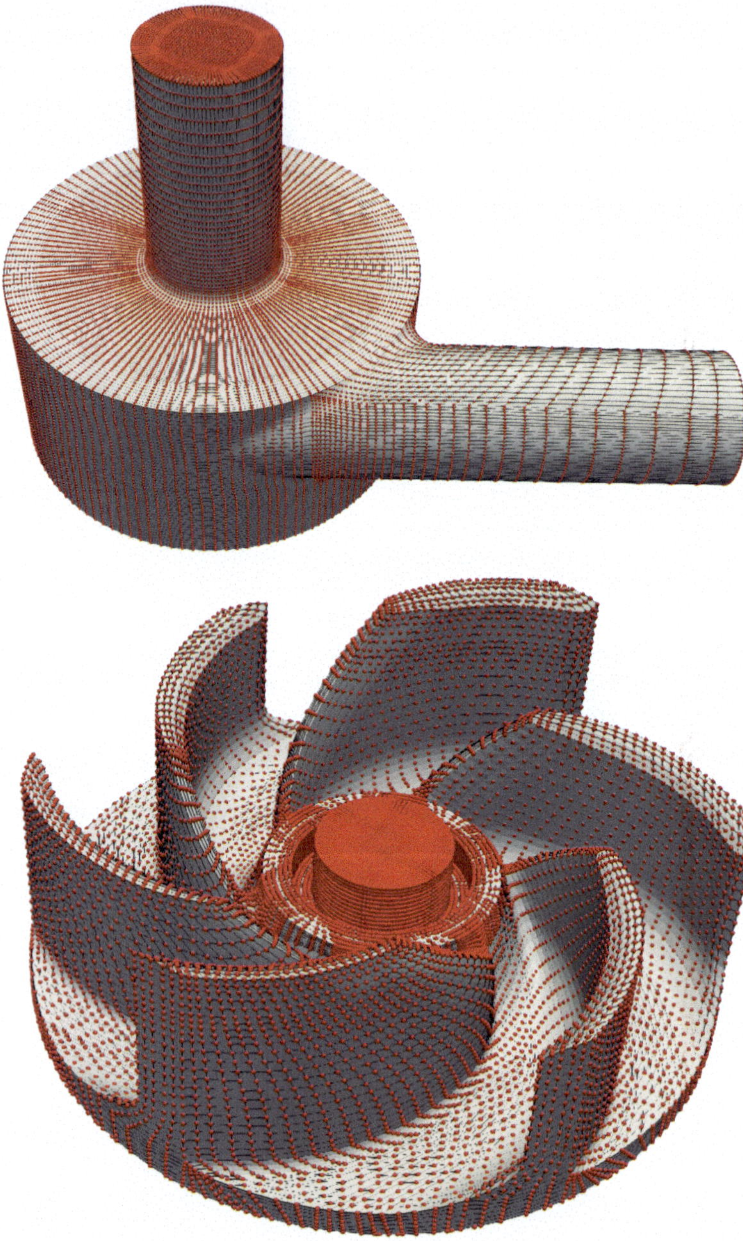

Fig. 10.32 3D ST-SI-IGA computation of fluid–cable interaction with one-way-dependence. Flow-driven string dynamics in a pump. Mesh. *Red* circles are the control points

Fig. 10.33 3D ST-SI-IGA computation of fluid–cable interaction with one-way-dependence. Flow-driven string dynamics in a pump. Inlet (*red*), outlet (*blue*), and SI (*olive*)

Fig. 10.34 3D ST-SI-IGA computation of fluid–cable interaction with one-way-dependence. Flow-driven string dynamics in a pump. Isosurfaces corresponding to a positive value of the second invariant of the velocity gradient tensor, colored by the velocity magnitude, at six instants. Color range from *blue* to *red* corresponds to the velocity range from low to high

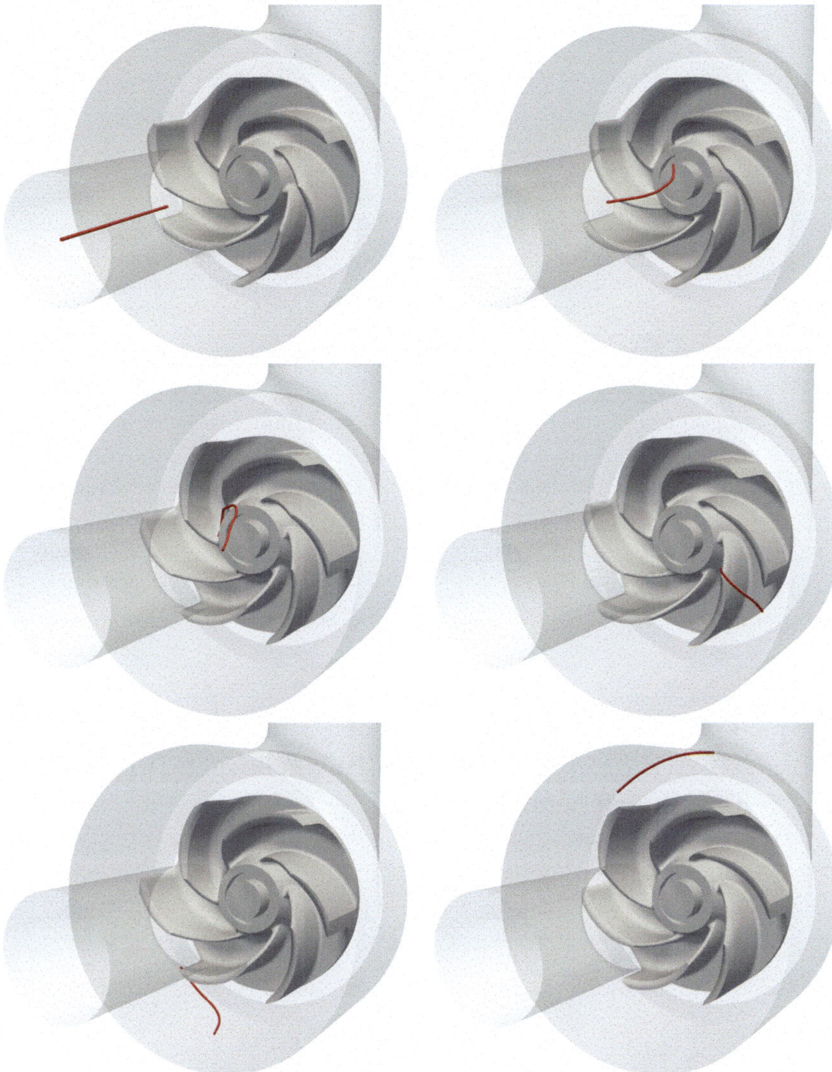

Fig. 10.35 3D ST-SI-IGA computation of fluid–cable interaction with one-way-dependence. Flow-driven string dynamics in a pump. One of the 51 cases of string dynamics computations

Chapter 11
Methods and Solutions in 2019–2021

11.1 First 3D Computation of Tire Aerodynamics with Near-Actual Tire Geometry, Road Contact, Tire Deformation, Road Roughness, Fluid Film, Actual-Contact and High-Resolution Boundary-Layer Representations, and IGA Discretization in Space, 2019

The first 3D computation of tire aerodynamics with near-actual tire geometry, road contact, tire deformation, road roughness, fluid film, actual-contact and high-resolution boundary-layer representations, and IGA discretization in space was reported in 2019 [173]. The computational framework is an enhanced version of the one in Sect. 10.13; the ST-SI-TC-IGA can, as described in Sect. 10.15, emulate the Reynolds-equation model. Flow in fluid films can be represented with just one quadratic NURBS element across the film thickness, with accuracy and computational cost comparable to that of the Reynolds-equation model.

The tire model and deformation pattern are the same as in Sect. 10.13. In representing the road roughness, it is understood that the mesh resolution on the tire surface would be higher than the mesh resolution on the road surface. Therefore, the dents containing the fluid films are represented not on the road surface but on the tire surface. Figure 11.1 illustrates this concept. The paths needed for deisolating the fluid trapped in the dents are created on the tire surface. The road roughness has random pattern. The computation was with the ST-SUPS. The mesh, generated manually, is made of 342,296 quadratic NURBS elements. Figure 11.2 shows the mesh. Figure 11.3 shows the flow patterns. Figure 11.4 shows the fluid friction on the tire surface.

T. E. Tezduyar, K. Takizawa, *Space–Time Computational Flow Analysis*,
Modeling and Simulation in Science, Engineering and Technology,
https://doi.org/10.1007/978-3-031-88727-7_11

Fig. 11.1 3D computation of tire aerodynamics with near-actual tire geometry, road contact, tire deformation, road roughness, fluid film, actual-contact and high-resolution boundary-layer representations, and IGA discretization. Road-roughness representation. Dents containing the fluid films are represented not on the road surface (*top*) but on the tire surface (*bottom*)

11.2 First 3D Bioprosthetic-Heart-Valve Flow Analysis with Realistic Leaflet Geometry and Motion, Leaflet Contact, Actual-Contact and High-Resolution Boundary-Layer Representations, and IGA Discretization in Space, 2019

The first 3D bioprosthetic-heart-valve flow analysis with realistic leaflet geometry and motion, leaflet contact, actual-contact and high-resolution boundary-layer representations, and IGA discretization in space was reported in 2019 [174]. The computation was with the ST-SUPS and ST-SI-TC-IGA. The bioprosthetic-heart-valve (BHV) and arterial-surface motion are from the mixed ALE-VMS IMGA FSI computation [175] in the framework of the FSITICT. The mesh, generated with three SIs and in three parts, is made of 289,452 quadratic NURBS elements. Figure 11.5 shows the BHV model, and Fig. 11.6 shows the SIs and the mesh. Part 1 is facing the SIs. It contains the elements that collapse and are reborn as the leaflets are moving. Its motion comes from a special-purpose mesh moving method that takes into account the contact. Part 2 is not changing during the leaflet motion. Part 3 is the

Fig. 11.2 3D computation of tire aerodynamics with near-actual tire geometry, road contact, tire deformation, road roughness, fluid film, actual-contact and high-resolution boundary-layer representations, and IGA discretization. Mesh. The mesh with the *yellow–gray* pattern is fixed. The mesh with the other color patterns is rotating with the tire. The elements in the *red–gray* pattern are the special elements for the fluid film gaps in the contact area. The elements in the *green–gray* pattern are those in the grooves. The *blue–gray* pattern indicates the rest of the elements

rest of the mesh, between Part 1, Part 2, and the artery surface. It is changing during the computation. Its motion is determined with the EBMR and MJBS. The motion is driven by the motion of Part 1, the valve motion, and the motion of the artery surface. Figure 11.7 shows the flow patterns, and Fig. 11.8 shows the corresponding valve WSS.

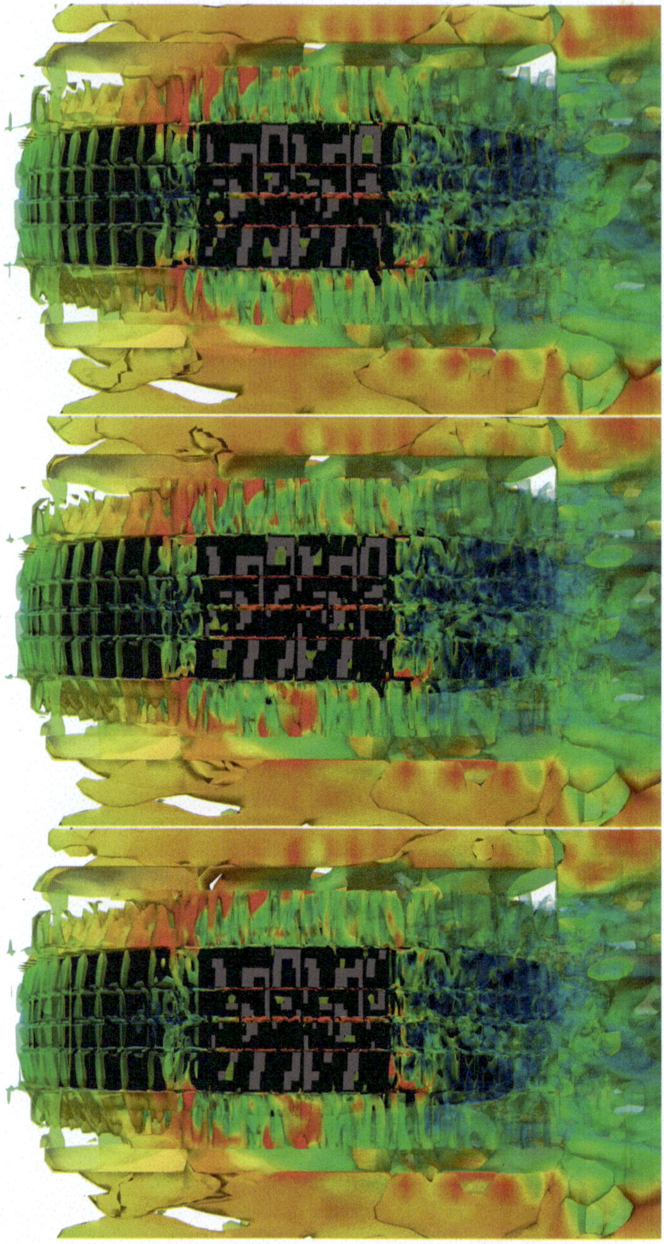

Fig. 11.3 3D computation of tire aerodynamics with near-actual tire geometry, road contact, tire deformation, road roughness, fluid film, actual-contact and high-resolution boundary-layer representations, and IGA discretization. Isosurfaces corresponding to a positive value of the second invariant of the velocity gradient tensor, colored by the velocity magnitude, viewed from the bottom, at three instants. Color range from *blue* to *red* corresponds to the velocity range from low to high. The *gray* indicates the full-contact zones

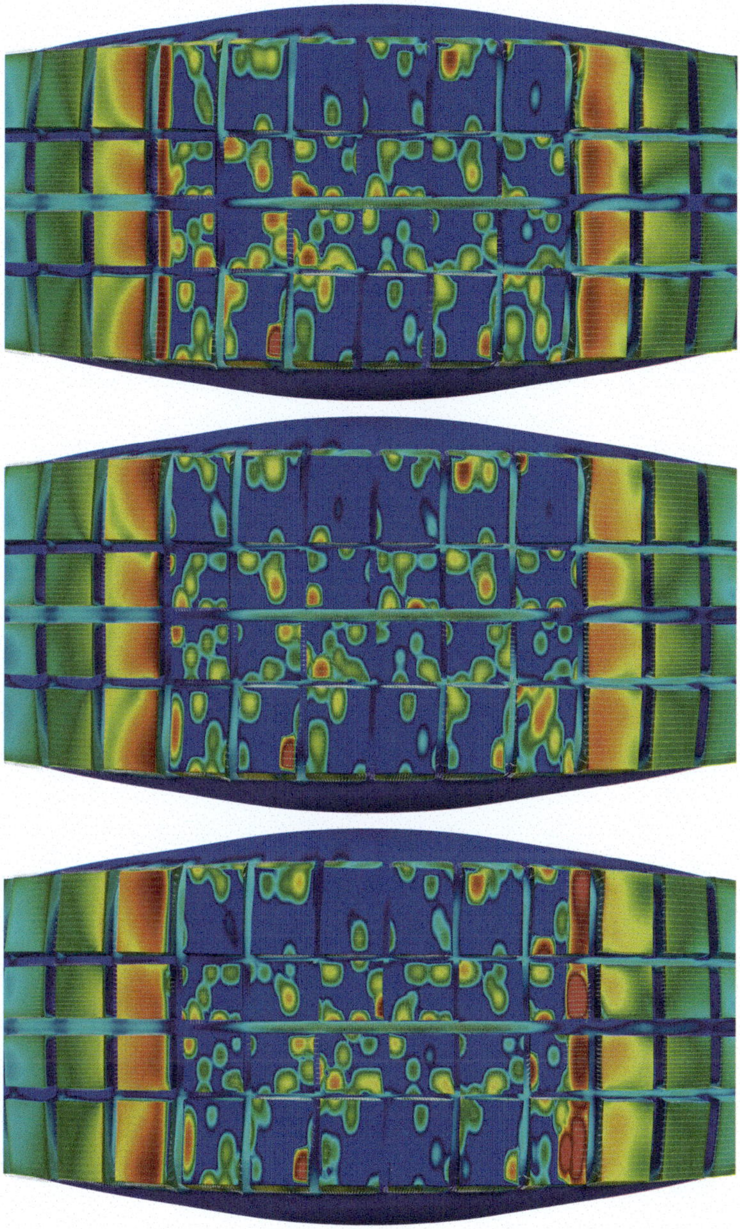

Fig. 11.4 3D computation of tire aerodynamics with near-actual tire geometry, road contact, tire deformation, road roughness, fluid film, actual-contact and high-resolution boundary-layer representations, and IGA discretization. Fluid friction on the tire surface in terms of the positional-averaged WSS at three instants. Color range from *blue* to *red* corresponds to the WSS range from low to high. The vectors indicate just the direction of the WSS

Fig. 11.5 3D BHV flow analysis with realistic leaflet geometry and motion, leaflet contact, actual-contact and high-resolution boundary-layer representations, and IGA discretization in space. BHV model

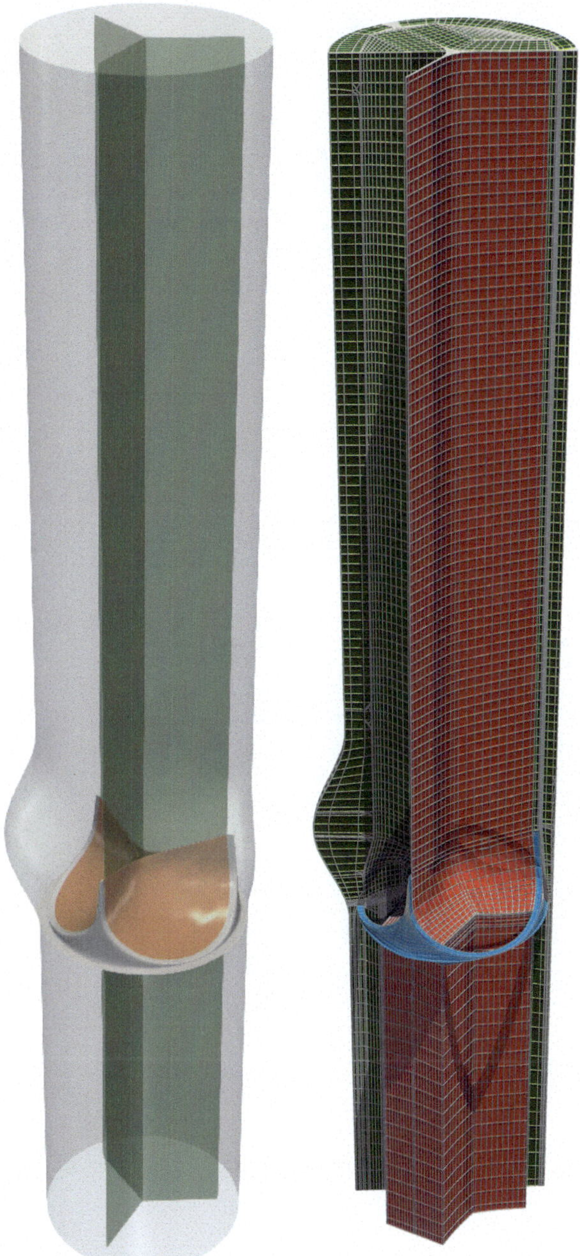

Fig. 11.6 3D BHV flow analysis with realistic leaflet geometry and motion, leaflet contact, actual-contact and high-resolution boundary-layer representations, and IGA discretization in space. The three SIs and the three mesh parts: Part 1 (*red*), Part 2 (*blue*), and half of Part 3 (*green*)

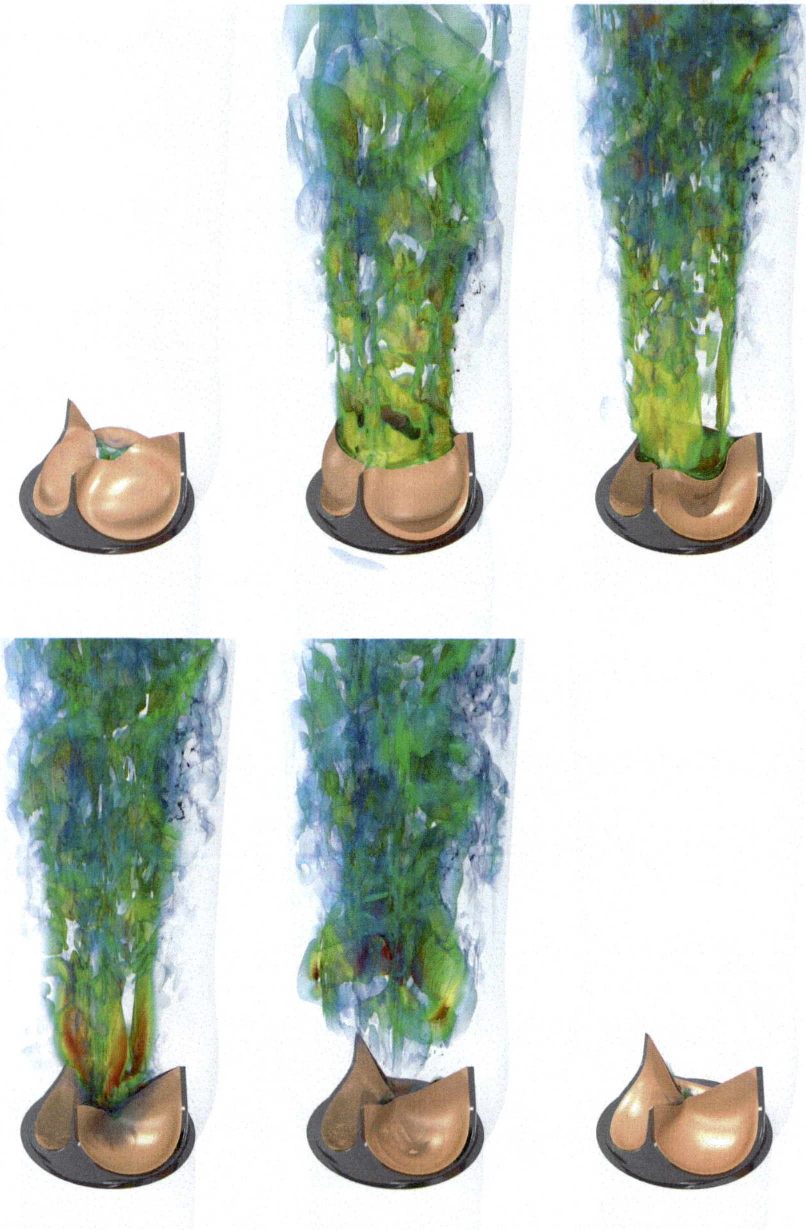

Fig. 11.7 3D BHV flow analysis with realistic leaflet geometry and motion, leaflet contact, actual-contact and high-resolution boundary-layer representations, and IGA discretization in space. Isosurfaces corresponding to a positive value of the second invariant of the velocity gradient tensor, colored by the velocity magnitude. Color range from *blue* to *red* corresponds to the velocity range from low to high

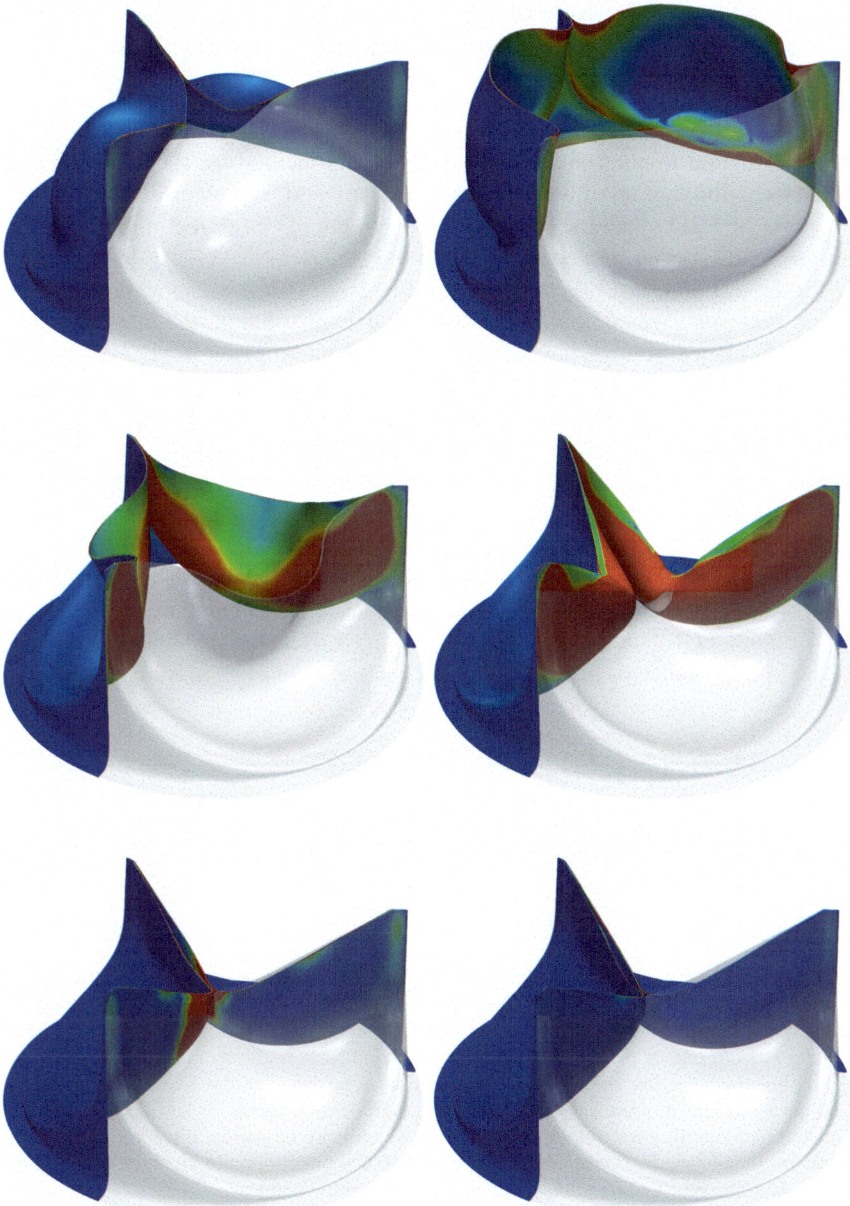

Fig. 11.8 3D BHV flow analysis with realistic leaflet geometry and motion, leaflet contact, actual-contact and high-resolution boundary-layer representations, and IGA discretization in space. WSS at the same instants as in Fig. 11.7. Color range from *blue* to *red* corresponds to the WSS range from low to high

11.3 Constrained-Flow-Profile (CFP) Traction: A Method for Specifying Traction Boundary Condition at an Inflow Boundary, 2019

The CFP Traction was introduced in 2019 [61] as a method for specifying traction boundary condition at an inflow boundary. The method is explained in Fig. 11.9 in the context of a 2D quadratic NURBS mesh. In the 3D case, the number of basis functions in the special-purpose element would be 27.

With the test platform of a 2D channel flow from [176], we show how the CFP Traction performs. We compare that to the analytical solution. We also compare it to the performance of applying the outflow stabilization given in [177] also to the inflow, and the abbreviation "OS" will denote that approach. There is a parameter "β" involved in that outflow stabilization, and it is selected as $\beta \geq \frac{1}{2}$ [24]. At the outflow, in all test cases, we use the OS with $\beta = 0.5$. At the inflow, we test the OS with both $\beta = 0.5$ and $\beta = 1.0$. The test setup is given in Fig. 11.10. The quadratic NURBS meshes are shown in Fig. 11.11. The computations were performed with the ST-VMS. Figure 11.12 shows the velocity magnitude for the OS (with $\beta = 0.5$ and $\beta = 1.0$) and CFP. The analytical solution parabolic velocity profile has a maximum of 4.84×10^{-2} m/s, and the Reynolds number based on that and the channel width is 157. Figure 11.13 shows the pressure averaged over the channel cross-section. The good performance of the CFP Traction is clear from Figs. 11.12 and 11.13.

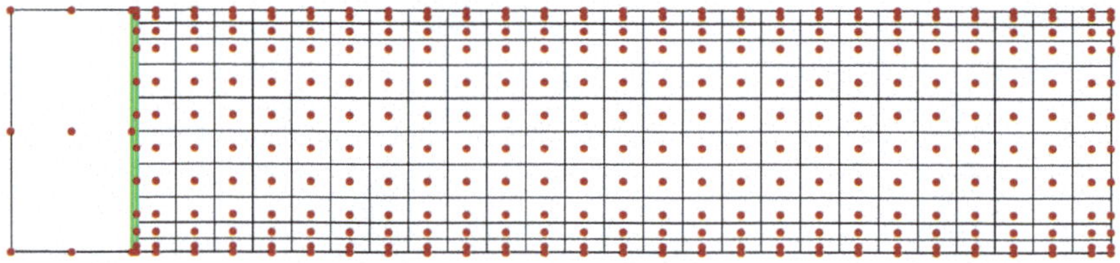

Fig. 11.9 CFP Traction. The method explained in the context of a 2D quadratic NURBS mesh. The inflow boundary is the *left* edge. The control points are denoted by the *red* circles and the element boundaries by the *black* lines. The special-purpose element on the *left*, with 9 basis functions, and the rest of the mesh on the *right* have an SI between them, denoted by the pair of *green* lines, connecting the parts of the solution obtained over the two mesh zones. The special-purpose element produces a constrained, quadratic flow profile. It has at the inflow only one unspecified control-point velocity component, which is in the normal direction. The solution obtained for the unspecified component and the quadratic velocity profile yield the flow rate that corresponds to the specified tractions at the inflow and outflow boundaries

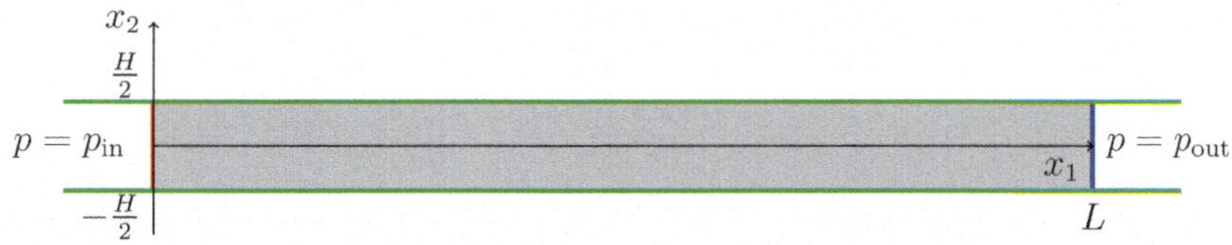

Fig. 11.10 CFP Traction. Test setup for the 2D channel flow. Test parameters: ρ = 1050 kg/m^3, μ = 4.2 × 10^{-3} Pa·s, L = 1.365 × 10^{-1} m, H = 1.3 × 10^{-2} m, p_{in} = 1.313 Pa, and $p_{\text{out}} = 0$. Channel walls: no-slip. Inflow and outflow: traction conditions coming from p_{in} and p_{out}

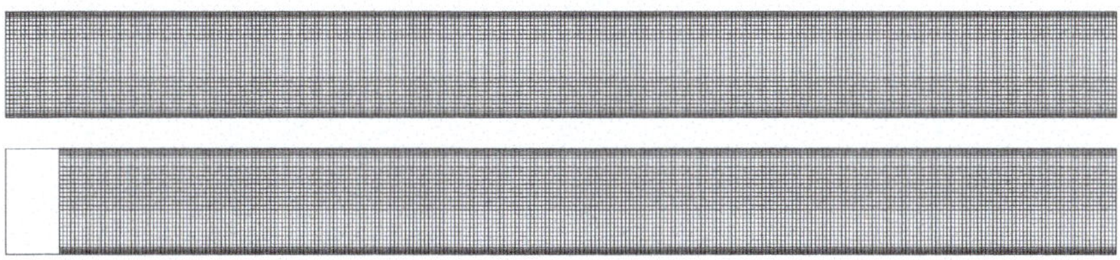

Fig. 11.11 CFP Traction. Meshes in the test computations with the OS and CFP. The lines mark the element boundaries. The two meshes are different only in the part of the domain where we have the special-purpose element of the CFP. Number of control points: 9472 (OS), 9033 (CFP). Number of elements: 8820 (OS), 8401 (CFP)

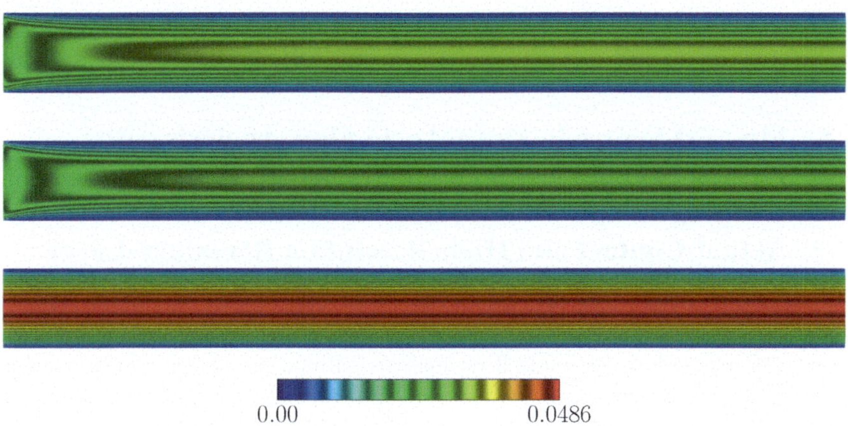

Fig. 11.12 CFP Traction. Velocity magnitude (m/s). OS ($\beta = 0.5$), OS ($\beta = 1.0$), and CFP

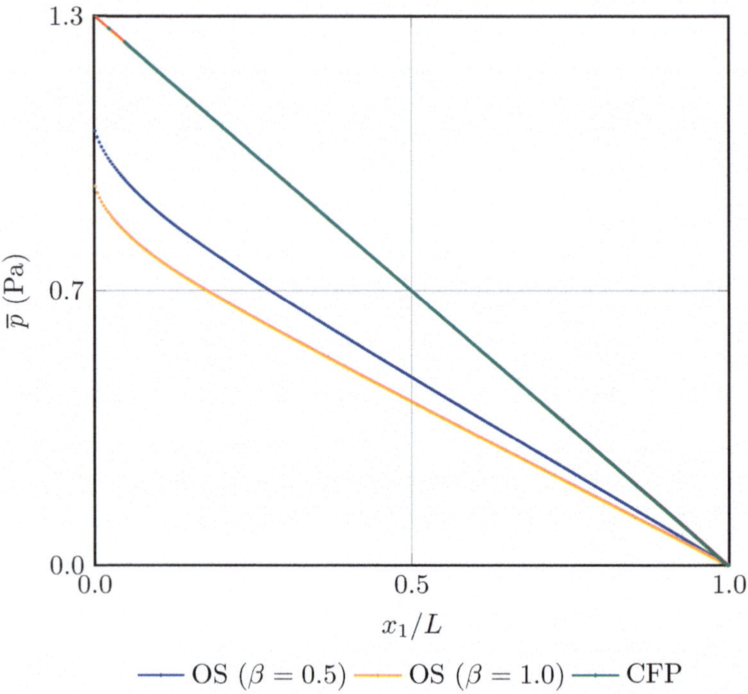

Fig. 11.13 CFP Traction. Pressure averaged over the channel cross-section. The *red* line is the analytical solution

11.4 First 3D Ventricle-Valve-Aorta Flow Analysis with Patient-Specific Aorta, Realistic Ventricle and Leaflet Geometries and Motion, Leaflet Contact, Actual-Contact and High-Resolution Boundary-Layer Representations, and IGA Discretization in Space, 2019

The first 3D ventricle-valve-aorta flow analysis with patient-specific aorta, realistic ventricle and leaflet geometries and motion, leaflet contact, actual-contact and high-resolution boundary-layer representations, and IGA discretization in space was reported in 2019 [61]. The computation was with the ST-SUPS and ST-SI-TC-IGA. Figure 11.14 shows the model for the left ventricle (LV), valve, and aorta sequence. The model does not include the mitral valve; it ends where the LV meets the left atrium, and that is the inflow boundary. The boundary condition at the inflow is zero-velocity when the mitral valve is closed, and the CFP Traction (see Sect. 11.3) when it is open. The LV shrinks and expands, and the valve opens and closes. The aorta is rigid. The mesh, generated separately for the LV, valve, and aorta, is made of 572,556 quadratic NURBS elements.

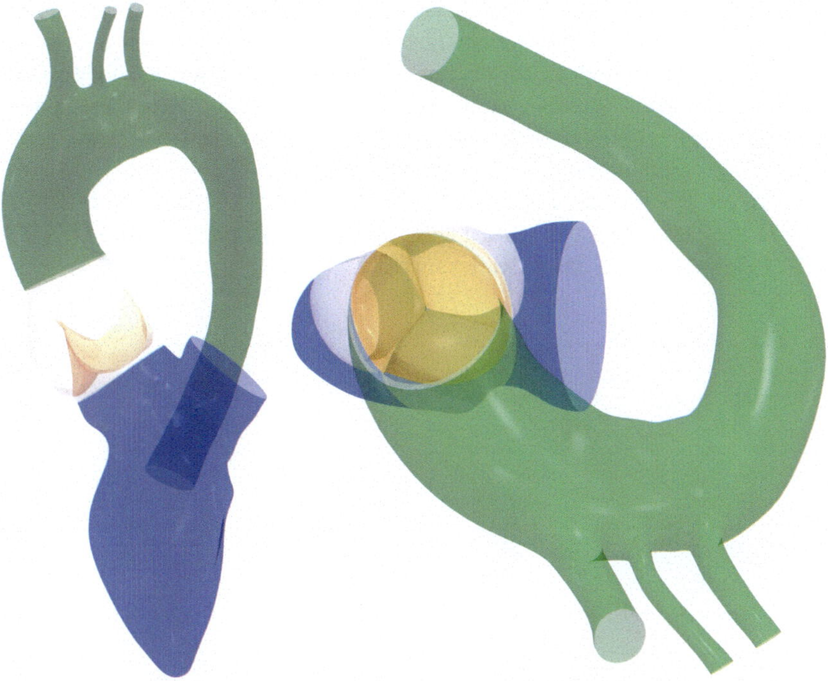

Fig. 11.14 3D ventricle-valve-aorta flow analysis with patient-specific aorta, realistic ventricle and leaflet geometry and motion, leaflet contact, actual-contact and high-resolution boundary-layer representations, and IGA discretization in space. Model. LV (*blue*), valve with the leaflets (*orange*) and sinuses (*transparent*), and aorta (*green*). Front view and view along the valve axis

The LV geometry and motion were built from CT scans of the LV at one instant in the cardiac cycle and other anatomically realistic data. The starting flow computation mesh was generated with the CGIMG. The cardiac-cycle representation of the mesh motion, conforming to the LV motion, was built by using cubic B-splines in time, the ST-C, and the procedure [178] that assures cycle periodicity. The computation was with the linear-elasticity mesh moving method with MJBS. The valve leaflet motion was produced from the shell structural mechanics computation in [179], which was performed with a time-periodic, spatially uniform pressure difference between the upper and lower surfaces of the leaflets. The time profile of the pressure was similar to the one used in [180]. The valve mesh motion was created, by a transformation, from the mesh motion of the valve ST-SI-TC-IGA computation in [179], where the leaflet contact was represented as an actual contact and the boundary layers were represented with high-resolution meshes. The aorta geometry, represented by cubic T-splines, is based on a set of CT scans different than the ones for the LV. Given that geometry, the mesh was generated with the CGIMG.

Figure 11.15 shows the mesh, seven SIs of the mesh, and the boundary conditions. The special-purpose element has 27 basis functions and is connected to the rest of the mesh with an SI. One of the seven SIs is between the LV segment and the special-purpose element, facilitating the CFP Traction. The other six are for mesh generation flexibility: two between the valve and the other two mesh segments, three between the valve mesh segments containing each leaflet, and one between the two LV mesh segments. Figures 11.16, 11.17, and 11.18 show the mesh at six instants in the cardiac cycle. Figure 11.19 shows the flow patterns, and Figs. 11.20 and 11.21 show the WSS on the leaflet surfaces.

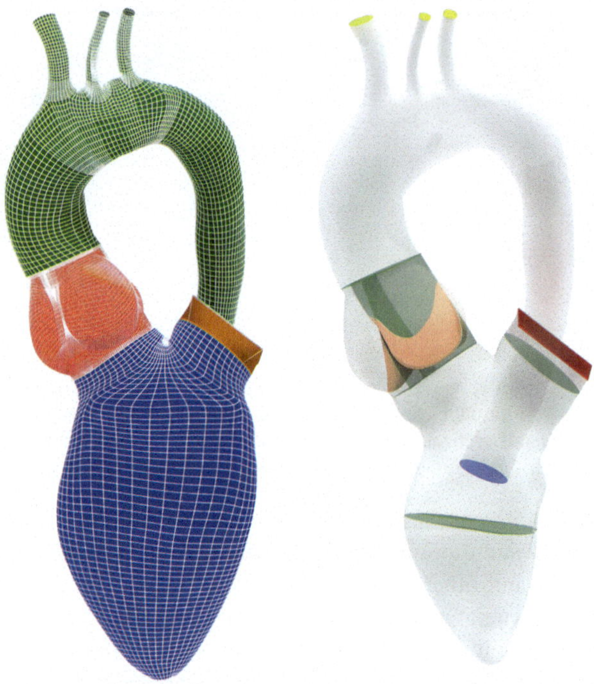

Fig. 11.15 3D ventricle-valve-aorta flow analysis with patient-specific aorta, realistic ventricle and leaflet geometry and motion, leaflet contact, actual-contact and high-resolution boundary-layer representations, and IGA discretization in space. *Left:* Mesh composed of the LV, valve, and aorta segments, plus the special-purpose element (*brown*) for the CFP Traction. The lines are the element boundaries. *Right:* The seven SIs (*green*) and boundary conditions. *Red*: CFP Traction inflow, *blue*: traction-specified outflow, *yellow*: velocity-specified outflow

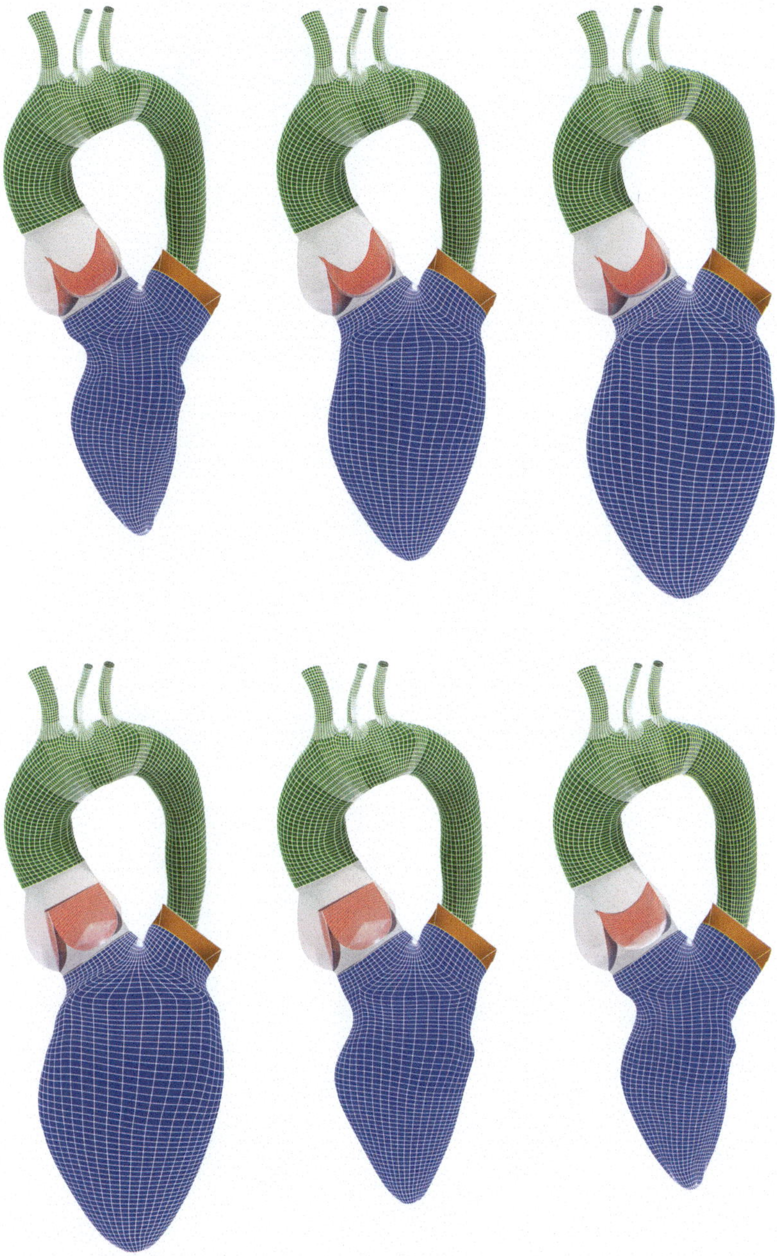

Fig. 11.16 3D ventricle-valve-aorta flow analysis with patient-specific aorta, realistic ventricle and leaflet geometry and motion, leaflet contact, actual-contact and high-resolution boundary-layer representations, and IGA discretization in space. Mesh at six instants

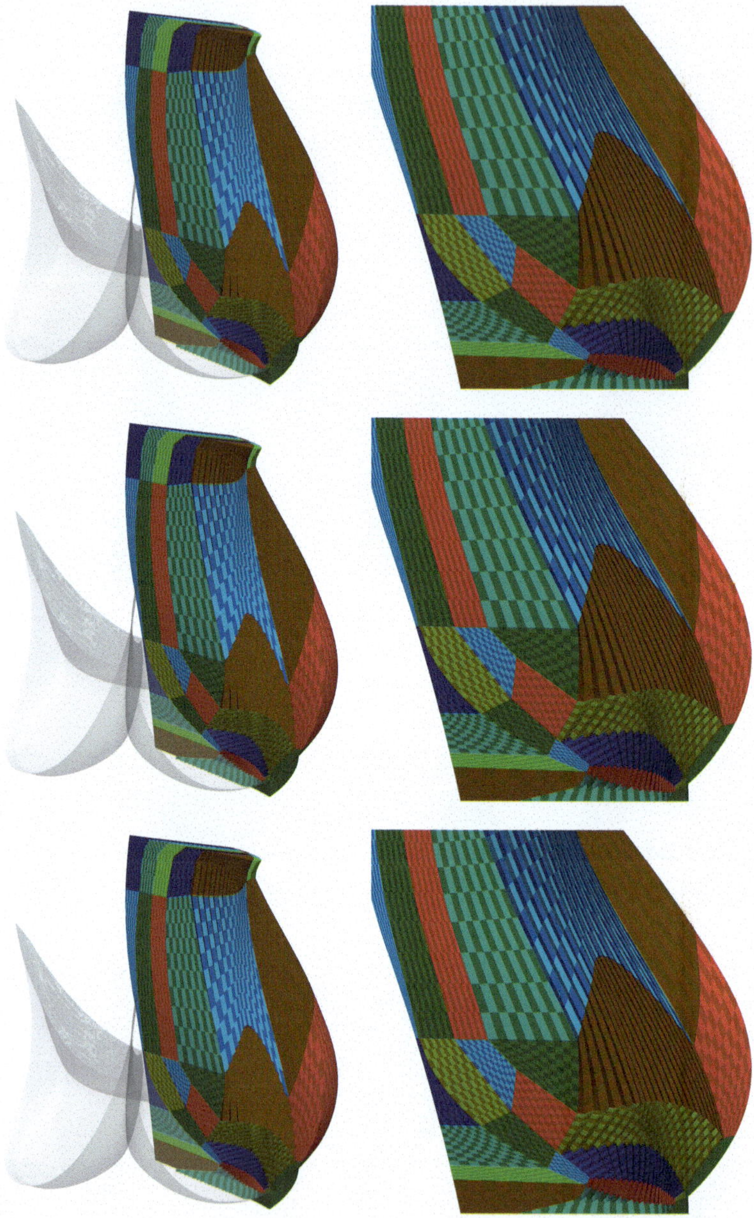

Fig. 11.17 3D ventricle-valve-aorta flow analysis with patient-specific aorta, realistic ventricle and leaflet geometry and motion, leaflet contact, actual-contact and high-resolution boundary-layer representations, and IGA discretization in space. Mesh in the valve at the cardiac-cycle instants of the first three pictures in Fig. 11.16. The *right* pictures are the zoomed views. The checkerboard pattern is for differentiating between the NURBS elements, and the colors are for differentiating between the NURBS patches

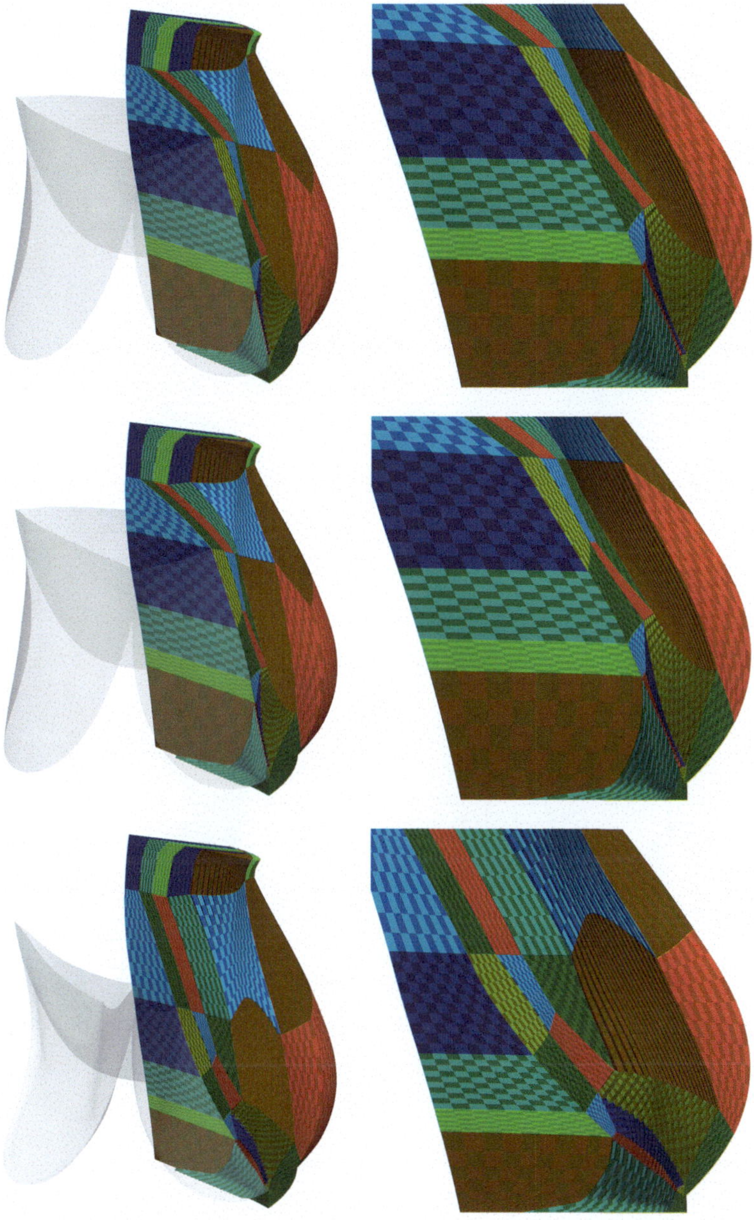

Fig. 11.18 3D ventricle-valve-aorta flow analysis with patient-specific aorta, realistic ventricle and leaflet geometry and motion, leaflet contact, actual-contact and high-resolution boundary-layer representations, and IGA discretization in space. Mesh in the valve at the cardiac-cycle instants of the last three pictures in Fig. 11.16. The *right* pictures are the zoomed views. The checkerboard pattern is for differentiating between the NURBS elements, and the colors are for differentiating between the NURBS patches

Fig. 11.19 3D ventricle-valve-aorta flow analysis with patient-specific aorta, realistic ventricle and leaflet geometry and motion, leaflet contact, actual-contact and high-resolution boundary-layer representations, and IGA discretization in space. Isosurfaces corresponding to a positive value of the second invariant of the velocity gradient tensor, colored by the velocity magnitude, at the same instants as in Fig. 11.16. Color range from *blue* to *red* corresponds to the velocity range from low to high

Fig. 11.20 3D ventricle-valve-aorta flow analysis with patient-specific aorta, realistic ventricle and leaflet geometry and motion, leaflet contact, actual-contact and high-resolution boundary-layer representations, and IGA discretization in space. WSS on the leaflet lower (*left*) and upper (*right*) surfaces. The frames are for the same instants as in the first three pictures of Fig. 11.16. Color range from *blue* to *red* corresponds to the WSS range from low to high

Fig. 11.21 3D ventricle-valve-aorta flow analysis with patient-specific aorta, realistic ventricle and leaflet geometry and motion, leaflet contact, actual-contact and high-resolution boundary-layer representations, and IGA discretization in space. WSS on the leaflet lower (*left*) and upper (*right*) surfaces. The frames are for the same instants as in the last three pictures of Fig. 11.16. Color range from *blue* to *red* corresponds to the WSS range from low to high

11.5 A Linear-Elasticity Mesh Moving Method with No Cycle-to-Cycle Accumulated Distortion: Back-Cycle-Based Mesh Moving (BCBMM), 2019

The BCBMM was introduced in 2019 [61] as a linear-elasticity mesh moving method with no cycle-to-cycle accumulated distortion. In computing the mesh motion from t_n to t_{n+1} with a linear-elasticity mesh moving method, if the nodal displacements are computed from the configuration at t_n, the mesh moving method is path-dependent. In flow computations where we expect cyclic or near-cyclic results, a path-dependent mesh moving method leads to non-cyclic results and cycle-to-cycle accumulated mesh distortion. Nonlinear-elasticity mesh moving methods, such as the EBMR, give us the path-independence we want. Nevertheless, linear-elasticity mesh moving methods where the displacements are not computed from the configuration at t_0, and other methods where the displacements are not computed from the configuration at t_0, are still widely used. The BCBMM enables their continued use without cycle-to-cycle accumulated distortion.

In the BCBMM, in computing the mesh motion from t_n and t_{n+1} with the linear-elasticity equations, in any cycle, the nodal displacements are computed from the configurations in the first cycle. It was pointed out in [61] that in later cycles, as needed, the configurations the nodal displacements are computed from can be changed from the first cycle to a higher cycle. The need may arise if, prior to the solution becoming cyclic or near-cyclic, there are significant differences between the solutions at different cycles. The half-cycle-based mesh moving (HCBMM) method was introduced in [181] as a special-case version of the BCBMM. It is for cases where the boundary or interface motion in the second half of the cycle consists of reversing the steps in the first half and we prefer the mesh motion to have the same reversal pattern. This is accomplished by modifying how the displacements are computed in the second half of the first cycle. In the second half, at a given time level, the displacements are computed from the configuration at the corresponding time level in the first half. In later cycles, the displacements are computed as in the BCBMM.

Detailed 2D and 3D test computations with finite element meshes were presented in [181], using the mesh motion associated with wing pitching as the test case, to demonstrate how the BCBMM and HCBMM function as mesh moving methods with no cycle-to-cycle accumulated mesh distortion. The BCBMM and HCBMM were introduced and tested in the context of a linear-elasticity mesh moving method. However, it is clear that they are applicable also to other mesh moving methods, such as those based on the Poisson's equation, where the displacements are not computed from the configuration at t_0.

11.6 A Low-Distortion Mesh Moving and Mesh Relaxation Method: Fiber-Reinforced Hyperelasticity Mesh Update Method (FRHEMUM), 2020

The FRHEMUM was introduced in 2020 [82]. The method name and acronym were coined in [182]. It is a low-distortion mesh moving method that is also used for mesh relaxation to improve the quality after the initial creation of a mesh. It is based on the fiber-reinforced hyperelasticity and optimized ZSS. The method was introduced targeting IGA discretization, but it is, of course, also applicable to finite element discretization. To reduce distortion during the mesh deformation, the element is stiffened in multiple directions by placing reinforcement fibers in those directions. The ZSS is optimized by seeking, with mesh relaxation, orthogonality of the parametric directions and making the ZSS time-dependent as needed. With the mesh relaxation, after the initial creation of a mesh, we improve its quality and have an equilibrium state with the optimized ZSS, boundary conditions, and constitutive law.

An extensive set of 2D and 3D mesh relaxation and mesh moving tests were presented in [82]. Here, we provide an overview of the 3D tests, where a square-section block is undergoing torsion. For comparison purposes, we also show the results from the linear-elasticity mesh moving with MJBS and from mesh relaxation and mesh moving with the plain neo-Hookean hyperelasticity. Figure 11.22 shows

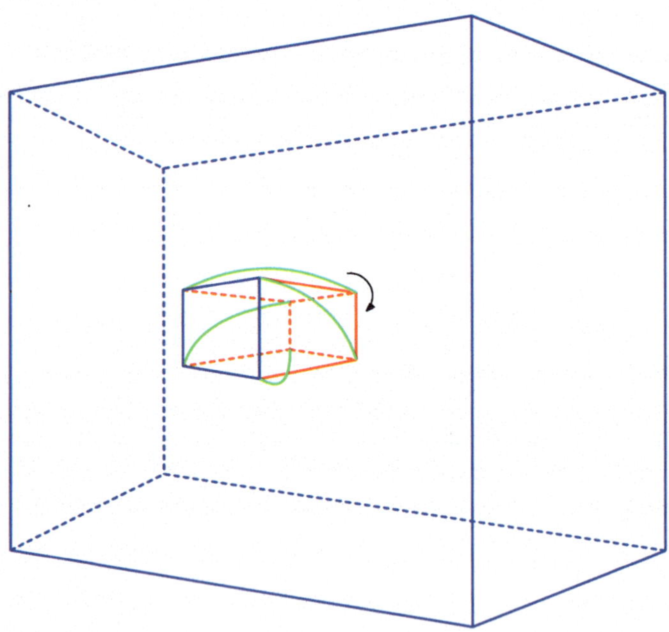

Fig. 11.22 3D mesh moving tests with a square-section block undergoing torsion. Box-shaped domain used in the tests, with the square-section block in its initial (*red*) and twisted (*green*) configurations. One end of the block is fixed centrally on the front plane of the box. The other end is at the midpoint between the front and back planes and is undergoing a 90° rotation about its own center

the box-shaped domain used in the test computations, with the square-section block in its initial and twisted configurations. The deformation of the block, as given in Fig. 11.22, defines the specified boundary positions in computing the positions of the rest of the points in the box-shaped domain. In all three mesh moving methods, the first two layers of elements around the block and the rest of the elements have different material properties; the inner elements are stiffer.

Fig. 11.23 3D mesh moving tests with a square-section block undergoing torsion. The initial mesh $\mathbf{x}_0$. The checkerboard pattern is for differentiating between the NURBS elements. The colors are for differentiating between the NURBS patches

The mesh is made of 136,000 quadratic NURBS elements. The initial mesh, $\mathbf{x}_0$, is shown in Fig. 11.23. Figure 11.24 shows the mesh after the relaxation. Table 11.1 shows the mesh orthogonality measures for the initial and post-relaxation meshes. In visualizing the deformed-mesh configurations, we use the portion of $\mathbf{x}_0$ shown in

Fig. 11.24 3D mesh moving tests with a square-section block undergoing torsion. Mesh after the relaxation. Neo-Hookean (*top*) and fiber-reinforced (*bottom*) hyperelasticity. The checkerboard pattern is for differentiating between the NURBS elements. The colors are for differentiating between the NURBS patches

Table 11.1 3D mesh moving tests with a square-section block undergoing torsion. Mesh orthogonality measures for the initial and post-relaxation meshes, with the neo-Hookean ("NHHE") and fiber-reinforced ("FRHE") hyperelasticity. Deviation of the mesh-line angles from $\pi/2$, represented in terms of the L_2 and max norms

Method	L_2 norm	Max norm
$\mathbf{x}_0$	4.8°	43.1°
NHHE	0.4°	75.8°
FRHE	0.1°	35.6°

Fig. 11.25. Of that portion, because of the symmetry, we visualize only half of the NURBS patches. Figure 11.26 shows the post-torsion meshes. Table 11.2 shows the mesh orthogonality measures for the post-torsion meshes.

Fig. 11.25 3D mesh moving tests with a square-section block undergoing torsion. The portion of $\mathbf{x}_0$ used in visualizing the deformed-mesh configurations. The checkerboard pattern is for differentiating between the NURBS elements. The colors are for differentiating between the NURBS patches

Fig. 11.26 3D mesh moving tests with a square-section block undergoing torsion. Post-torsion meshes. Linear-elasticity mesh moving with MJBS (*top*) and neo-Hookean (*middle*) and fiber-reinforced (*bottom*) hyperelasticity. The *right* frame is the zoomed view. The checkerboard pattern is for differentiating between the NURBS elements. The colors are for differentiating between the NURBS patches

Table 11.2 3D mesh moving tests with a square-section block undergoing torsion. Mesh orthogonality measures for the post-torsion meshes. Linear-elasticity mesh moving with MJBS ("LE-MJBS") and neo-Hookean ("NHHE") and fiber-reinforced ("FRHE") hyperelasticity. Deviation of the mesh-line angles from $\pi/2$, represented in terms of the L_2 and max norms

Method	L_2 norm	Max norm
LE-MJBS	8.4°	78.1°
NHHE	1.2°	88.5°
FRHE	1.1°	64.2°

11.7 First 3D Wind Turbine Computation with the ST-SI-IGA and CGIMG: Aerodynamics of a Tsunami-Shelter VAWT, 2020

The first 3D wind turbine computation with the ST-SI-IGA and CGIMG was reported in 2020 [183]. The computation was for the aerodynamics of a tsunami-shelter VAWT. The model is the same as in Fig. 9.10. The mesh is made of 955,477 quadratic NURBS elements. There are nine SIs in total, three with actual slip and six for mesh generation flexibility. Figures 11.27, 11.28, and 11.29 show

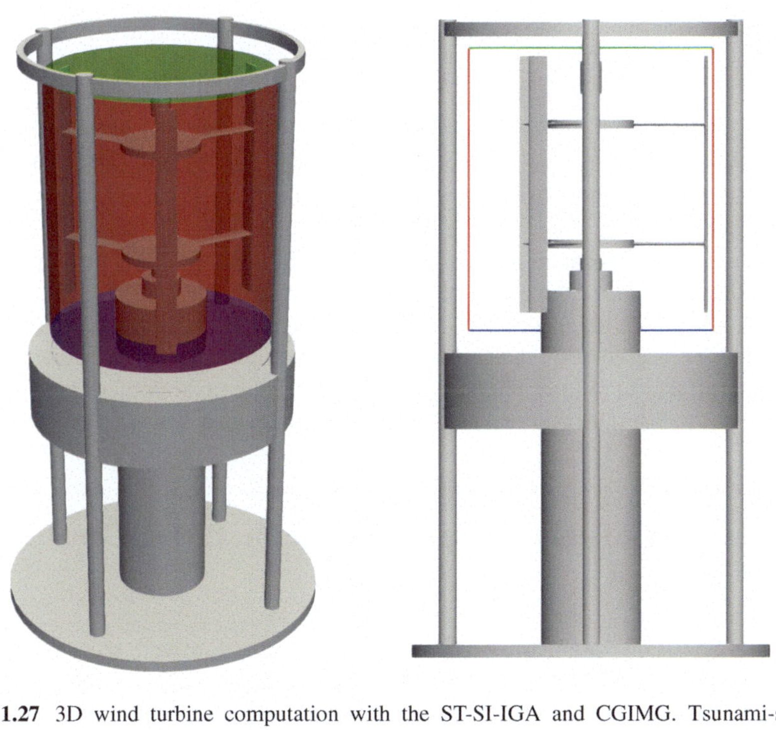

Fig. 11.27 3D wind turbine computation with the ST-SI-IGA and CGIMG. Tsunami-shelter VAWT. The three SIs with actual slip, marked in different colors

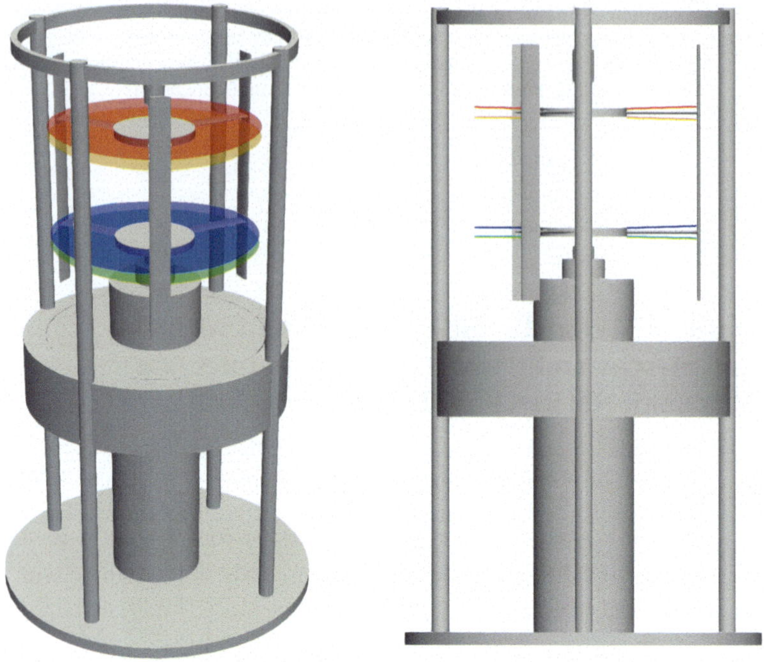

Fig. 11.28 3D wind turbine computation with the ST-SI-IGA and CGIMG. Tsunami-shelter VAWT. The four SIs in the rotating mesh that are for mesh generation flexibility, marked in different colors

the SIs. After the initial creation of the mesh with the CGIMG, its quality was improved by mesh relaxation with the FRHEMUM. The meshes after the initial creation and after the mesh relaxation are shown in Figs. 11.30 and 11.31. We see noticeable improvement in the mesh-line orthogonality, with little effect on local mesh resolution or the element aspect ratio. With the ST-IGA with IGA basis functions in time, more specifically quadratic NURBS basis functions in time, we achieve exact representation of the rotor and mesh motions. At each time step, which has four nonlinear iterations, the ST-SUPS is used in the first two iterations and the ST-VMS in the last two. Figure 11.32 shows the flow patterns near a blade.

11.8 First ST-SI-IGA Computations of the Taylor–Couette Flow, 2021

The first ST-SI-IGA computations of the Taylor–Couette flow were reported in 2021 [184]. The computations were for combinations of different Taylor numbers, full-domain and rotational-periodicity representations of the flow field (see Fig. 11.33 for representation of the circular geometry by using quadratic NURBS patches with

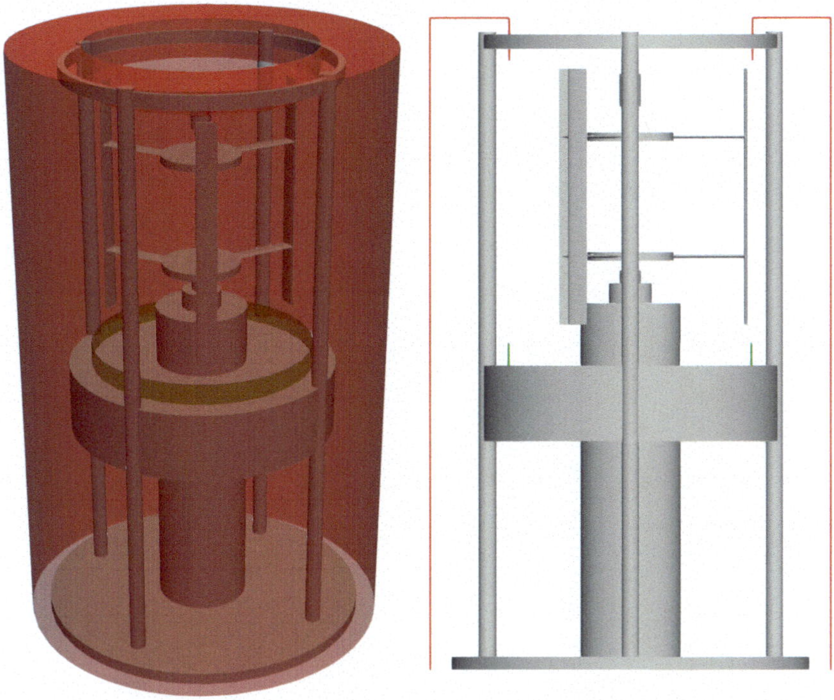

Fig. 11.29 3D wind turbine computation with the ST-SI-IGA and CGIMG. Tsunami-shelter VAWT. The two SIs in the stationary mesh that are for mesh generation flexibility, marked in different colors

the minimum number of elements), different mesh refinements, and inertial and noninertial reference frames. The Taylor numbers considered were 0 (Couette flow), 3681 (axisymmetric Taylor vortex flow), 5300 (axisymmetric Taylor vortex flow), and 8281 (wavy Taylor vortex flow). The rotational-periodicity representations are with quarter, one-third, and half domains. The no-slip conditions are enforced strongly or weakly (fluid–solid SI), depending on the test computation, and the periodicity is enforced with the ST-SI (fluid–fluid SI). The meshes are made of quadratic NURBS elements. With the ST-IGA with IGA basis functions in time, more specifically quadratic NURBS basis functions in time, we achieve exact representation of the mesh motions. The computations were with the ST-VMS. We summarize the computation for Taylor number 8281, inertial reference frame, half-domain rotational-periodicity, and mesh with $16 \times 48 \times 128$ elements in the radial, circumferential, and axial directions. Figure 11.34 shows the flow patterns.

Fig. 11.30 3D wind turbine computation with the ST-SI-IGA and CGIMG. Tsunami-shelter VAWT. Mesh in horizontal cut plane. After the initial creation (*top*) and after the mesh relaxation (*bottom*)

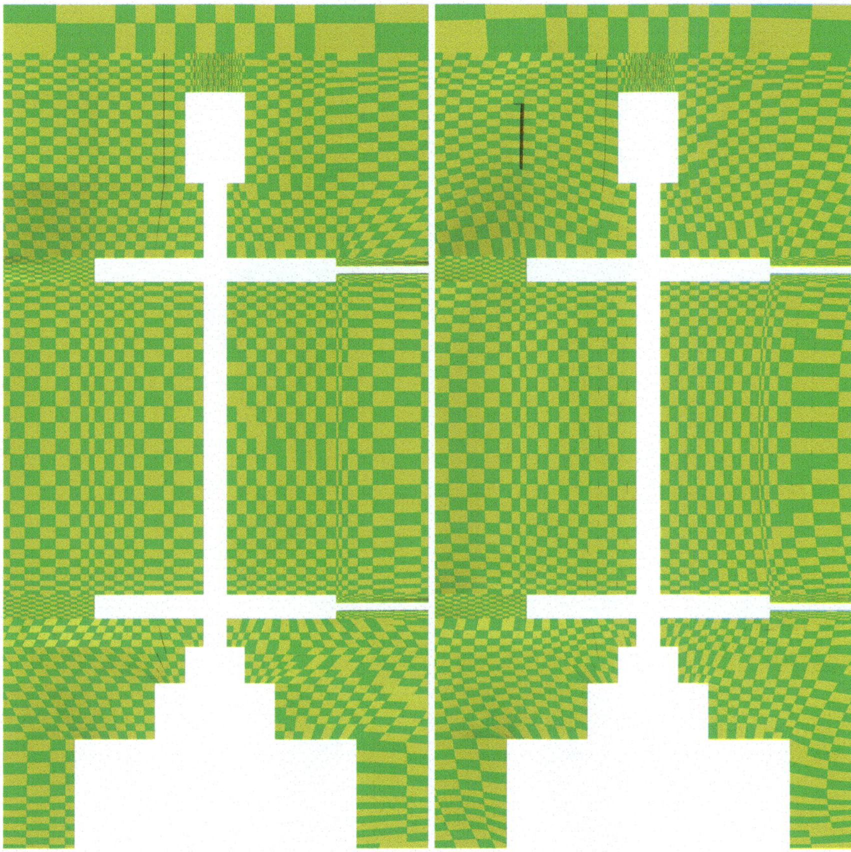

Fig. 11.31 3D wind turbine computation with the ST-SI-IGA and CGIMG. Tsunami-shelter VAWT. Mesh in vertical cut plane. After the initial creation (*left*) and after the mesh relaxation (*right*)

11.9 First 3D ST-IGA Computations of Long-Wake Flows: Wind Turbine Wake, 2021

The first 3D ST-IGA computations of long-wake flows were reported in 2021 [154], for a wind turbine wake. The computations were with the MDM, where the long wake is computed over a sequence of subdomains, instead of a single, long domain, thus somewhat reducing the computational cost. Furthermore, with the MDM, the computation over a downstream subdomain can start several turbine rotations after the computation over the upstream subdomain starts, thus reducing the computational cost even more. In the computations, the velocity data on the inflow plane comes from a previous wind turbine computation [185], extracted by projection from a plane located 10 m downstream of the turbine, which has a diameter of

Fig. 11.32 3D wind turbine computation with the ST-SI-IGA and CGIMG. Tsunami-shelter VAWT. Isosurfaces corresponding to a positive value of the second invariant of the velocity gradient tensor, colored by the velocity magnitude, for four different positions of the blade separated by 90°. Color range from *blue* to *red* corresponds to the velocity range from low to high

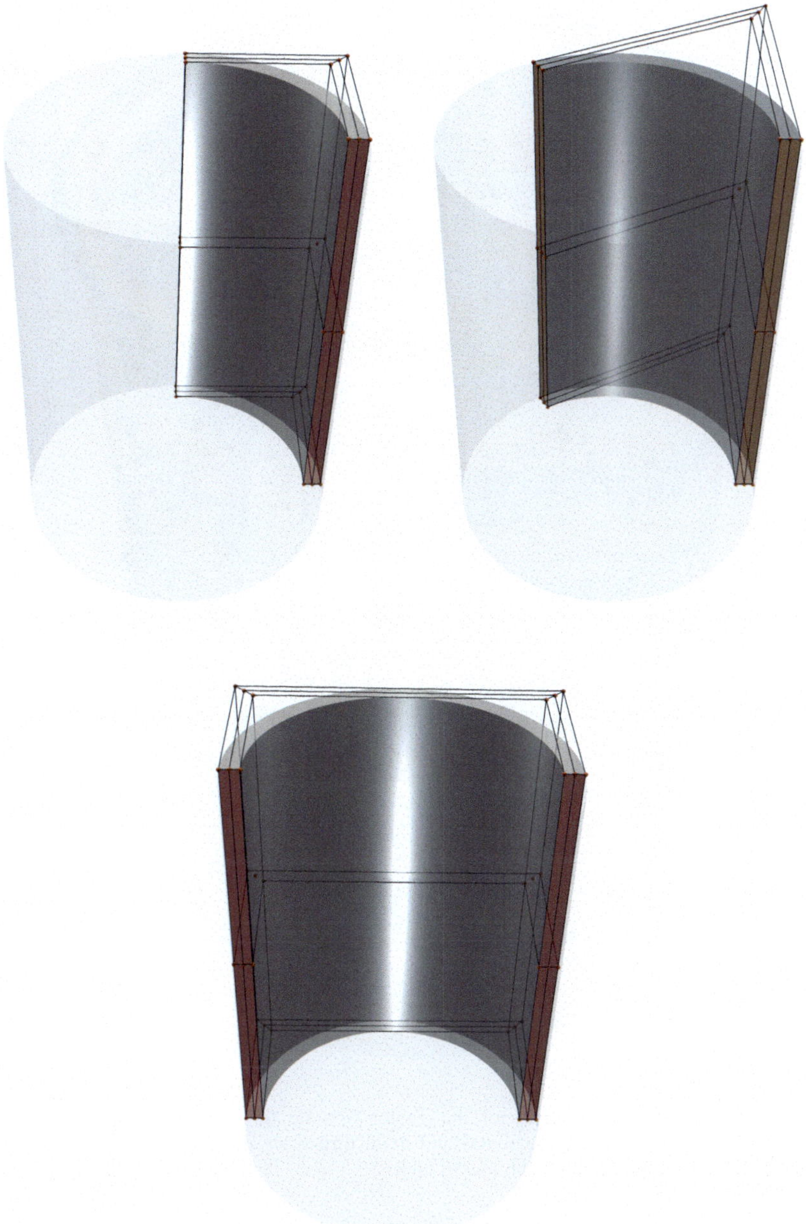

Fig. 11.33 ST-SI-IGA computations of the Taylor–Couette flow. Representation of the circular geometry by using quadratic NURBS patches with the minimum number of elements. Patches that have quarter, one-third and half-domain sizes

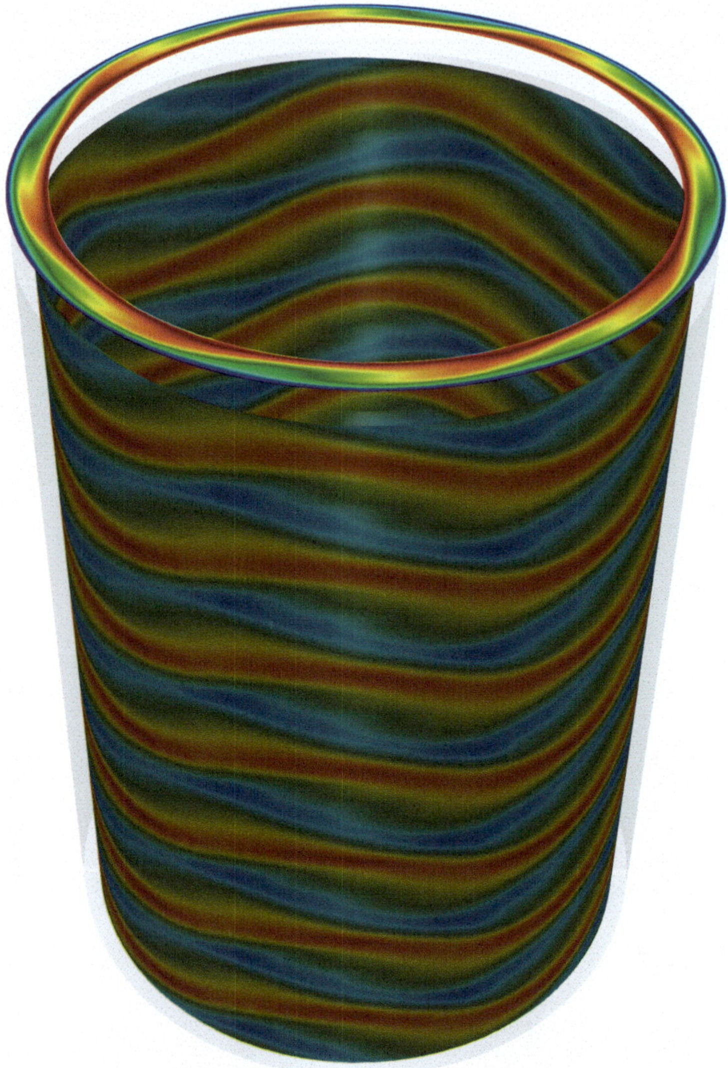

Fig. 11.34 ST-SI-IGA computations of the Taylor–Couette flow. Velocity magnitude on the sections at the midpoint along the radial and axial directions. Color range from *blue* to *red* corresponds to the velocity range from low to high

126 m. Using a method introduced in [154], the velocity data sourced from [185] is converted to data that has full temporal periodicity. The temporally-periodic data is represented with cubic B-splines in time, using the ST-C. Computations were carried out with a single, full domain and the combination of two subdomains with a total length essentially equal to the full-domain length. The computation over the downstream subdomain was under two scenarios: starting the computation at

the same actual time as the computation over the upstream subdomain starts, and five turbine rotations after. The meshes are made of quadratic NURBS elements. The number of elements in the streamwise, horizontal, and vertical directions are $116 \times 80 \times 52$ for the full domain, $58 \times 80 \times 52$ for the upstream subdomain, and $59 \times 80 \times 52$ for the downstream subdomain. The computations were with the ST-VMS. Figures 11.35 and 11.36 show the flow patterns for the two-subdomain computation with delayed start for the downstream subdomain and for the full-domain computation.

11.10 NURBS Surface-to-Volume Guided Mesh Generation (NSVGMG) Method, 2021

The NSVGMG was introduced in 2021 [62]. Like the CGIMG, it is a general-purpose IGA mesh generation method, introduced to increase the scope of IGA in computing complex-geometry problems. In the NSVGMG, NURBS patch surface meshes serve as guides in generating the patch volume meshes. The interior control points are determined independent of each other, with only a small subset of the surface control points playing a role in determining each interior point $\mathbf{x}_{\mathrm{c}}$:

$$\left(\sum_{\alpha=1}^{n_{\mathrm{pd}}} \mathbf{t}^{\alpha}\mathbf{t}^{\alpha}\right) \cdot \mathbf{x}_{\mathrm{c}} = \sum_{\alpha=1}^{n_{\mathrm{pd}}} \mathbf{t}^{\alpha}\mathbf{t}^{\alpha} \cdot \mathbf{c}^{\alpha}. \tag{11.1}$$

Here n_{pd} is the number of parametric dimensions, $\mathbf{t}^{\alpha}$ is the unit vector along the inner guide associated with the parametric direction α, and $\mathbf{c}^{\alpha}$ is the origin calculated along that guide. The NSVGMG was first used in ST-IGA computation of car and tire aerodynamics with road contact and tire deformation, reported in [62]. In the updated version of the NSVGMG introduced in [182], in the process of determining the location of an interior point in a parametric direction, more weight is given to the closer guides, with the closeness measured along the guides in the other parametric directions. Tests with 2D and 3D shapes reported in [182] show the effectiveness of the NSVGMG in generating good quality meshes, and the robustness of the updated NSVGMG even in mesh generation for complex shapes with distorted boundaries (see Fig. 11.37).

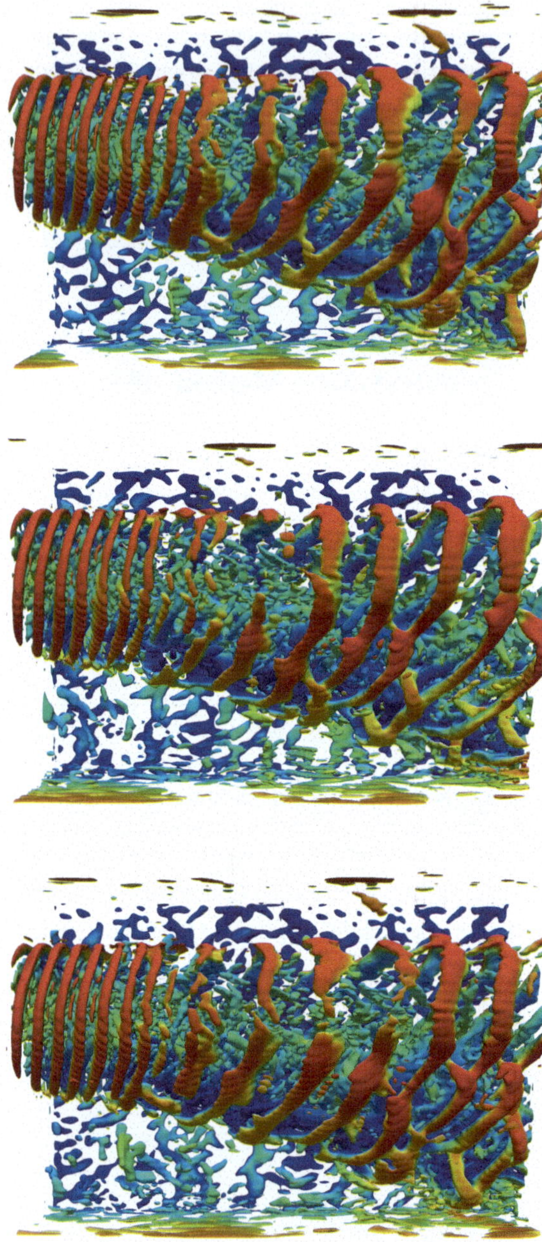

Fig. 11.35 3D ST-IGA computations of long-wake flows. Wind turbine wake. Two-subdomain computation with five-rotation delayed start for the downstream subdomain. Isosurfaces corresponding to a positive value of the second invariant of the velocity gradient tensor, colored by the velocity magnitude, at three instants. Color range from *blue* to *red* corresponds to the velocity range from low to high. Top view

Fig. 11.36 3D ST-IGA computations of long-wake flows. Wind turbine wake. Superimposed isosurfaces corresponding to a positive value of the second invariant of the velocity gradient tensor, for the full-domain computation (*red*) and two-subdomain computation with five-rotation delayed start for the downstream subdomain (*blue*). Top view

11.11 First 3D Computation of Car and Tire Aerodynamics with Near-Actual Car Body and Tire Geometries, Road Contact, Tire Deformation, Actual-Contact and High-Resolution Boundary-Layer Representations, and IGA Discretization in Space, 2021

The first 3D computation of car and tire aerodynamics with near-actual car body and tire geometries, road contact, tire deformation, actual-contact and high-resolution boundary-layer representations, and IGA discretization in space was reported in 2021 [62]. The car body and tire models are shown in Fig. 11.38. The tire model includes longitudinal grooves. A steady-state structural mechanics computation was carried out to obtain the tire deformation, resulting in the shape shown in Fig. 11.38. The computation was with the ST-VMS and ST-SI-TC-IGA. The mesh, made of 707,372 quadratic NURBS elements, was generated over five domain parts, separately, and the mesh parts are connected with SIs. Both the NSVGMG and special-purpose mesh generation methods were used. The FRHEMUM was also used for improving the quality after the initial creation of the mesh. Figure 11.39 shows the mesh. Figure 11.40 shows the mesh around the tire and wheel. Although the tire model includes only longitudinal grooves, the mesh around the tire is still

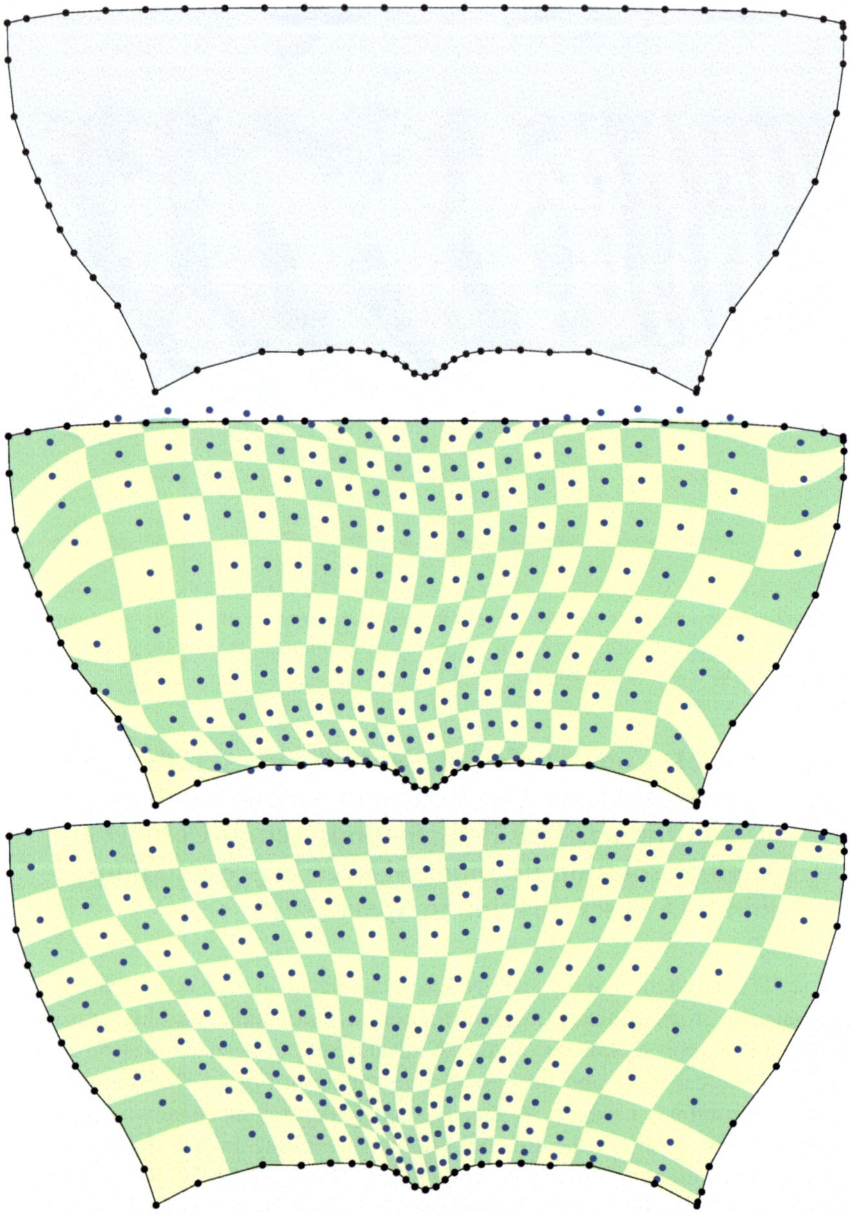

Fig. 11.37 NURBS Surface-to-Volume Guided Mesh Generation (NSVGMG) method. Tests with a complex 2D shape. Mesh along the edges (*top*). Meshes generated with the original NSVGMG (*middle*) and updated NSVGMG (*bottom*). The *blue* control points are those determined, and the checkerboard pattern is for differentiating between the elements

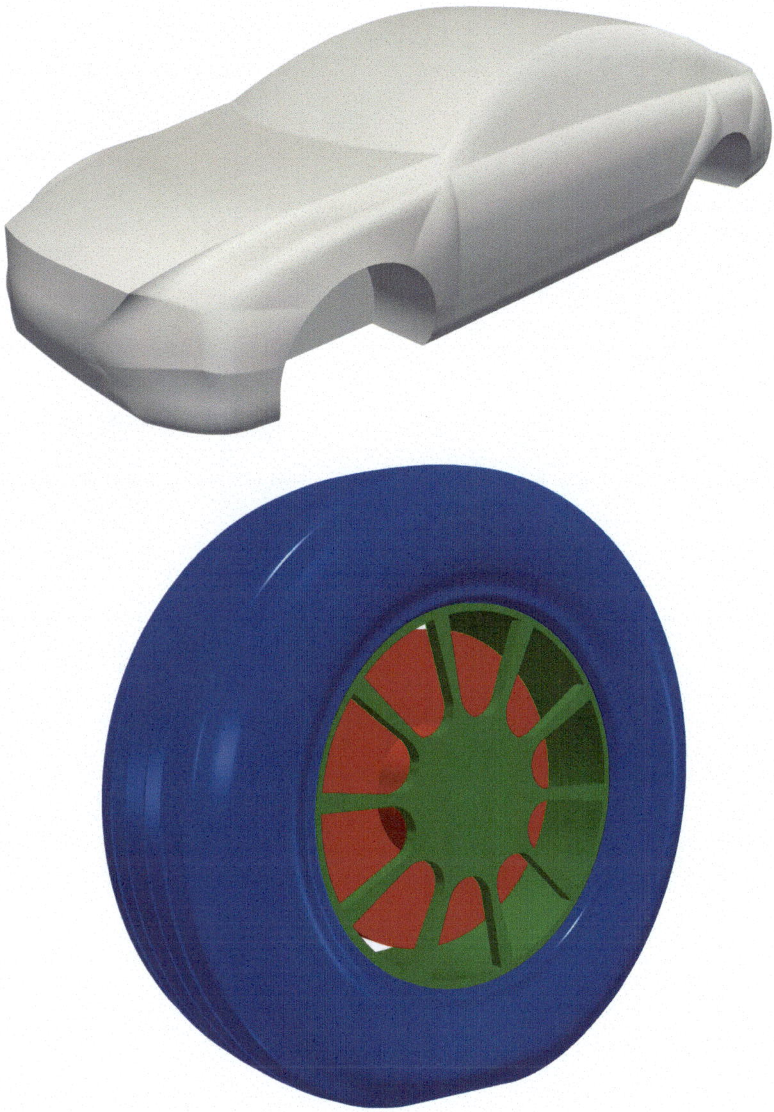

Fig. 11.38 3D computation of car and tire aerodynamics with near-actual car body and tire geometries, road contact, tire deformation, actual-contact and high-resolution boundary-layer representations, and IGA discretization in space. Car body and tire models. Tire (*blue*), wheel (*green*), and disk rotor (*red*) models

rotating and handled with the ST-TC mechanism. That is the only part of the mesh that is moving. Figure 11.41 shows the flow patterns at three instants. Figures 11.42 and 11.43 show the flow patterns around the tire–road contact areas at the same instants.

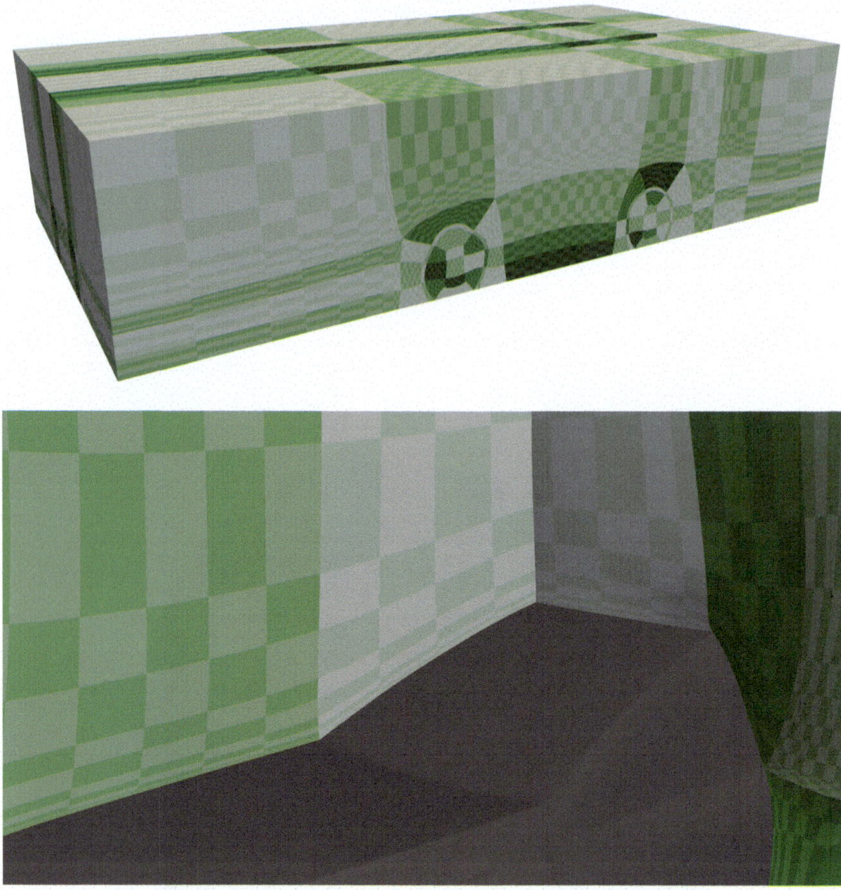

Fig. 11.39 3D computation of car and tire aerodynamics with near-actual car body and tire geometries, road contact, tire deformation, actual-contact and high-resolution boundary-layer representations, and IGA discretization in space. Mesh. Entire domain and cross-sections along the width and flow directions. *Gray* indicates the car body. The checkerboard pattern is for differentiating between the elements. The colors are for differentiating between the patches

Fig. 11.40 3D computation of car and tire aerodynamics with near-actual car body and tire geometries, road contact, tire deformation, actual-contact and high-resolution boundary-layer representations, and IGA discretization in space. Mesh around the tire and wheel. *Dark gray* indicates the tire and *light gray* indicates the wheel. The checkerboard pattern is for differentiating between the elements. The colors are for differentiating between the patches

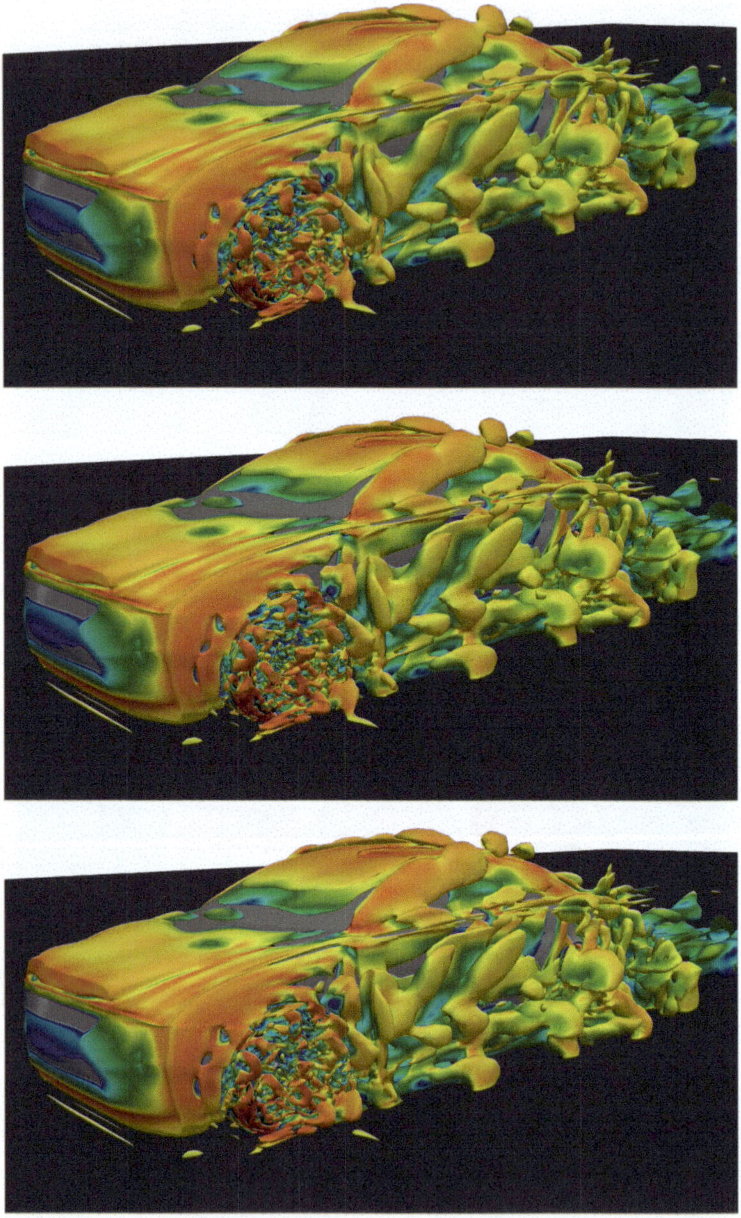

Fig. 11.41 3D computation of car and tire aerodynamics with near-actual car body and tire geometries, road contact, tire deformation, actual-contact and high-resolution boundary-layer representations, and IGA discretization in space. Isosurfaces corresponding to a positive value of the second invariant of the velocity gradient tensor, colored by the velocity magnitude, at 3 uniformly spaced instants. Color range from *blue* to *red* corresponds to the velocity range from low to high

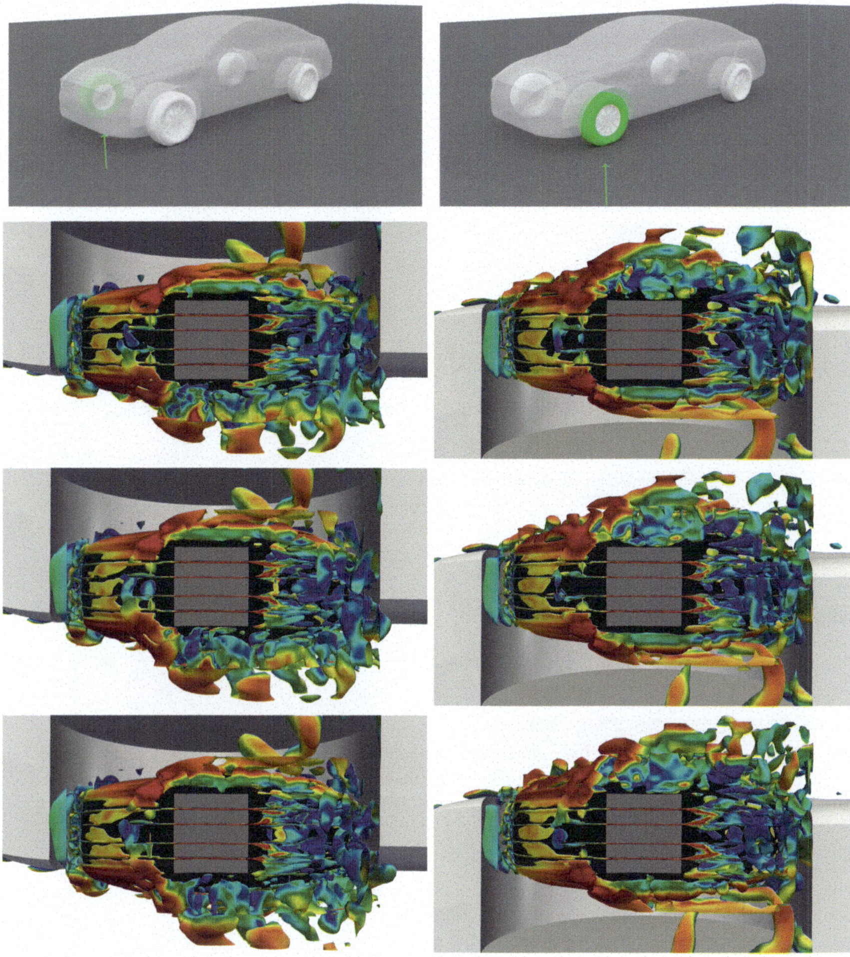

Fig. 11.42 3D computation of car and tire aerodynamics with near-actual car body and tire geometries, road contact, tire deformation, actual-contact and high-resolution boundary-layer representations, and IGA discretization in space. Isosurfaces corresponding to a positive value of the second invariant of the velocity gradient tensor, colored by the velocity magnitude, viewed from below the tires, at the same instants as in Fig. 11.41. Color range from *blue* to *red* corresponds to the velocity range from low to high. Front right and left tires. The free-stream flow is from *left* to *right*. The *dark gray* zones are the contact areas. The *top* frames show the tires viewed and the arrows indicate the viewing direction

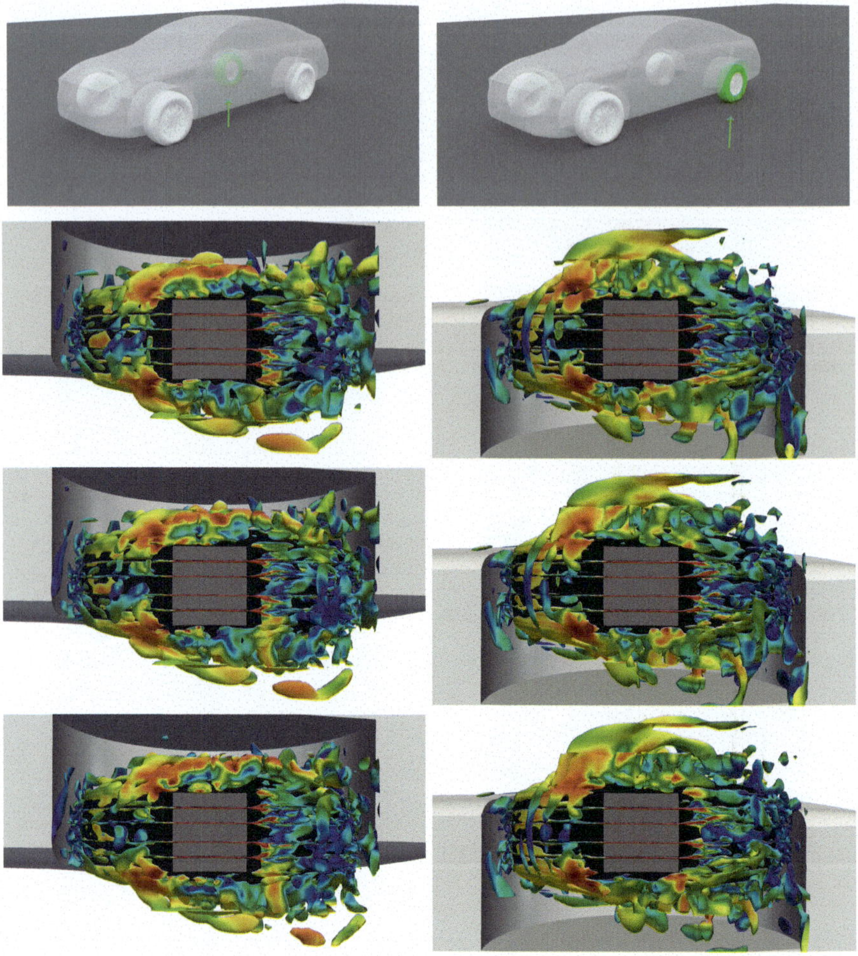

Fig. 11.43 3D computation of car and tire aerodynamics with near-actual car body and tire geometries, road contact, tire deformation, actual-contact and high-resolution boundary-layer representations, and IGA discretization in space. Isosurfaces corresponding to a positive value of the second invariant of the velocity gradient tensor, colored by the velocity magnitude, viewed from below the tires, at the same instants as in Fig. 11.41. Color range from *blue* to *red* corresponds to the velocity range from low to high. Rear right and left tires. The free-stream flow is from *left* to *right*. The *dark gray* zones are the contact areas. The *top* frames show the tires viewed and the arrows indicate the viewing direction

Chapter 12
Methods and Solutions in 2022–2024

12.1 ST-IGA Flow Computations with Higher-Order IGA Basis Functions in Time: 2D Flow Past a Cylinder, 2022

ST-IGA flow computations with higher-order IGA basis functions in time were reported in 2022 [186], for 2D incompressible flow past a cylinder at Reynolds number 100. The computations were with the ST-VMS. The spatial IGA discretization is with a mesh made of 6582 quadratic NURBS elements. The temporal IGA discretization is with B-splines elements with polynomial orders 1, 2, 3, and 4. The set of test cases computed were formed from combinations of these polynomial orders with a range of time-step sizes and other computational parameters. Here, we provide an overview of a computation with polynomial order 4. Figure 12.1 shows, at an instant in the vortex shedding cycle, the velocity magnitude.

12.2 First High-Resolution 3D Computation of Car and Tire Aerodynamics with Near-Actual Car Body and Tire Geometries, Road Contact, Tire Deformation, Actual-Contact and High-Resolution Boundary-Layer Representations, and IGA Discretization in Space, 2022

The first high-resolution 3D computation of car and tire aerodynamics with near-actual car body and tire geometries, road contact, tire deformation, actual-contact and high-resolution boundary-layer representations, and IGA discretization in space was reported in 2022 [149]. The aerodynamic influence of the car body on the tires and influence of the front-tire wake on the rear-tire aerodynamics were accounted for in the framework of the MDM. The car body and tire models are the same as

T. E. Tezduyar, K. Takizawa, *Space–Time Computational Flow Analysis*,
Modeling and Simulation in Science, Engineering and Technology,
https://doi.org/10.1007/978-3-031-88727-7_12

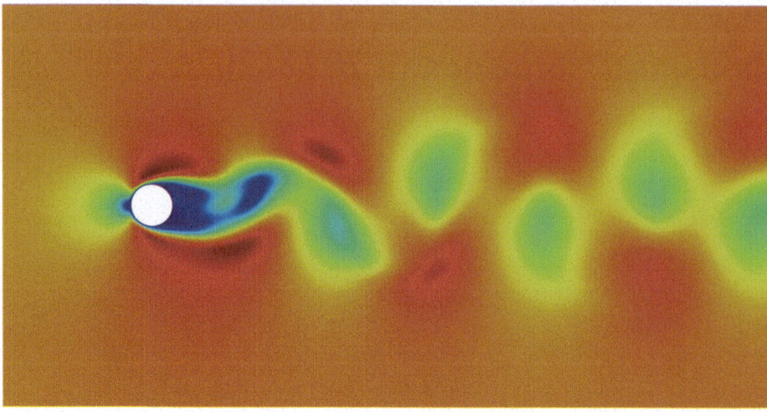

Fig. 12.1 ST-IGA flow computations with higher-order IGA basis functions in time. 2D flow past a cylinder. The discretization is with quadratic NURBS in space and quartic B-splines in time. Velocity magnitude. Color range from *blue* to *red* corresponds to the velocity range from low to high

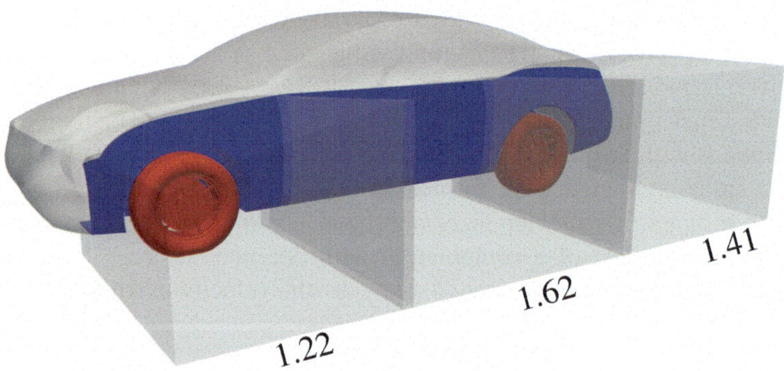

Fig. 12.2 High-resolution 3D computation of car and tire aerodynamics with near-actual car body and tire geometries, road contact, tire deformation, actual-contact and high-resolution boundary-layer representations, and IGA discretization in space. Sequence of three local domains. The dimensions are in m. *Blue* indicates the body and *red* indicates the tires, wheels, and disk rotors

in Sect. 11.11. Taking the data from the computation described in Sect. 11.11 as the global-computation data, high-resolution computations for the left set of tires were carried out over the sequence of three local domains shown in Fig. 12.2. Figure 12.3 shows the boundary conditions for the three local domains. The prescribed-velocity boundary conditions are extracted from the global solution, except for the inflow planes of the second and third local domains. The velocities at those planes come from the solutions over the first and second local domains. The stress boundary conditions are extracted from the global solution. The global data used as the extraction source is stored with the ST-C, using cubic B-splines in time.

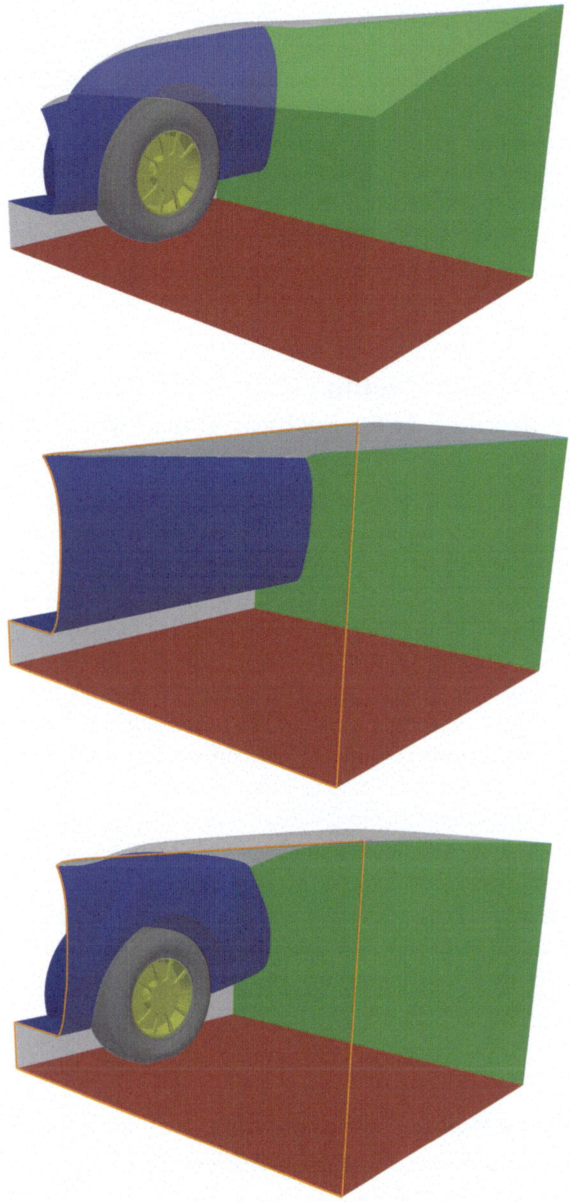

Fig. 12.3 High-resolution 3D computation of car and tire aerodynamics with near-actual car body and tire geometries, road contact, tire deformation, actual-contact and high-resolution boundary-layer representations, and IGA discretization in space. Boundary conditions for the three local domains. *Blue*: car body, *red*: ground, *green*: stress boundary condition, *dark gray*: tire, *yellow*: wheel, and *transparent*: prescribed-velocity boundaries. The two planes with *orange* frames are the inflow planes for those subdomains

The computations were with the ST-VMS and ST-SI-TC-IGA. The meshes over the first, second, and third local domains have 1,487,632, 845,000, and 1,576,232 elements. The meshes over the first and third local domains were generated over four parts each, and the meshes are connected with SIs. Both the CGIMG and special-purpose mesh generation methods were used. The FRHEMUM was also used for improving the quality after the initial creation of the mesh. Figure 12.4 shows the meshes over the three local domains. Figure 12.5 shows the local mesh around the tire and wheel. Figure 12.6 shows the flow patterns at three instants. Figure 12.7 shows the flow patterns around the tire–road contact areas at the same instants.

12.3 Carrier-Domain Method (CDM) for High-Resolution Computation of Time-Periodic Long-Wake Flows, 2022

The was introduced in 2022 [187] for high-resolution computation of time-periodic long-wake flows. The cost-effectiveness of the method makes such high-resolution computations practical in 3D. The CDM is a descendant of the MDM, originally intended also for cost-effective computation of long-wake flows. In the CDM, a short segment of the wake domain, the moves in the free-stream direction, from the beginning of the long wake domain to the end. The motion is with a velocity that sustains the outflow nature of the CD outflow plane. The data at the moving inflow plane comes from the time-periodic data computed at an earlier position of the CD. With the high mesh resolution that can easily be afforded over the short domain segment, the wake flow patterns can be carried, with superior accuracy, far downstream. Computing the long-wake flow with a high-resolution moving mesh that covers a short segment of the wake domain at any instant during the computation would certainly be far more cost-effective than computing it with a high-resolution fixed mesh that covers the entire length. A CDM version where the CD moves in a discrete fashion was also introduced in [187]. The abbreviations and identify the versions with continuous and discrete motion of the CD (see Figs. 12.8 and 12.9). In the figures, T is the period, U is the free-stream velocity, and $L = UT$.

12.4 First Long-Wake Flow Computation with the CDM and ST-IGA: 2D Flow Past a Cylinder, 2022

The first long-wake flow computation with the CDM and ST-IGA was reported in 2022 [187], in the same article where the CDM was introduced. The computation was for 2D flow past a cylinder at Reynolds number 100, with the long-wake flow computed up to 350 diameters downstream of the cylinder, far enough to see the secondary vortex street. For comparison, the wake was computed also with the MDM. The computations were with the ST-VMS. The meshes are made of quadratic

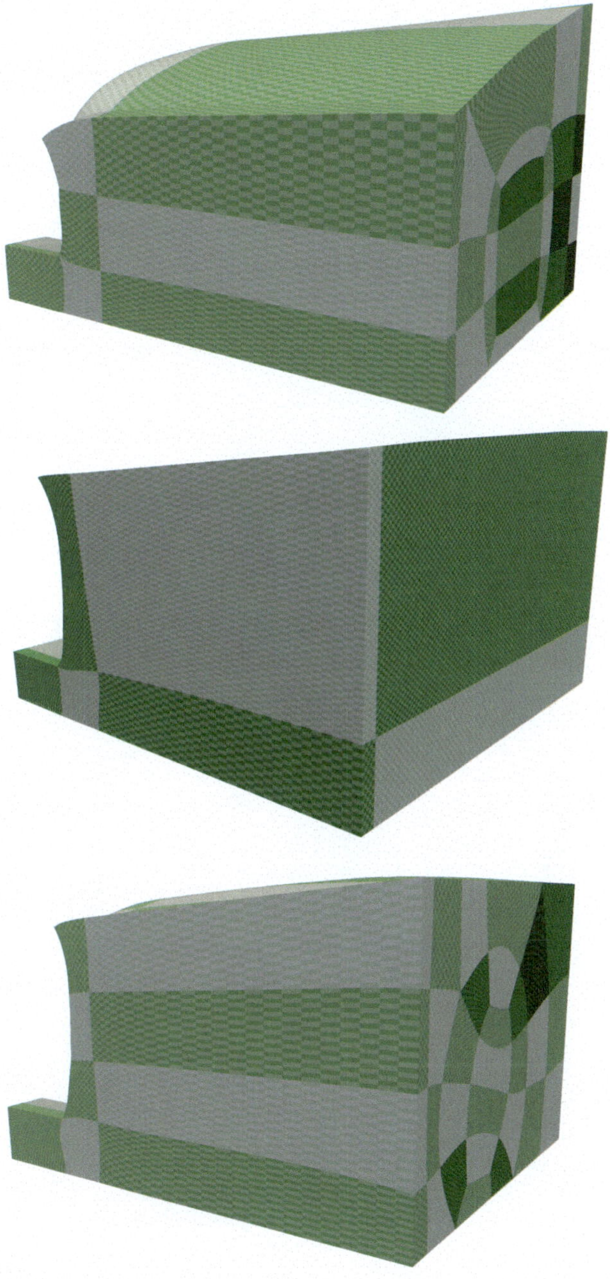

Fig. 12.4 High-resolution 3D computation of car and tire aerodynamics with near-actual car body and tire geometries, road contact, tire deformation, actual-contact and high-resolution boundary-layer representations, and IGA discretization in space. Meshes over the three local domains. The checkerboard pattern is for differentiating between the elements. The colors are for differentiating between the patches

Fig. 12.5 High-resolution 3D computation of car and tire aerodynamics with near-actual car body and tire geometries, road contact, tire deformation, actual-contact and high-resolution boundary-layer representations, and IGA discretization in space. Local mesh around the tire and wheel. *Dark gray* indicates the tire and *light gray* indicates the wheel. The checkerboard pattern is for differentiating between the elements. The colors are for differentiating between the patches

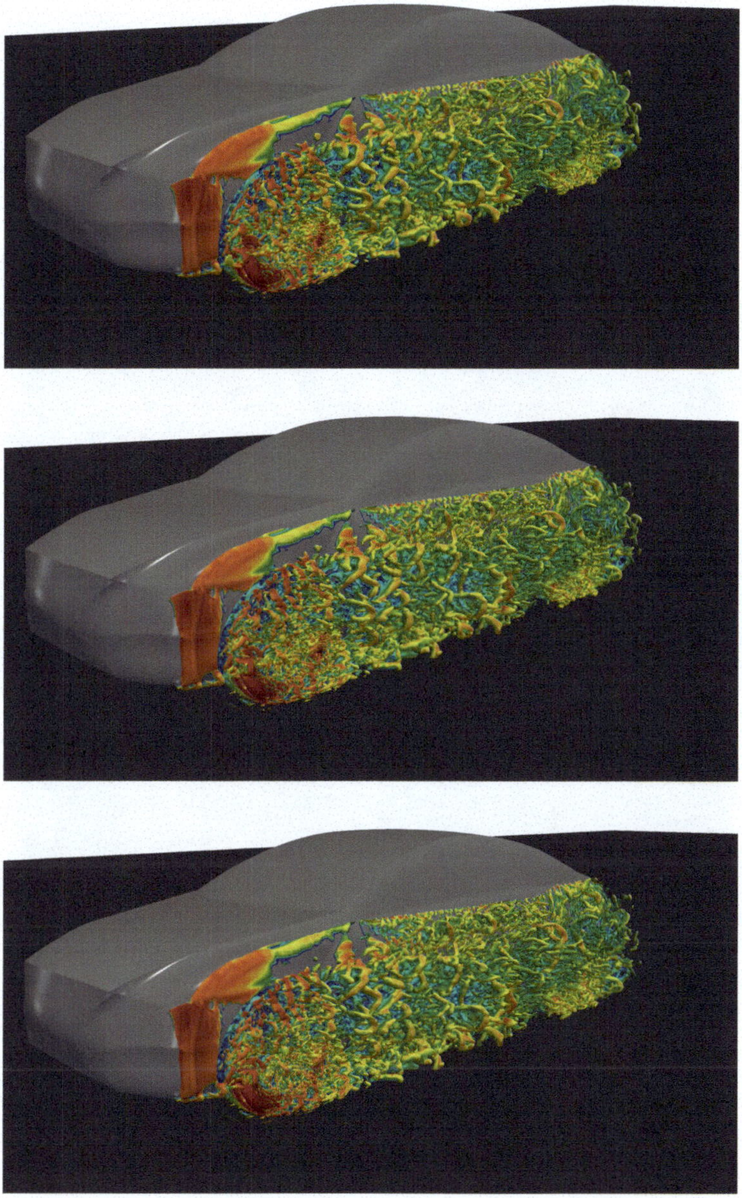

Fig. 12.6 High-resolution 3D computation of car and tire aerodynamics with near-actual car body and tire geometries, road contact, tire deformation, actual-contact and high-resolution boundary-layer representations, and IGA discretization in space. Isosurfaces corresponding to a positive value of the second invariant of the velocity gradient tensor, colored by the velocity magnitude, at 3 uniformly spaced instants. Color range from *blue* to *red* corresponds to the velocity range from low to high

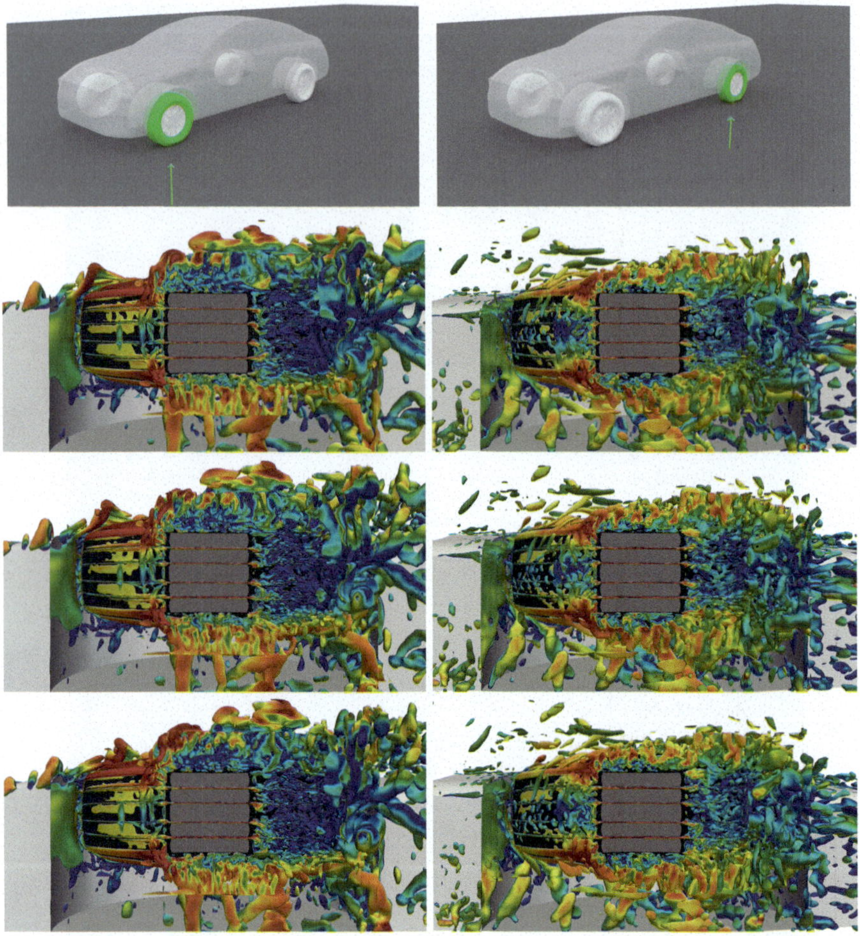

Fig. 12.7 High-resolution 3D computation of car and tire aerodynamics with near-actual car body and tire geometries, road contact, tire deformation, actual-contact and high-resolution boundary-layer representations, and IGA discretization in space. Isosurfaces corresponding to a positive value of the second invariant of the velocity gradient tensor, colored by the velocity magnitude, viewed from below the tires, at the same instants as in Fig. 12.6. Color range from *blue* to *red* corresponds to the velocity range from low to high. The free-stream flow is from *left* to *right*. The *dark gray* zones are the contact areas. The top frames show the tires viewed and the arrows indicate the viewing direction

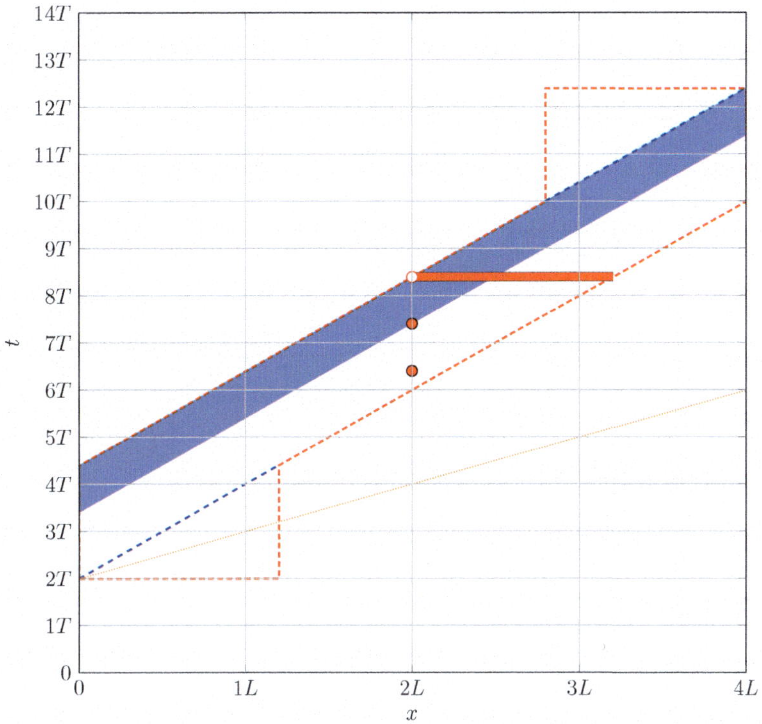

Fig. 12.8 CDM-C. The CD (*red*), the ST domain swept by that (enclosed by the *red* dashed lines), candidate data extraction instants going back in time one period at a time (*black* circles with *red* filling), and reporting ST domain (shaded in *blue*)

NURBS elements. First the cylinder short-wake flow was computed to establish the wake-flow time period and to generate the source data for the inflow velocity of the long-wake flow. The mesh for that has 6582 elements. Figure 12.10 shows the flow patterns for the short-wake flow. Using the time-periodic data from that, the long-wake flow was computed, first with a single, second subdomain ("SD-2") in the framework of the MDM, and then with the CDM-C, using the carrier domain CD. In the MDM terminology, the domain used in computation of the cylinder short-wake flow becomes "SD-1." The mesh for SD-2 has 1392×64 elements, and the mesh for CD has 68×64 elements. Figure 12.11 shows the flow patterns for the long-wake flow from the two computations. They are in full agreement.

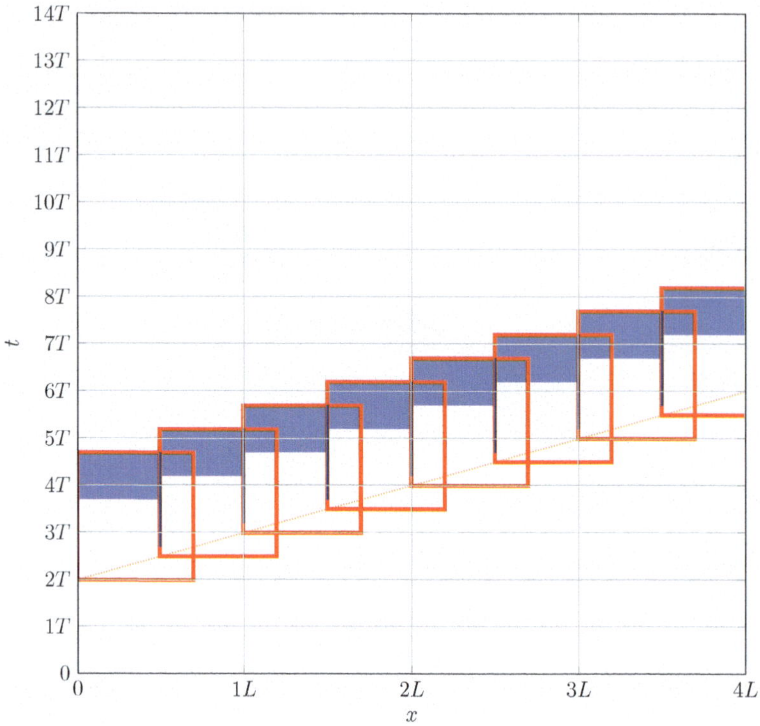

Fig. 12.9 CDM-D. The ST subdomains (*red*) and reporting ST domain (shaded in *blue*)

Fig. 12.10 Long-wake flow computation with the CDM and ST-IGA. 2D flow past a cylinder. Short-wake flow. Vorticity. Color range from *blue* to *red* corresponds to the vorticity range from low to high

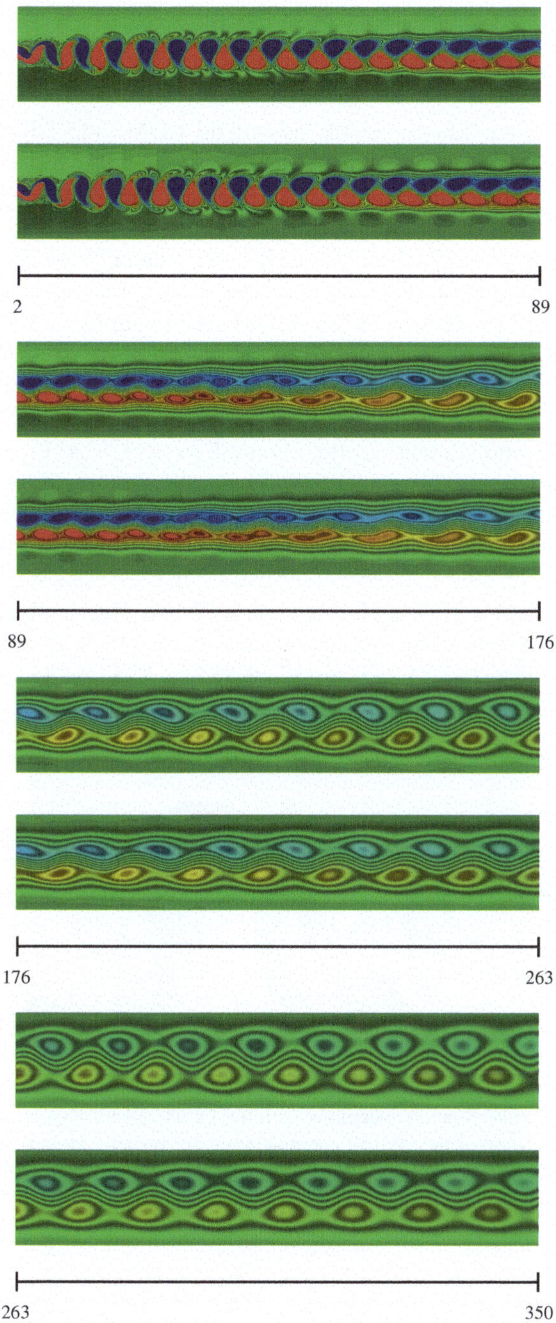

Fig. 12.11 Long-wake flow computation with the CDM and ST-IGA. 2D flow past a cylinder. Vorticity. MDM (*top*) and CDM-C (*bottom*). Color range from *blue* to *red* corresponds to the vorticity range from low to high

12.5 First 3D Long-Wake Flow Computations with the CDM and ST-IGA, 2023

The first 3D long-wake flow computations with the CDM and ST-IGA were reported in 2023 [188], for flow past a cylinder at Reynolds number 100 and for the wind turbine wake flow discussed in Sect. 11.9. The computations were with the ST-VMS and CDM-C. The meshes are made of quadratic NURBS elements. The computation for the cylinder long-wake flow is essentially the 3D version of the CDM computation in Sect. 12.4, and the computation for the wind turbine long-wake flow is the CDM version of the MDM computations in Sect. 11.9. For the cylinder wake, first the short-wake flow was computed to establish the wake-flow time period and to generate the source data for the inflow velocity of the long-wake flow. The mesh for that has 6582×16 elements, with the mesh in the 2D plane having the same structure as the 2D mesh in Sect. 12.4. Figure 12.12 shows the flow patterns for the short-wake flow. Using the time-periodic data from that, the long-wake flow was computed with the CDM-C, where the mesh for CD has $60 \times 64 \times 16$ elements. Figure 12.13 shows the flow patterns for the long-wake flow. For the wind turbine wake, for comparison, first the wake was computed with the full wake domain (FWD), and then with CDM-C. The number of elements in the streamwise, horizontal, and vertical directions are $118 \times 80 \times 52$ for the FWD and $14 \times 80 \times 52$ for the CD. Figures 12.14 and 12.15 show the flow patterns for the CDM-C and FWD computations.

Fig. 12.12 3D long-wake flow computations with the CDM and ST-IGA. Flow past a cylinder. Short-wake flow. 3D view of the vertical component of the vorticity at an instant after reaching time-periodicity. Color range from *blue* to *red* corresponds to the vorticity range from low to high

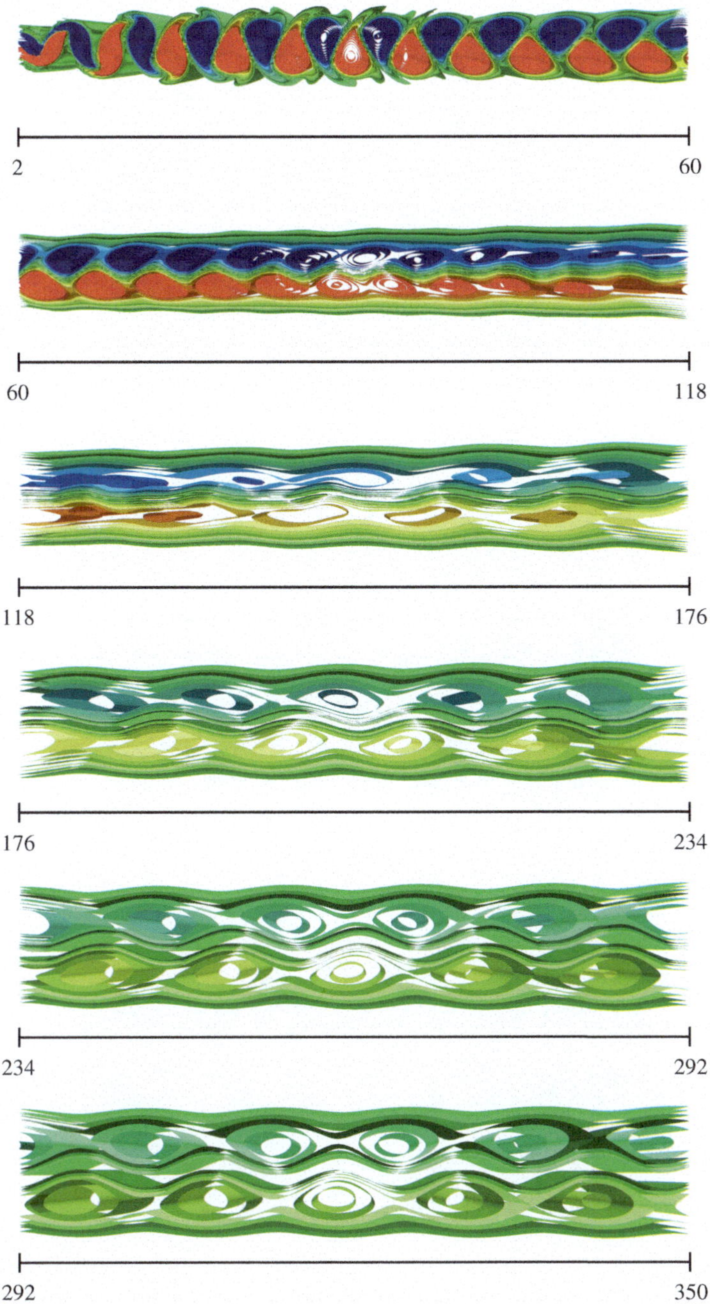

Fig. 12.13 3D long-wake flow computations with the CDM and ST-IGA. Flow past a cylinder. CDM-C. 3D view of the vertical component of the vorticity. Color range from *blue* to *red* corresponds to the vorticity range from low to high

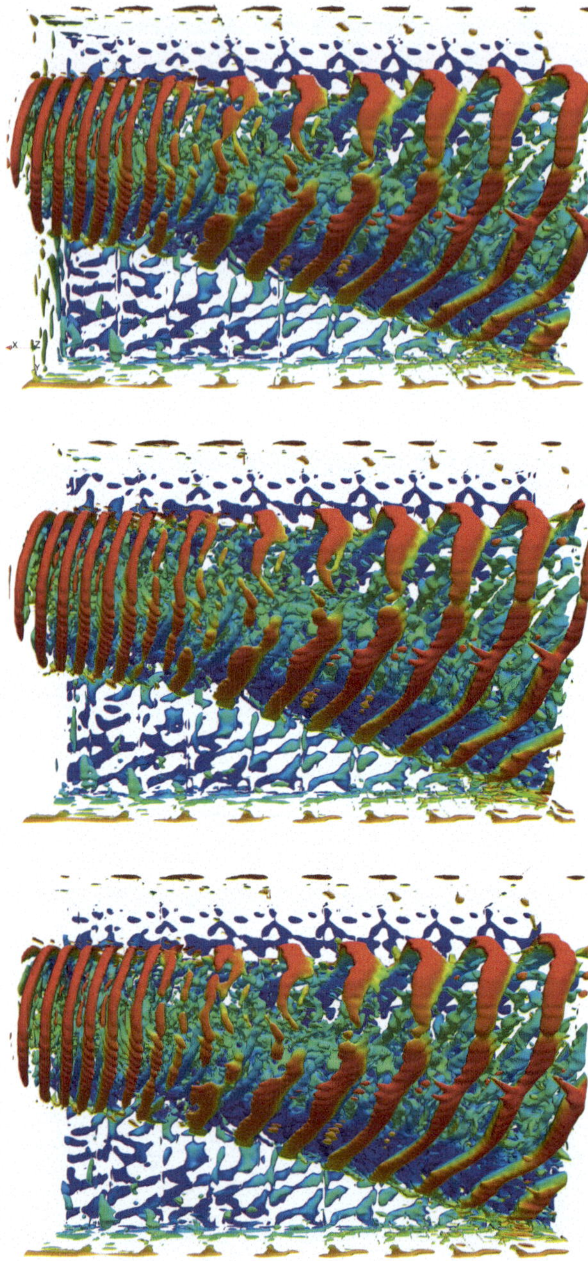

Fig. 12.14 3D long-wake flow computations with the CDM and ST-IGA. Wind turbine wake. CDM-C. Isosurfaces corresponding to a positive value of the second invariant of the velocity gradient tensor, colored by the velocity magnitude, at three instants. Color range from *blue* to *red* corresponds to the velocity range from low to high. Top view

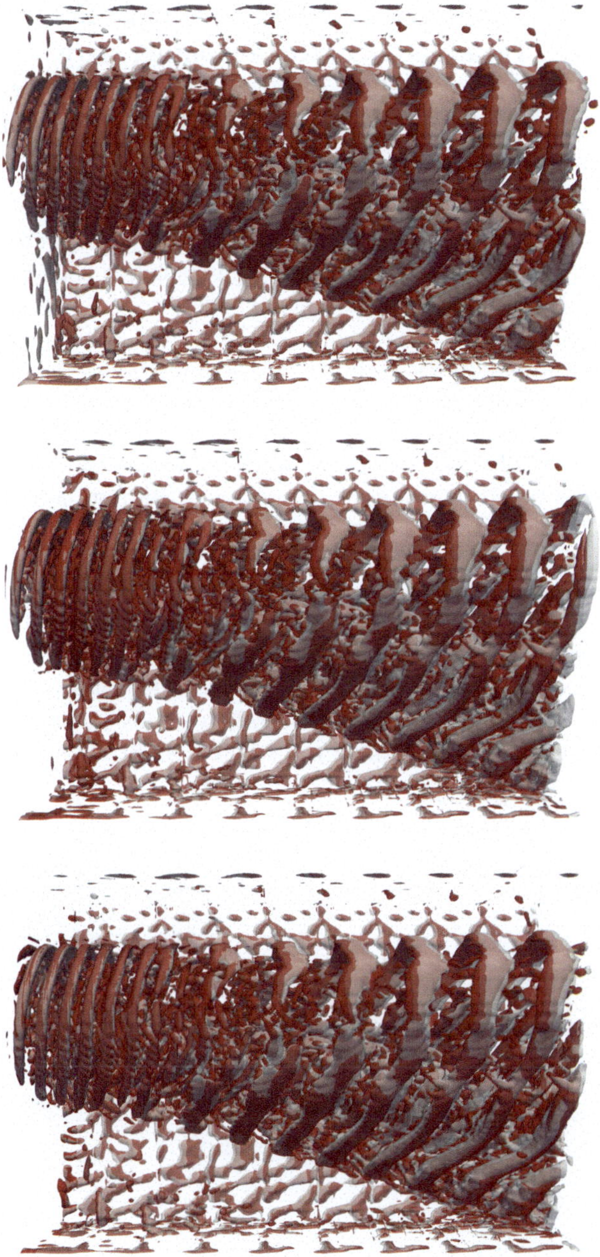

Fig. 12.15 3D long-wake flow computations with the CDM and ST-IGA. Wind turbine wake. FWD and CDM-C. Superimposed isosurfaces corresponding to a positive value of the second invariant of the velocity gradient tensor, for the FWD (*gray*) and CDM-C (*red*), at the same instants as in Fig. 12.14. Top view

12.6 First 3D Computation of Tire Aerodynamics with Complex Tread Pattern, Road Contact, Tire Deformation, Actual-Contact and High-Resolution Boundary-Layer Representations, and IGA Discretization in Space, 2024

The first 3D computation of tire aerodynamics with complex tread pattern, road contact, tire deformation, actual-contact and high-resolution boundary-layer representations, and IGA discretization in space was reported in 2024 [189]. The tire model has the same diameter and width as in Sect. 11.11. This time, however, the tire has a complex, asymmetric tread pattern, and that brings the model closer to the actual tire. The tread pattern has 28-fold rotational periodicity. Figure 12.16 shows the tread pattern. Figure 12.17 shows the mesh inside the SI enclosing the undeformed tire, to illustrate the mesh structure conceptually. The deformation of the treads is obtained by using the deformation data from Sect. 11.11 and the mapping from the undeformed geometry. Figure 12.18 shows the tire deformation pattern. The mesh is made of 656,573 quadratic NURBS elements. The CGIMG was used in generating the tire surface mesh. The NSVGMG was used in generating the mesh inside the SI enclosing the deformed tire. After the initial creation of the mesh, its quality was improved by mesh relaxation with the FRHEMUM. The mesh inside the SI is created for each time step in the first 1/28 rotation of the tire, and the meshes created are reused in the subsequent 1/28 rotations. Figure 12.19 shows the mesh at two instants. The computation was with the ST-SUPS and ST-SI-TC-IGA. Figures 12.20, 12.21, and 12.22 show the flow patterns.

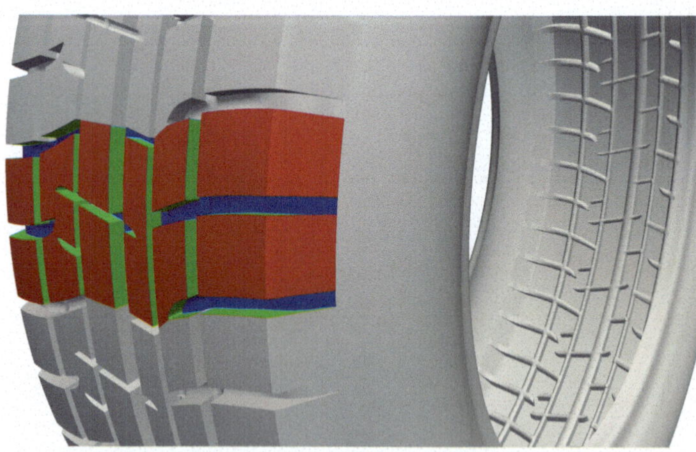

Fig. 12.16 3D computation of tire aerodynamics with complex tread pattern, road contact, tire deformation, actual-contact and high-resolution boundary-layer representations, and IGA discretization in space. Tire surface rotational-periodicity cell of the tread pattern. *Red*, *blue*, and *green* indicate the tread surface, groove bottom, and groove wall

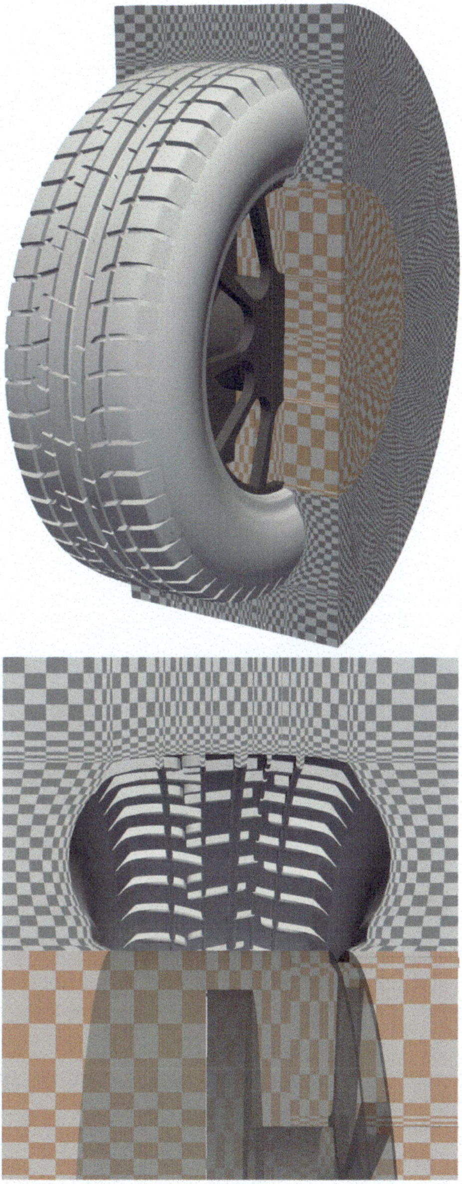

Fig. 12.17 3D computation of tire aerodynamics with complex tread pattern, road contact, tire deformation, actual-contact and high-resolution boundary-layer representations, and IGA discretization in space. The mesh inside the SI enclosing the undeformed tire. This mesh is just for illustrating the mesh structure conceptually. *Gray* indicates the mesh between the tire and the SI, and *orange* indicates the mesh between the wheel and the SI. There is a second SI, which is between the *gray* and *orange* meshes, and that is for mesh generation flexibility. The checkerboard pattern is for differentiating between the elements. The *bottom* figure includes the inside view of the tire surface with treads and the transparent view of the wheel

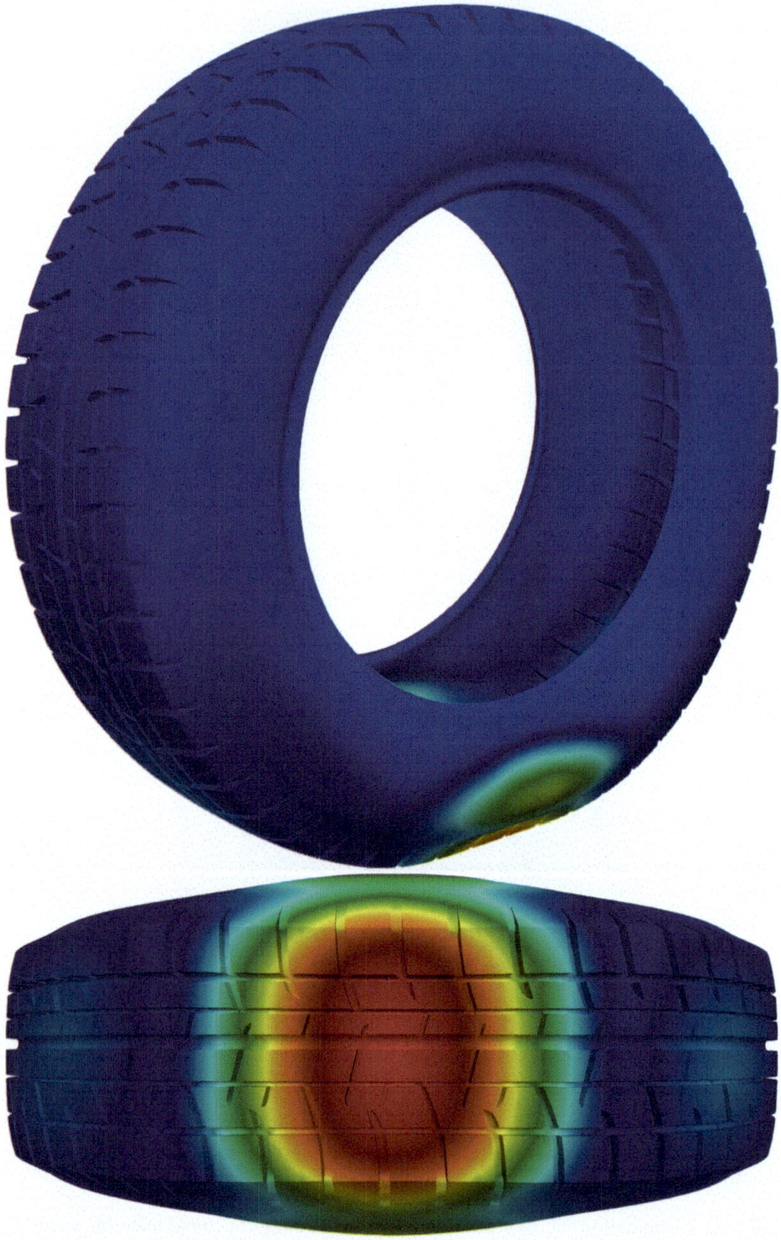

Fig. 12.18 3D computation of tire aerodynamics with complex tread pattern, road contact, tire deformation, actual-contact and high-resolution boundary-layer representations, and IGA discretization in space. Tire deformation pattern. Color range from *blue* to *red* corresponds to the deformation range from low to high

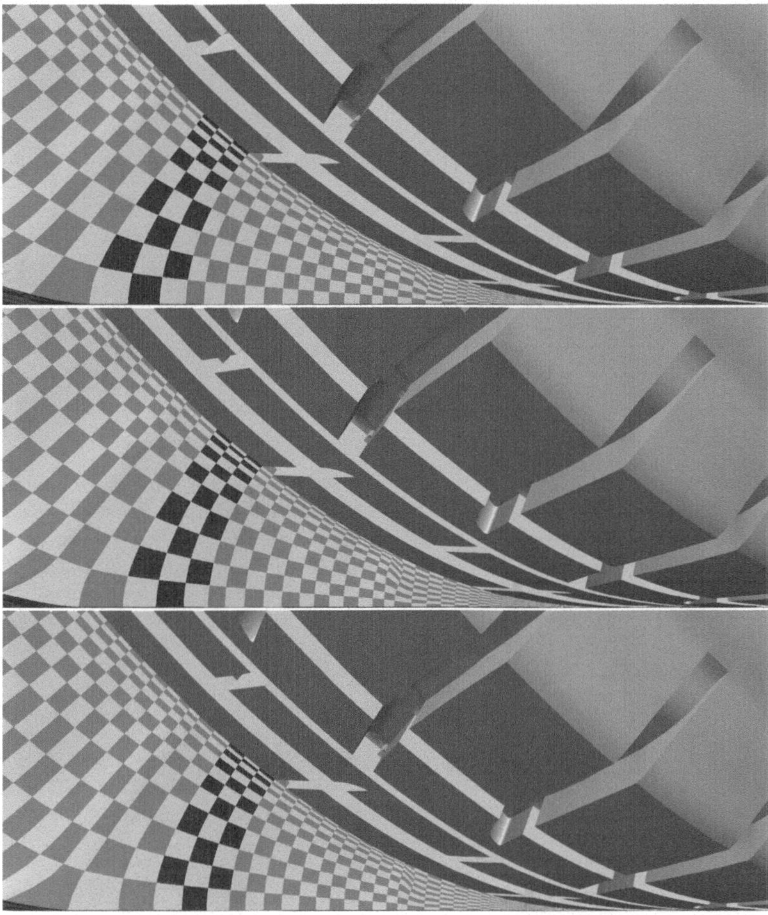

Fig. 12.19 3D computation of tire aerodynamics with complex tread pattern, road contact, tire deformation, actual-contact and high-resolution boundary-layer representations, and IGA discretization in space. The mesh at a cross-section placed centrally in the width direction and the tire surface, viewed near the contact area, at three instants. The checkerboard pattern is for differentiating between the elements. The coloring of a band of elements is for easier visualization of the mesh motion

12.7 First 3D Wind Turbine Rotor and Tower Aerodynamics Computation with the ST-SI-IGA and CGIMG, 2024

The first 3D wind turbine rotor and tower aerodynamics computation with the ST-SI-IGA and CGIMG was reported in 2024 [190], for the NREL 5MW offshore baseline wind turbine. Figure 12.23 shows the model. The ST-IGA with IGA basis functions in time is used for the temporal representation of the rotor motion, enabling us to represent the circular paths associated with that motion exactly and

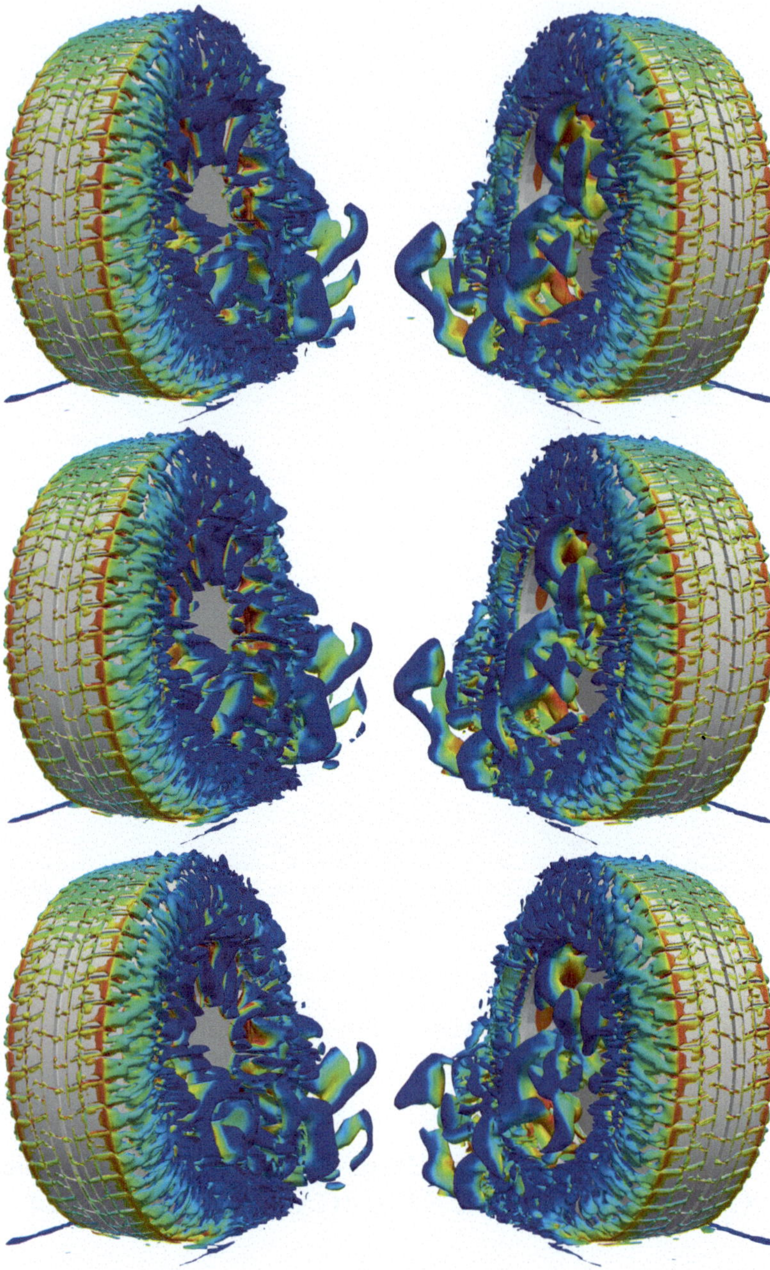

Fig. 12.20 3D computation of tire aerodynamics with complex tread pattern, road contact, tire deformation, actual-contact and high-resolution boundary-layer representations, and IGA discretization in space. Isosurfaces corresponding to a positive value of the second invariant of the velocity gradient tensor, colored by the magnitude of the velocity relative to the road, at three instants. Color range from *blue* to *red* corresponds to the velocity range from low to high. Showing the outer (*left*) and inner (*right*) sides of the tire, which is rolling toward the reader

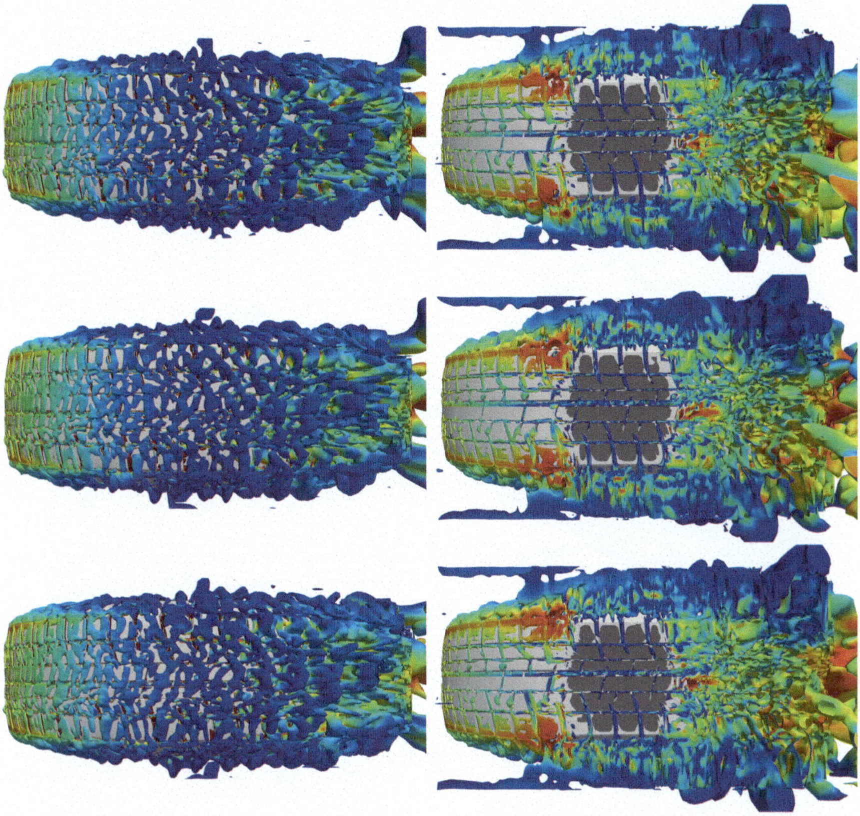

Fig. 12.21 3D computation of tire aerodynamics with complex tread pattern, road contact, tire deformation, actual-contact and high-resolution boundary-layer representations, and IGA discretization in space. Isosurfaces corresponding to a positive value of the second invariant of the velocity gradient tensor, colored by the magnitude of the velocity relative to the road, at the same instants as in Fig. 12.20. Color range from *blue* to *red* corresponds to the velocity range from low to high. Viewed from above (*left*) and below (*right*) the tire. The *gray* zones are the contact areas. The tire is rolling to the left

specify a constant angular velocity corresponding to the invariant speeds along those paths. The mesh is made of 571,158 quadratic NURBS elements. It was generated with the CGIMG. The mesh structure around the rotor and the SIs are shown in Figs. 12.24 and 12.25. The blade SIs are for mesh generation flexibility, and the disk-shaped rotor SI is where we have slip between the rotating disk and the rest of the mesh. Figure 12.26 shows the blade surface mesh and the cross-section mesh inside the blade SI. Figure 12.27 shows the mesh on the inner and outer surfaces of the blade SI. Figure 12.28 shows the cross-sections of the mesh between the blade SIs and rotor SI. Figure 12.29 shows the mesh on the inner surface of the rotor SI and the tower surface mesh. The three mesh zones between the blades and blade SIs have total 51,882 elements, and the zone between the blade SIs and rotor SI has

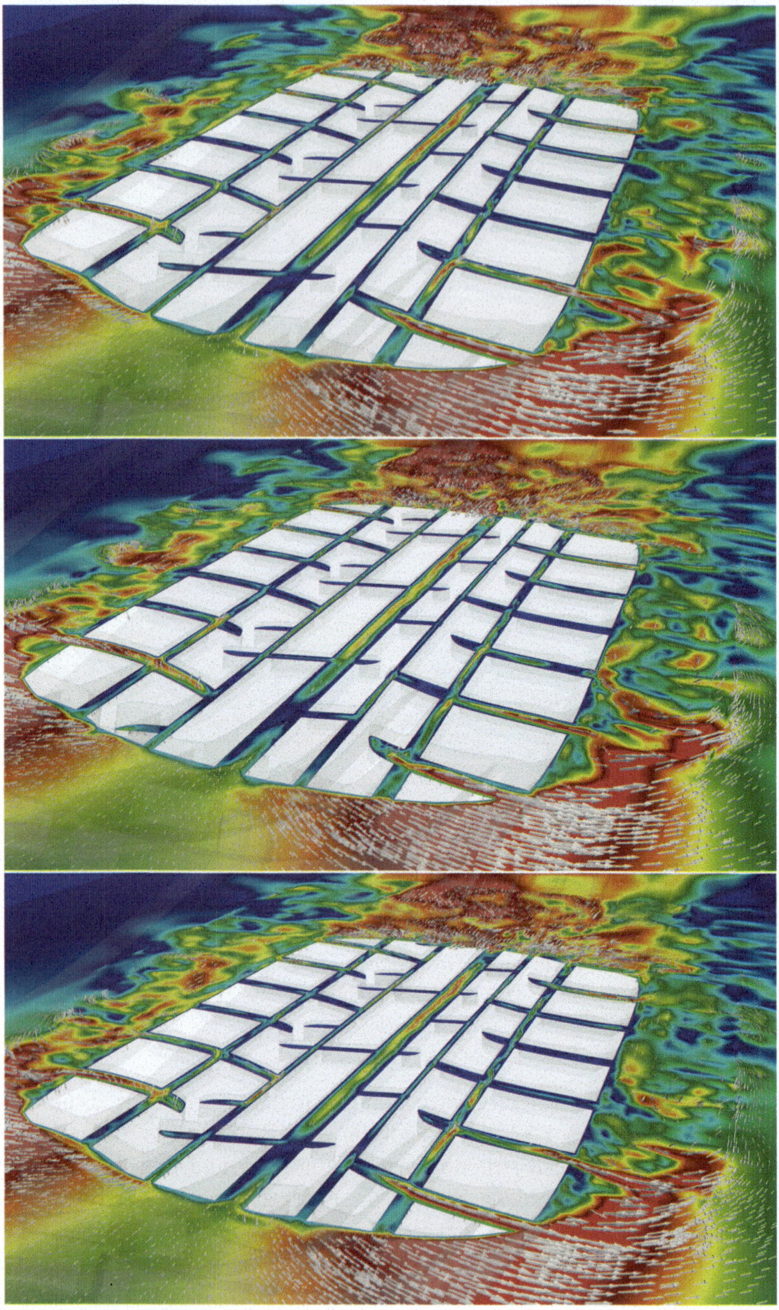

Fig. 12.22 3D computation of tire aerodynamics with complex tread pattern, road contact, tire deformation, actual-contact and high-resolution boundary-layer representations, and IGA discretization in space. Magnitude of the velocity relative to the road, on a plane 5.9 mm above the road, at the same instants as in Fig. 12.20. The velocity vectors are also relative to the road. The tire is rolling toward the reader

Fig. 12.23 3D wind turbine rotor and tower aerodynamics computation with the ST-SI-IGA and CGIMG. Model

89,748 elements. Figure 12.30 shows the entire mesh, together with the blade and rotor SIs. Figure 12.31 shows the mesh on the inner and outer surfaces of the rotor SI. The computation was with the ST-VMS. Figure 12.32 shows the flow patterns at an instant. The torque generated is shown in Fig. 12.33. Figure 12.34 shows the pressure coefficient over the rotor. Figures 12.35 and 12.36 show the velocity magnitude over different cross-sections.

Fig. 12.24 3D wind turbine rotor and tower aerodynamics computation with the ST-SI-IGA and CGIMG. Mesh structure around the rotor. Blade SIs (*green*) and the rotor SI (*gray*)

Fig. 12.25 3D wind turbine rotor and tower aerodynamics computation with the ST-SI-IGA and CGIMG. Blade SIs (*green*)

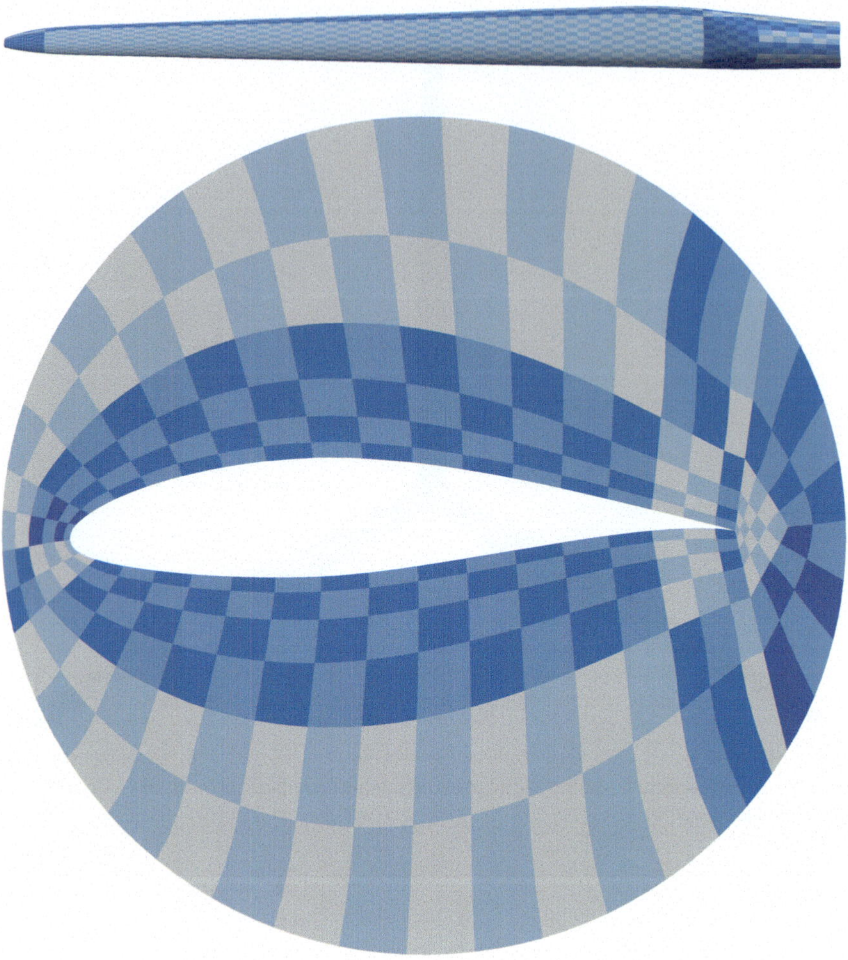

Fig. 12.26 3D wind turbine rotor and tower aerodynamics computation with the ST-SI-IGA and CGIMG. Blade surface mesh and the cross-section mesh inside the blade SI at approximately $0.95R$. The checkerboard pattern is for differentiating between the NURBS elements, and the color shades are for differentiating between the NURBS patches

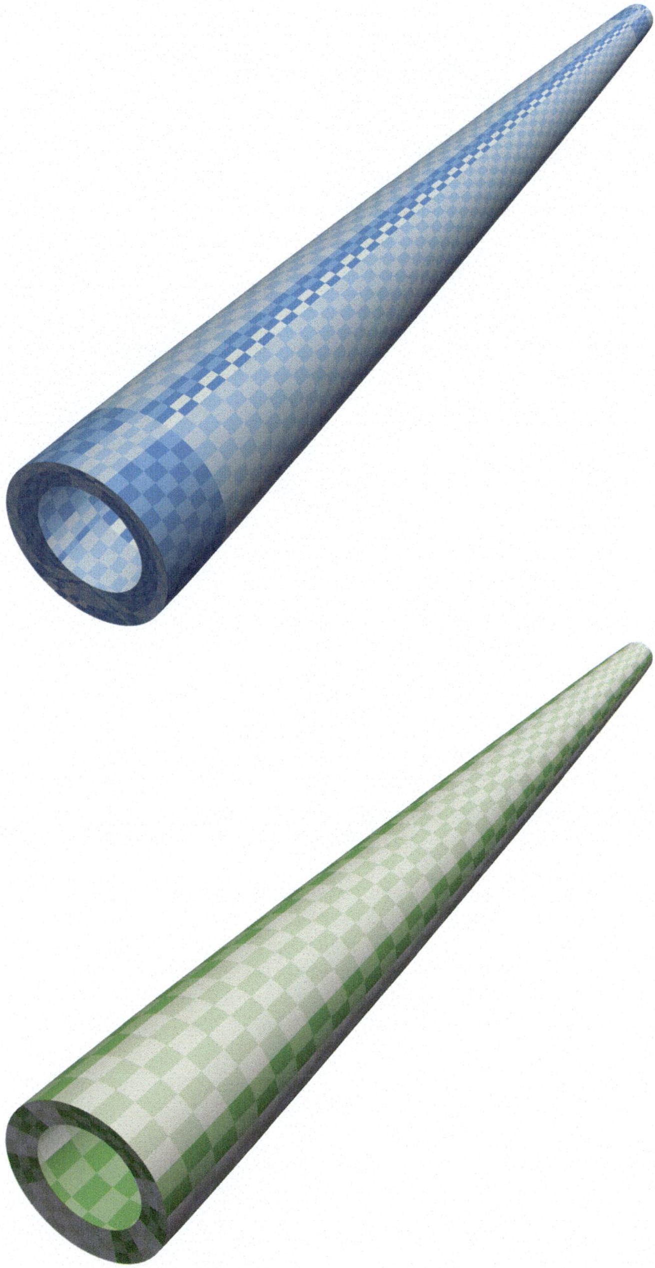

Fig. 12.27 3D wind turbine rotor and tower aerodynamics computation with the ST-SI-IGA and CGIMG. Nonmatching meshes on the inner (*blue*) and outer (*green*) surfaces of a blade SI. The checkerboard pattern is for differentiating between the NURBS elements, and the color shades are for differentiating between the NURBS patches

Fig. 12.28 3D wind turbine rotor and tower aerodynamics computation with the ST-SI-IGA and CGIMG. Cross-sections of the mesh between the blade SIs and rotor SI. The circular cross-section cuts the blades centrally. The checkerboard pattern is for differentiating between the NURBS elements, and the color shades are for differentiating between the NURBS patches

Fig. 12.29 3D wind turbine rotor and tower aerodynamics computation with the ST-SI-IGA and CGIMG. Mesh on the inner surface of the rotor SI and the tower surface mesh. The checkerboard pattern is for differentiating between the NURBS elements, and the color shades are for differentiating between the NURBS patches

Fig. 12.30 3D wind turbine rotor and tower aerodynamics computation with the ST-SI-IGA and CGIMG. Cross-sections of the mesh. The checkerboard pattern is for differentiating between the NURBS elements, and the color shades are for differentiating between the NURBS patches

Fig. 12.31 3D wind turbine rotor and tower aerodynamics computation with the ST-SI-IGA and CGIMG. Nonmatching meshes on the inner (*green*) and outer (*blue*) surfaces of the rotor SI. The checkerboard pattern is for differentiating between the NURBS elements, and the color shades are for differentiating between the NURBS patches

Fig. 12.32 3D wind turbine rotor and tower aerodynamics computation with the ST-SI-IGA and CGIMG. Isosurfaces corresponding to a positive value of the second invariant of the velocity gradient tensor, colored by the velocity magnitude. Color range from *light blue* to *dark blue* corresponds to the velocity range from low to high

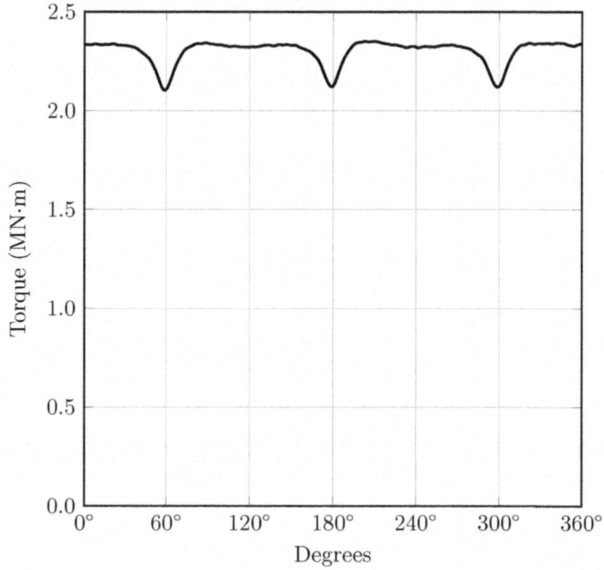

Fig. 12.33 3D wind turbine rotor and tower aerodynamics computation with the ST-SI-IGA and CGIMG. Torque generated during one rotation

Fig. 12.34 3D wind turbine rotor and tower aerodynamics computation with the ST-SI-IGA and CGIMG. Pressure coefficient. Pressure and suction sides

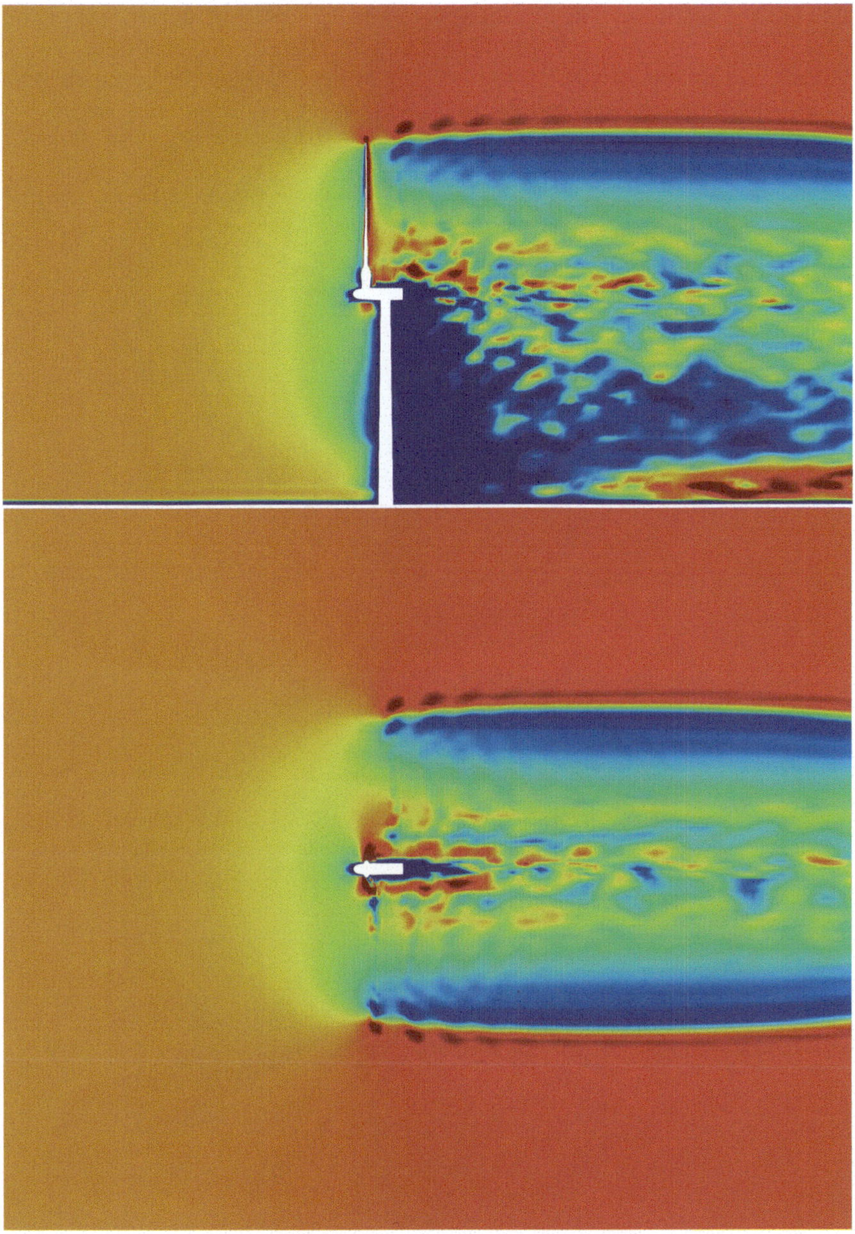

Fig. 12.35 3D wind turbine rotor and tower aerodynamics computation with the ST-SI-IGA and CGIMG. Velocity magnitude over the vertical and horizontal cross-sections passing through the rotor axis. Color range from *blue* to *red* corresponds to the velocity range from low to high

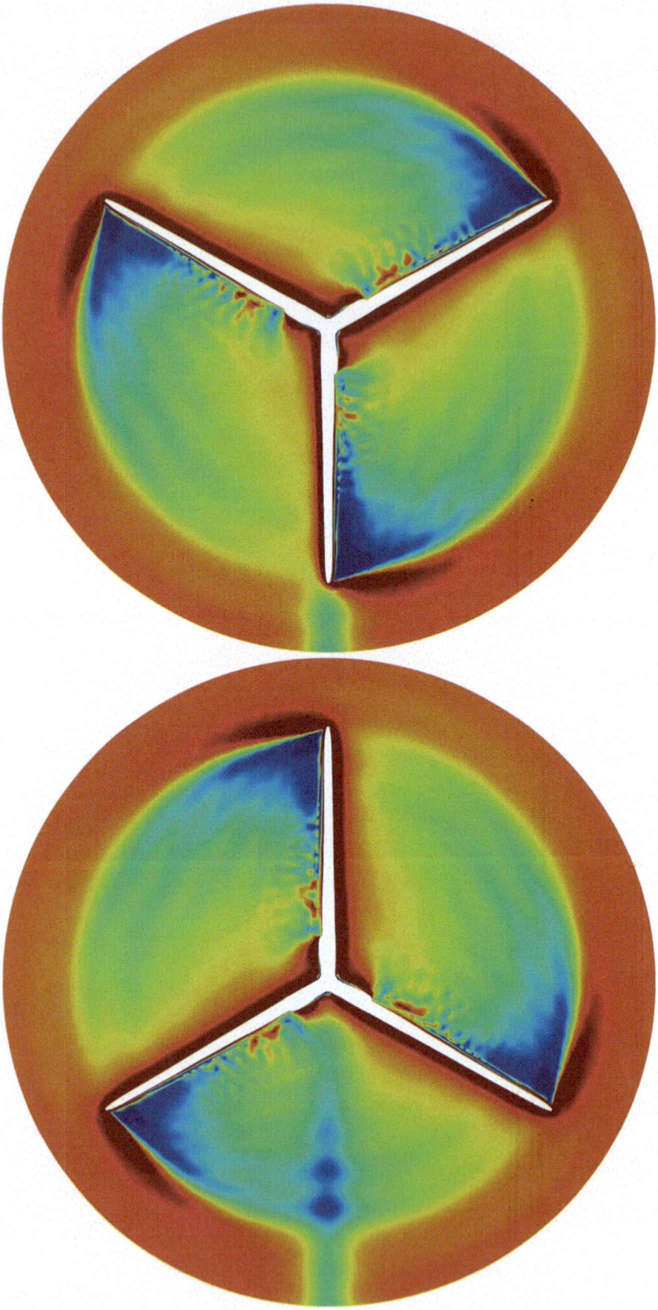

Fig. 12.36 3D wind turbine rotor and tower aerodynamics computation with the ST-SI-IGA and CGIMG. Velocity magnitude over the blade central plane, at two instants. Color range from *blue* to *red* corresponds to the velocity range from low to high

Chapter 13
Concluding Remarks

This book serves as a chronological catalog of the methods and solutions in the STCFA. We presented the material in two parts. In *Part I*, we focused on the methods and solutions in finite element analysis, and in *Part II*, the methods and solutions in isogeometric analysis.

The STCFA started in 1990 with the inception of the DSD/SST, which was introduced as a moving-mesh method for flows with MBI and which is now called ST-SUPS, reflecting its stabilization components. Flows with MBI is a wide class of problems that includes fluid–particle interactions, FSI, and free-surface and multi-fluid flows. As the STCFA evolved, advanced mesh moving methods, FSI coupling methods, and problem-class-specific methods were introduced to increase its scope and the ST-VMS was introduced to upgrade its stabilization components to the VMS. Many of these methods were introduced in the context of finite element analysis, and we covered them in *Part I*. Complementary general-purpose methods developed in the evolution of the STCFA include the ST-TC, which enables high-fidelity moving-mesh computations even in the presence of topology changes in the fluid mechanics domain, such as an actual contact between moving solid surfaces. We covered that too in *Part I*.

The methods we covered in *Part II* include the ST-IGA, ST-SI, and the methods that are a synthesis of the complementary general-purpose methods, with the most comprehensive one being the ST-SI-TC-IGA. The ST-IGA delivers superior accuracy through IGA basis functions not only in space but also in time. The ST-SI enables high-fidelity moving-mesh computations even over meshes made of patches with nonmatching meshes at the interfaces between those patches. That includes cases where the inner mesh around a rotating solid surface, such as a turbine rotor, rotates with the solid surface while the outer mesh remains unaffected by that rotation. The ST-SI-TC-IGA combines the ST-IGA, ST-SI, and ST-TC, and their desirable features, in a well-integrated fashion.

Many of the methods covered in *Part I* would also work in isogeometric analysis. Similarly, some of the methods covered in *Part II* would also work in finite element

T. E. Tezduyar, K. Takizawa, *Space–Time Computational Flow Analysis*,
Modeling and Simulation in Science, Engineering and Technology,
https://doi.org/10.1007/978-3-031-88727-7_13

analysis. With all these and other unconventional methods, the STCFA brought first-of-its-kind solutions in many classes of problems, ranging from fluid–particle interactions in particle-laden flows to FSI in parachute aerodynamics, flapping-wing aerodynamics of an actual locust to ventricle-valve-aorta flow analysis to car and tire aerodynamics with near-actual geometries, road contact, and tire deformation. These and the other first-of-its-kind solutions covered in the book show how the STCFA brought solutions in so many classes of challenging flow problems.

Appendix A
Compressible-Flow Expressions for U, $\mathbf{F}_i$, $\mathbf{E}_i$, $\mathbf{A}_i$, $\mathbf{K}_{ij}$, and S

We provide the expressions for $\mathbf{U}$, $\mathbf{F}_i$, $\mathbf{E}_i$, $\mathbf{A}_i$, $\mathbf{K}_{ij}$, and $\mathbf{S}$. The vector $\mathbf{U}$ is given as

$$\mathbf{U} = \rho \begin{bmatrix} 1 \\ \mathbf{u} \\ e \end{bmatrix}, \tag{A.1}$$

where ρ is the density, $\mathbf{u}$ is the velocity, and e is the total energy per unit mass, composed of the specific internal and kinetic energies: $e = e_{\text{INTE}} + e_{\text{KINE}}$. The Euler fluxes are

$$\mathbf{F}_i = u_i \mathbf{U} + p \begin{bmatrix} 0 \\ \mathbf{e}_i \\ u_i \end{bmatrix} \tag{A.2}$$

and the viscous fluxes are

$$\mathbf{E}_i = \begin{bmatrix} 0 \\ \mathbf{e}_i \cdot \mathbf{T} \\ \mathbf{e}_i \cdot (\mathbf{T} \cdot \mathbf{u} - \mathbf{q}) \end{bmatrix}, \tag{A.3}$$

where p is the pressure, $\mathbf{e}_i$ are the basis vectors, and the viscous-stress tensor and heat flux vectors are

$$\mathbf{T} = \lambda \nabla \cdot \mathbf{u}\, \mathbf{I} + 2\mu \boldsymbol{\varepsilon}(\mathbf{u}), \tag{A.4}$$

$$\mathbf{q} = -\kappa \nabla \theta. \tag{A.5}$$

T. E. Tezduyar, K. Takizawa, *Space–Time Computational Flow Analysis*,
Modeling and Simulation in Science, Engineering and Technology,
https://doi.org/10.1007/978-3-031-88727-7

Here λ is the bulk viscosity, $\mathbf{I}$ is the identity tensor, μ is the viscosity, $\boldsymbol{\varepsilon}(\mathbf{u}) = \left(\boldsymbol{\nabla}\mathbf{u} + (\boldsymbol{\nabla}\mathbf{u})^{\mathsf{T}}\right)/2$ is the strain-rate tensor, κ is the thermal conductivity, and θ is the temperature. For ideal gases, the pressure can be written as

$$p = (\gamma - 1)\rho e_{\text{INTE}} = (\gamma - 1)\rho(e - e_{\text{KINE}}) = (\gamma - 1)\left(\rho e - \frac{1}{2\rho}\|\rho\mathbf{u}\|^2\right), \tag{A.6}$$

and the absolute temperature can be written as

$$\theta = \frac{e_{\text{INTE}}}{c_{\text{V}}} = \frac{e - e_{\text{KINE}}}{c_{\text{V}}} = \frac{1}{\rho c_{\text{V}}}\left(\rho e - \frac{1}{2\rho}\|\rho\mathbf{u}\|^2\right), \tag{A.7}$$

where γ and c_{V} are the ratio of the specific heats and constant-volume heat capacity. With Eqs. (A.6) and (A.7), we express p and θ in terms of the components of the conservation-variables vector, and in 3D those expressions are

$$p = (\gamma - 1)\left(U_5 - \frac{1}{2U_1}\left(U_2^2 + U_3^2 + U_4^2\right)\right), \tag{A.8}$$

$$\theta = \frac{1}{c_{\text{V}}}\left(\frac{U_5}{U_1} - \frac{1}{2U_1^2}\left(U_2^2 + U_3^2 + U_4^2\right)\right). \tag{A.9}$$

The coefficient matrices $\mathbf{A}_i$ and $\mathbf{K}_{ij}$ are derived from Eq. (2.2). The matrices $\mathbf{A}_i$ are written as

$$\mathbf{A}_i = \begin{bmatrix} 0 & \mathbf{e}_i & 0 \\ -u_i\mathbf{u} & u_i\mathbf{I} + \mathbf{u}\mathbf{e}_i & \mathbf{0} \\ -\left(e + \frac{p}{\rho}\right)u_i & \left(e + \frac{p}{\rho}\right)\mathbf{e}_i & u_i \end{bmatrix} + (\gamma - 1)\begin{bmatrix} 0 & \mathbf{0} & 0 \\ e_{\text{KINE}}\mathbf{e}_i & -\mathbf{e}_i\mathbf{u} & \mathbf{e}_i \\ e_{\text{KINE}}u_i & -u_i\mathbf{u} & u_i \end{bmatrix}. \tag{A.10}$$

The second term is specifically for ideal gases, and that comes from

$$\frac{\partial p}{\partial \mathbf{U}} = (\gamma - 1)\begin{bmatrix} e_{\text{KINE}} \\ -\mathbf{u} \\ 1 \end{bmatrix}. \tag{A.11}$$

The matrices $\mathbf{K}_{ij}$ are written as

$$\mathbf{K}_{ij} = \frac{1}{\rho} \begin{bmatrix} 0 & \mathbf{0} & 0 \\ -\lambda u_j \mathbf{e}_i - \mu \left(u_i \mathbf{e}_j + \delta_{ij} \mathbf{u} \right) & \lambda \mathbf{e}_i \mathbf{e}_j + \mu \left(\mathbf{e}_j \mathbf{e}_i + \delta_{ij} \mathbf{I} \right) & \mathbf{0} \\ -(\lambda + \mu) u_i u_j - 2\mu e_{\text{KINE}} \delta_{ij} & \lambda u_i \mathbf{e}_j + \mu \left(u_j \mathbf{e}_i + \delta_{ij} \mathbf{u} \right) & 0 \end{bmatrix}$$

$$+ \frac{\kappa}{\rho c_{\text{V}}} \delta_{ij} \begin{bmatrix} 0 & \mathbf{0} & 0 \\ \mathbf{0} & \mathbf{0} & \mathbf{0} \\ (2e_{\text{KINE}} - e) & -\mathbf{u} & 1 \end{bmatrix} . \tag{A.12}$$

The second term is specifically for ideal gases, and that comes from

$$\frac{\partial q_i}{\partial \mathbf{U}_{,j}} = -\frac{\kappa}{\rho c_{\text{V}}} \delta_{ij} \begin{bmatrix} 2e_{\text{KINE}} - e \\ -\mathbf{u} \\ 1 \end{bmatrix} , \tag{A.13}$$

where

$$\mathbf{U}_{,j} = \frac{\partial \mathbf{U}}{\partial x_j} . \tag{A.14}$$

The source vector is

$$\mathbf{S} = \rho \begin{bmatrix} 0 \\ \mathbf{f} \\ \mathbf{f} \cdot \mathbf{u} + s_{\text{HEAT}} \end{bmatrix} , \tag{A.15}$$

where s_{HEAT} is the heat source per unit mass.

In the subsections, we provide the coefficient matrices in 3D, 2D, and 1D.

A.1 Coefficient Matrices in 3D

From Eqs. (A.10) and (A.12), the matrices in 3D are

$$\mathbf{A}_1 = \begin{bmatrix} 0 & 1 & 0 & 0 & 0 \\ -u_1 u_1 & 2u_1 & 0 & 0 & 0 \\ -u_1 u_2 & u_2 & u_1 & 0 & 0 \\ -u_1 u_3 & u_3 & 0 & u_1 & 0 \\ -\left(e + \frac{p}{\rho} \right) u_1 & e + \frac{p}{\rho} & 0 & 0 & u_1 \end{bmatrix}$$

$$
+ (\gamma - 1) \begin{bmatrix} 0 & 0 & 0 & 0 & 0 \\ e_{\text{KINE}} & -u_1 & -u_2 & -u_3 & 1 \\ 0 & 0 & 0 & 0 & 0 \\ 0 & 0 & 0 & 0 & 0 \\ e_{\text{KINE}} u_1 & -u_1 u_1 & -u_1 u_2 & -u_1 u_3 & u_1 \end{bmatrix}, \tag{A.16}
$$

$$
\mathbf{A}_2 = \begin{bmatrix} 0 & 0 & 1 & 0 & 0 \\ -u_2 u_1 & u_2 & u_1 & 0 & 0 \\ -u_2 u_2 & 0 & 2u_2 & 0 & 0 \\ -u_2 u_3 & 0 & u_3 & u_2 & 0 \\ -\left(e + \frac{p}{\rho}\right) u_2 & 0 & e + \frac{p}{\rho} & 0 & u_2 \end{bmatrix}
$$

$$
+ (\gamma - 1) \begin{bmatrix} 0 & 0 & 0 & 0 & 0 \\ 0 & 0 & 0 & 0 & 0 \\ e_{\text{KINE}} & -u_1 & -u_2 & -u_3 & 1 \\ 0 & 0 & 0 & 0 & 0 \\ e_{\text{KINE}} u_2 & -u_2 u_1 & -u_2 u_2 & -u_2 u_3 & u_2 \end{bmatrix}, \tag{A.17}
$$

$$
\mathbf{A}_3 = \begin{bmatrix} 0 & 0 & 0 & 1 & 0 \\ -u_3 u_1 & u_3 & 0 & u_1 & 0 \\ -u_3 u_2 & 0 & u_3 & u_2 & 0 \\ -u_3 u_3 & 0 & 0 & 2u_3 & 0 \\ -\left(e + \frac{p}{\rho}\right) u_3 & 0 & 0 & e + \frac{p}{\rho} & u_3 \end{bmatrix}
$$

$$
+ (\gamma - 1) \begin{bmatrix} 0 & 0 & 0 & 0 & 0 \\ 0 & 0 & 0 & 0 & 0 \\ 0 & 0 & 0 & 0 & 0 \\ e_{\text{KINE}} & -u_1 & -u_2 & -u_3 & 1 \\ e_{\text{KINE}} u_3 & -u_3 u_1 & -u_3 u_2 & -u_3 u_3 & u_3 \end{bmatrix}, \tag{A.18}
$$

$$
\mathbf{K}_{11} = \frac{1}{\rho} \begin{bmatrix} 0 & 0 & 0 & 0 & 0 \\ -(\lambda + 2\mu) u_1 & \lambda + 2\mu & 0 & 0 & 0 \\ -\mu u_2 & 0 & \mu & 0 & 0 \\ -\mu u_3 & 0 & 0 & \mu & 0 \\ -(\lambda + \mu) u_1 u_1 - 2\mu e_{\text{KINE}} & (\lambda + 2\mu) u_1 & \mu u_2 & \mu u_3 & 0 \end{bmatrix}
$$

$$
+ \frac{\kappa}{\rho c_{\text{v}}} \begin{bmatrix} 0 & 0 & 0 & 0 & 0 \\ 0 & 0 & 0 & 0 & 0 \\ 0 & 0 & 0 & 0 & 0 \\ 0 & 0 & 0 & 0 & 0 \\ (2e_{\text{KINE}} - e) & -u_1 & -u_2 & -u_3 & 1 \end{bmatrix}, \tag{A.19}
$$

$$\mathbf{K}_{12} = \frac{1}{\rho}\begin{bmatrix} 0 & 0 & 0 & 0 & 0 \\ -\lambda u_2 & 0 & \lambda & 0 & 0 \\ -\mu u_1 & \mu & 0 & 0 & 0 \\ 0 & 0 & 0 & 0 & 0 \\ -(\lambda+\mu)\,u_1u_2 & \mu u_2 & \lambda u_1 & 0 & 0 \end{bmatrix}, \tag{A.20}$$

$$\mathbf{K}_{13} = \frac{1}{\rho}\begin{bmatrix} 0 & 0 & 0 & 0 & 0 \\ -\lambda u_3 & 0 & 0 & \lambda & 0 \\ 0 & 0 & 0 & 0 & 0 \\ -\mu u_1 & \mu & 0 & 0 & 0 \\ -(\lambda+\mu)\,u_1u_3 & \mu u_3 & 0 & \lambda u_1 & 0 \end{bmatrix}, \tag{A.21}$$

$$\mathbf{K}_{21} = \frac{1}{\rho}\begin{bmatrix} 0 & 0 & 0 & 0 & 0 \\ -\mu u_2 & 0 & \mu & 0 & 0 \\ -\lambda u_1 & \lambda & 0 & 0 & 0 \\ 0 & 0 & 0 & 0 & 0 \\ -(\lambda+\mu)\,u_2u_1 & \lambda u_2 & \mu u_1 & 0 & 0 \end{bmatrix}, \tag{A.22}$$

$$\mathbf{K}_{22} = \frac{1}{\rho}\begin{bmatrix} 0 & 0 & 0 & 0 & 0 \\ -\mu u_1 & \mu & 0 & 0 & 0 \\ -(\lambda+2\mu)\,u_2 & 0 & \lambda+2\mu & 0 & 0 \\ -\mu u_3 & 0 & 0 & \mu & 0 \\ -(\lambda+\mu)\,u_2u_2 - 2\mu e_{\text{KINE}} & \mu u_1 & (\lambda+2\mu)\,u_2 & \mu u_3 & 0 \end{bmatrix}$$

$$+\frac{\kappa}{\rho c_{\mathrm{V}}}\begin{bmatrix} 0 & 0 & 0 & 0 & 0 \\ 0 & 0 & 0 & 0 & 0 \\ 0 & 0 & 0 & 0 & 0 \\ 0 & 0 & 0 & 0 & 0 \\ (2e_{\text{KINE}} - e) & -u_1 & -u_2 & -u_3 & 1 \end{bmatrix}, \tag{A.23}$$

$$\mathbf{K}_{23} = \frac{1}{\rho}\begin{bmatrix} 0 & 0 & 0 & 0 & 0 \\ 0 & 0 & 0 & 0 & 0 \\ -\lambda u_3 & 0 & 0 & \lambda & 0 \\ -\mu u_2 & 0 & \mu & 0 & 0 \\ -(\lambda+\mu)\,u_2u_3 & 0 & \mu u_3 & \lambda u_2 & 0 \end{bmatrix}, \tag{A.24}$$

$$\mathbf{K}_{31} = \frac{1}{\rho}\begin{bmatrix} 0 & 0 & 0 & 0 & 0 \\ -\mu u_3 & 0 & 0 & \mu & 0 \\ 0 & 0 & 0 & 0 & 0 \\ -\lambda u_1 & \lambda & 0 & 0 & 0 \\ -(\lambda+\mu)\,u_3u_1 & \lambda u_3 & 0 & \mu u_1 & 0 \end{bmatrix}, \tag{A.25}$$

$$\mathbf{K}_{32} = \frac{1}{\rho}\begin{bmatrix} 0 & 0 & 0 & 0 & 0 \\ 0 & 0 & 0 & 0 & 0 \\ -\mu u_3 & 0 & 0 & \mu & 0 \\ -\lambda u_2 & 0 & \lambda & 0 & 0 \\ -(\lambda+\mu) u_3 u_2 & 0 & \lambda u_3 & \mu u_2 & 0 \end{bmatrix}, \tag{A.26}$$

$$\mathbf{K}_{33} = \frac{1}{\rho}\begin{bmatrix} 0 & 0 & 0 & 0 & 0 \\ -\mu u_1 & \mu & 0 & 0 & 0 \\ -\mu u_2 & 0 & \mu & 0 & 0 \\ -(\lambda+2\mu) u_3 & 0 & 0 & \lambda+2\mu & 0 \\ -(\lambda+\mu) u_3 u_3 - 2\mu e_{\text{KINE}} & \mu u_1 & \mu u_2 & (\lambda+2\mu) u_3 & 0 \end{bmatrix}$$
$$+ \frac{\kappa}{\rho c_{\text{V}}}\begin{bmatrix} 0 & 0 & 0 & 0 & 0 \\ 0 & 0 & 0 & 0 & 0 \\ 0 & 0 & 0 & 0 & 0 \\ 0 & 0 & 0 & 0 & 0 \\ (2e_{\text{KINE}} - e) & -u_1 & -u_2 & -u_3 & 1 \end{bmatrix}. \tag{A.27}$$

A.2 Coefficient Matrices in 2D

From Eqs. (A.10) and (A.12), the matrices in 2D are

$$\mathbf{A}_1 = \begin{bmatrix} 0 & 1 & 0 & 0 \\ -u_1 u_1 & 2u_1 & 0 & 0 \\ -u_1 u_2 & u_2 & u_1 & 0 \\ -\left(e+\frac{p}{\rho}\right) u_1 & e+\frac{p}{\rho} & 0 & u_1 \end{bmatrix}$$
$$+ (\gamma - 1)\begin{bmatrix} 0 & 0 & 0 & 0 \\ e_{\text{KINE}} & -u_1 & -u_2 & 1 \\ 0 & 0 & 0 & 0 \\ e_{\text{KINE}} u_1 & -u_1 u_1 & -u_1 u_2 & u_1 \end{bmatrix}, \tag{A.28}$$

$$\mathbf{A}_2 = \begin{bmatrix} 0 & 0 & 1 & 0 \\ -u_2 u_1 & u_2 & u_1 & 0 \\ -u_2 u_2 & 0 & 2u_2 & 0 \\ -\left(e+\frac{p}{\rho}\right) u_2 & 0 & e+\frac{p}{\rho} & u_2 \end{bmatrix}$$
$$+ (\gamma - 1)\begin{bmatrix} 0 & 0 & 0 & 0 \\ 0 & 0 & 0 & 0 \\ e_{\text{KINE}} & -u_1 & -u_2 & 1 \\ e_{\text{KINE}} u_2 & -u_2 u_1 & -u_2 u_2 & u_2 \end{bmatrix}, \tag{A.29}$$

$$
\mathbf{K}_{11} = \frac{1}{\rho} \begin{bmatrix} 0 & 0 & 0 & 0 \\ -(\lambda + 2\mu)\, u_1 & \lambda + 2\mu & 0 & 0 \\ -\mu u_2 & 0 & \mu & 0 \\ -(\lambda + \mu)\, u_1 u_1 - 2\mu e_{\text{KINE}} & (\lambda + 2\mu)\, u_1 & \mu u_2 & 0 \end{bmatrix}
$$

$$
+ \frac{\kappa}{\rho c_{\text{v}}} \begin{bmatrix} 0 & 0 & 0 & 0 \\ 0 & 0 & 0 & 0 \\ 0 & 0 & 0 & 0 \\ (2e_{\text{KINE}} - e) & -u_1 & -u_2 & 1 \end{bmatrix}, \tag{A.30}
$$

$$
\mathbf{K}_{12} = \frac{1}{\rho} \begin{bmatrix} 0 & 0 & 0 & 0 \\ -\lambda u_2 & 0 & \lambda & 0 \\ -\mu u_1 & \mu & 0 & 0 \\ -(\lambda + \mu)\, u_1 u_2 & \mu u_2 & \lambda u_1 & 0 \end{bmatrix}, \tag{A.31}
$$

$$
\mathbf{K}_{21} = \frac{1}{\rho} \begin{bmatrix} 0 & 0 & 0 & 0 \\ -\mu u_2 & 0 & \mu & 0 \\ -\lambda u_1 & \lambda & 0 & 0 \\ -(\lambda + \mu)\, u_2 u_1 & \lambda u_2 & \mu u_1 & 0 \end{bmatrix}, \tag{A.32}
$$

$$
\mathbf{K}_{22} = \frac{1}{\rho} \begin{bmatrix} 0 & 0 & 0 & 0 \\ -\mu u_1 & \mu & 0 & 0 \\ -(\lambda + 2\mu)\, u_2 & 0 & \lambda + 2\mu & 0 \\ -(\lambda + \mu)\, u_2 u_2 - 2\mu e_{\text{KINE}} & \mu u_1 & (\lambda + 2\mu)\, u_2 & 0 \end{bmatrix}
$$

$$
+ \frac{\kappa}{\rho c_{\text{v}}} \begin{bmatrix} 0 & 0 & 0 & 0 \\ 0 & 0 & 0 & 0 \\ 0 & 0 & 0 & 0 \\ (2e_{\text{KINE}} - e) & -u_1 & -u_2 & 1 \end{bmatrix}. \tag{A.33}
$$

A.3 Coefficient Matrices in 1D

From Eqs. (A.10) and (A.12), the matrices in 1D are

$$
\mathbf{A}_1 = \begin{bmatrix} 0 & 1 & 0 \\ -u_1 u_1 & 2u_1 & 0 \\ -\left(e + \frac{p}{\rho}\right) u_1 & e + \frac{p}{\rho} & u_1 \end{bmatrix}
$$

$$
+ (\gamma - 1) \begin{bmatrix} 0 & 0 & 0 \\ e_{\text{KINE}} & -u_1 & 1 \\ e_{\text{KINE}} u_1 & -u_1 u_1 & u_1 \end{bmatrix}, \tag{A.34}
$$

$$\mathbf{K}_{11} = \frac{1}{\rho}\begin{bmatrix} 0 & 0 & 0 \\ -(\lambda + 2\mu)\,u_1 & \lambda + 2\mu & 0 \\ -(\lambda + \mu)\,u_1 u_1 - 2\mu e_{\text{KINE}} & (\lambda + 2\mu)\,u_1 & 0 \end{bmatrix}$$

$$+ \frac{\kappa}{\rho c_{\text{v}}}\begin{bmatrix} 0 & 0 & 0 \\ 0 & 0 & 0 \\ (2e_{\text{KINE}} - e) & -u_1 & 1 \end{bmatrix}. \tag{A.35}$$

Appendix B
Advection–Diffusion ST-SUPG Method

The advection–diffusion ST-SUPG formulation is written, from [24, 115], as

$$\begin{aligned}
&\int_{Q_n} w^h \left(\frac{\partial \phi^h}{\partial t} + \mathbf{u}^h \cdot \boldsymbol{\nabla} \phi^h - f^h \right) \mathrm{d}Q + \int_{Q_n} \boldsymbol{\nabla} w^h \cdot \nu \boldsymbol{\nabla} \phi^h \mathrm{d}Q \\
&\quad - \int_{(P_n)_\mathrm{h}} w^h h^h \mathrm{d}P + \int_{\Omega_n} (w^h)_n^+ \left((\phi^h)_n^+ - (\phi^h)_n^- \right) \mathrm{d}\Omega \\
&\quad + \sum_{e=1}^{(n_\mathrm{el})_n} \int_{Q_n^e} \tau_\mathrm{SUPG} \left(\frac{\partial w^h}{\partial t} + \mathbf{u}^h \cdot \boldsymbol{\nabla} w^h \right) \\
&\quad \times \left(\frac{\partial \phi^h}{\partial t} + \mathbf{u}^h \cdot \boldsymbol{\nabla} \phi^h - \boldsymbol{\nabla} \cdot \nu \boldsymbol{\nabla} \phi^h - f^h \right) \mathrm{d}Q = 0, \qquad \text{(B.1)}
\end{aligned}$$

where ϕ is the scalar function governed by the advection–diffusion equation, ν is the diffusivity, f is the source term, w is the test function associated with ϕ, and h is the diffusive flux specified at the boundary.

T. E. Tezduyar, K. Takizawa, *Space–Time Computational Flow Analysis*,
Modeling and Simulation in Science, Engineering and Technology,
https://doi.org/10.1007/978-3-031-88727-7

Appendix C
Shallow-Water-Flow Expressions for U, $\mathbf{F}_i$, $\mathbf{E}_i$, $\mathbf{A}_i$, $\mathbf{K}_{ij}$, and S

We provide, from [95], the expressions for $\mathbf{U}$, $\mathbf{F}_i$, $\mathbf{E}_i$, $\mathbf{A}_i$, $\mathbf{K}_{ij}$, and $\mathbf{S}$. The vector $\mathbf{U}$ is given as

$$\mathbf{U} = \begin{bmatrix} U_1 \\ U_2 \\ U_3 \end{bmatrix} = \begin{bmatrix} H \\ Hu_1 \\ Hu_2 \end{bmatrix}, \tag{C.1}$$

where H is the total water height, and u_1 and u_2 are the components of the depth-averaged horizontal velocity. The Euler fluxes are

$$\mathbf{F}_1 = \begin{bmatrix} u_1 H \\ u_1 H u_1 + \frac{1}{2} g H^2 \\ u_1 H u_2 \end{bmatrix}, \tag{C.2}$$

$$\mathbf{F}_2 = \begin{bmatrix} u_2 H \\ u_2 H u_1 \\ u_2 H u_2 + \frac{1}{2} g H^2 \end{bmatrix}, \tag{C.3}$$

and the viscous fluxes are

$$\mathbf{E}_1 = \begin{bmatrix} 0 \\ 2\nu H \varepsilon_{11} \\ 2\nu H \varepsilon_{12} \end{bmatrix}, \tag{C.4}$$

T. E. Tezduyar, K. Takizawa, *Space–Time Computational Flow Analysis*,
Modeling and Simulation in Science, Engineering and Technology,
https://doi.org/10.1007/978-3-031-88727-7

$$\mathbf{E}_2 = \begin{bmatrix} 0 \\ 2\nu H \varepsilon_{21} \\ 2\nu H \varepsilon_{22} \end{bmatrix}. \tag{C.5}$$

Here g is the gravitational acceleration, and in $\boldsymbol{\varepsilon}(\mathbf{u}) = \left(\nabla\mathbf{u} + (\nabla\mathbf{u})^{\mathsf{T}}\right)/2$, the velocity $\mathbf{u}$ is interpreted as the depth-averaged velocity. The Euler fluxes are rewritten as

$$\mathbf{F}_1 = \begin{bmatrix} U_2 \\ \frac{U_2}{U_1} U_2 + \frac{1}{2} g {U_1}^2 \\ \frac{U_2}{U_1} U_3 \end{bmatrix}, \tag{C.6}$$

$$\mathbf{F}_2 = \begin{bmatrix} U_3 \\ \frac{U_3}{U_1} U_2 \\ \frac{U_3}{U_1} U_3 + \frac{1}{2} g {U_1}^2 \end{bmatrix}. \tag{C.7}$$

Then, the coefficient matrices are written as

$$\mathbf{A}_1 = \begin{bmatrix} 0 & 1 & 0 \\ g U_1 - \left(\frac{U_2}{U_1}\right)^2 & 2\frac{U_2}{U_1} & 0 \\ -\frac{U_2}{U_1}\frac{U_3}{U_1} & \frac{U_3}{U_1} & \frac{U_2}{U_1} \end{bmatrix}, \tag{C.8}$$

$$\mathbf{A}_2 = \begin{bmatrix} 0 & 0 & 1 \\ -\frac{U_3}{U_1}\frac{U_2}{U_1} & \frac{U_3}{U_1} & \frac{U_2}{U_1} \\ g U_1 - \left(\frac{U_3}{U_1}\right)^2 & 0 & 2\frac{U_3}{U_1} \end{bmatrix}, \tag{C.9}$$

$$\mathbf{K}_{11} = \begin{bmatrix} 0 & 0 & 0 \\ -2\nu\frac{U_2}{U_1} & 2\nu & 0 \\ -\nu\frac{U_3}{U_1} & 0 & \nu \end{bmatrix}, \tag{C.10}$$

$$\mathbf{K}_{12} = \begin{bmatrix} 0 & 0 & 0 \\ 0 & 0 & 0 \\ -\nu\frac{U_2}{U_1} & \nu & 0 \end{bmatrix}, \tag{C.11}$$

$$\mathbf{K}_{21} = \begin{bmatrix} 0 & 0 & 0 \\ -\nu\frac{U_3}{U_1} & 0 & \nu \\ 0 & 0 & 0 \end{bmatrix}, \tag{C.12}$$

$$\mathbf{K}_{22} = \begin{bmatrix} 0 & 0 & 0 \\ -\nu\frac{U_2}{U_1} & \nu & 0 \\ -\nu\frac{U_3}{U_1} & 0 & 2\nu \end{bmatrix}. \tag{C.13}$$

The vector **S** is given as

$$\mathbf{S} = \begin{bmatrix} 0 \\ gH\frac{\partial h}{\partial x_1} - 2\nu\left(\frac{\partial H}{\partial x_1}\varepsilon_{11} + \frac{\partial H}{\partial x_2}\varepsilon_{21}\right) + B_1 \\ gH\frac{\partial h}{\partial x_2} - 2\nu\left(\frac{\partial H}{\partial x_1}\varepsilon_{12} + \frac{\partial H}{\partial x_2}\varepsilon_{22}\right) + B_2 \end{bmatrix}, \tag{C.14}$$

where

$$\mathbf{B} = \begin{bmatrix} \frac{1}{\rho}\left((T_{31})_{\mathrm{s}} - (T_{31})_{\mathrm{b}}\right) \\ \frac{1}{\rho}\left((T_{32})_{\mathrm{s}} - (T_{32})_{\mathrm{b}}\right) \end{bmatrix}. \tag{C.15}$$

Here the T_{31} and T_{32} are the components of the shear stress, and the subscripts "s" and "b" refer to the surface and bottom.

Acronym Definitions and Method-Description Sections

Acronym	Definition	Section
ALE	Arbitrary Lagrangian–Eulerian	
ALE-SUPS	Arbitrary Lagrangian–Eulerian SUPG/PSPG	
ALE-VMS	Arbitrary Lagrangian–Eulerian Variational Multiscale	
BCBMM	Back-cycle-based mesh moving	11.5
BHV	Bioprosthetic heart valve	
CD	Carrier domain	12.3
CDM	Carrier-Domain Method	12.3
CDM-C	Carrier-Domain Method – Continuous carrier-domain motion	12.3
CDM-D	Carrier-Domain Method – Discrete carrier-domain motion	12.3
CFP	Constrained Flow Profile	11.3
CGIMG	Complex-Geometry IGA Mesh Generation	10.9
DSD/SST	Deforming-Spatial-Domain/Stabilized Space–Time	3.2
DSD/SST-VMST	DSD/SST with variational multiscale (turbulence) model	7.4
EBMR	Element-Based Mesh Relaxation	7.10
EDSTT	Enhanced-Discretization Space–Time Technique	5.3
EDSTT-MM	Enhanced-Discretization Space–Time Technique – Multi-Mesh	5.3
EDSTT-SM	Enhanced-Discretization Space–Time Technique – Single-Mesh	5.3
ETILT	Edge-Tracked Interface Locator Technique	5.2
FRHEMUM	Fiber-Reinforced Hyperelasticity Mesh Update Method	11.6
FSI	Fluid–structure interaction	
FSITICT	Fluid–Solid Interface-Tracking/Interface-Capturing Technique	6.8

(continued)

T. E. Tezduyar, K. Takizawa, *Space–Time Computational Flow Analysis*,
Modeling and Simulation in Science, Engineering and Technology,
https://doi.org/10.1007/978-3-031-88727-7

Acronym	Definition	Section
FSITICT-AP	Fluid–Solid Interface-Tracking/Interface-Capturing Technique – Adaptive Partitioning	6.8
FSITICT-FP	Fluid–Solid Interface-Tracking/Interface-Capturing Technique – Fixed Partitioning	6.8
FWD	Full wake domain	
HCBMM	Half-cycle-based mesh moving	11.5
HMGP	Homogenized Modeling of Geometric Porosity	6.5
HMGP-FG	HMGP with separate fabric and geometric porosities	6.5
IMGA	Immersogeometric Analysis	
LSIC	Stabilization based on least-squares on incompressibility constraint	2.2
LV	Left ventricle	
MDM	Multidomain Method	4.7
MBI	Moving boundaries and interfaces	
MIMICT	Mixed Interface-Meshing/Interface-Capturing Technique	5.2
MITICT	Mixed Interface-Tracking/Interface-Capturing Technique	5.2
MITIMT	Mixed Interface-Tracking/Interface-Meshing Technique	5.2
MJBS	Mesh-Jacobian-based stiffening	3.8
NSVGMG	NURBS Surface-to-Volume Guided Mesh Generation	11.10
NURBS	Non-Uniform Rational B-Splines	
OICL	Overinflation control line of the Orion spacecraft landing parachute	
PA	Orion spacecraft landing parachute earlier design configuration with all sails in place	
PM5	Orion spacecraft landing parachute modified-geometric-porosity design configuration with missing Sail 5	
PM11	Orion spacecraft landing parachute modified-geometric-porosity design configuration with missing Sail 11	
PSPG	Pressure-Stabilizing/Petrov-Galerkin	2.2
RBVMS	Residual-based variational multiscale	
SCFSI	Sequentially-Coupled Fluid–Structure Interaction	6.7
SEMMT	Solid-Extension Mesh Moving Technique	5.4
SEMMT-MD	Solid-Extension Mesh Moving Technique – Multiple Domain	5.4
SEMMT-SD	Solid-Extension Mesh Moving Technique – Single Domain	5.4
SENCT	Surface-Edge-Node Contact Tracking Technique	6.1
SENCT-D	Surface-Edge-Node Contact Tracking Technique – Displacement	6.1
SENCT-F	Surface-Edge-Node Contact Tracking Technique – Force	6.1
SENCT-FC	Surface-Edge-Node Contact Tracking Technique – Force-Conservative	6.1
SENCT-FC-M1	SENCT-FC based on the fluid mechanics ("1") mesh	6.1

(continued)

Acronym	Definition	Section
SHOC	Shock-capturing	
SSMUM	Shear–Slip Mesh Update Method	3.18
ST	Space–time	
ST-C	Space–Time Computation Techniques with Continuous Representation in Time	9.2
ST-IGA	Space–Time Isogeometric Analysis	8.1
ST-SI	Space–Time Slip Interface	9.5
ST-SI-IGA	Synthesis of ST-SI and ST-IGA	9.10
ST-SI-TC	Synthesis of ST-SI and ST-TC	10.2
ST-SI-TC-IGA	Synthesis of ST-SI-TC and ST-IGA	10.4
ST-SUPG	Space–Time SUPG	3.10
ST-SUPS	Space–Time SUPG/PSPG	3.2
ST-TC	Space–Time Topology Change	7.12
ST-VMS	Space–Time Variational Multiscale	7.4
STCFA	Space–Time Computational Flow Analysis	3.1
STCT	Space–Time Contact Technique	4.9
STNMUM	Space–Time/NURBS Mesh Update Method	8.3
SUPG	Streamline-Upwind/Petrov-Galerkin	2.2
$(\text{SUPG})_{82}$	1982 version of the compressible-flow SUPG method	2.1
SUPS	SUPG/PSPG	2.2
SWE	Shallow-water equations	
TC	Topology change	
VAWT	Vertical-axis wind turbine	
VMS	Variational multiscale	
WSS	Wall shear stress	
ZSS	Zero-stress state	

References

1. T.E. Tezduyar, Stabilized finite element formulations for incompressible flow computations. Adv. Appl. Mech. **28**, 1–44 (1992). https://doi.org/10.1016/S0065-2156(08)70153-4
2. T.E. Tezduyar, M. Behr, J. Liou, A new strategy for finite element computations involving moving boundaries and interfaces – the deforming-spatial-domain/space–time procedure: I. The concept and the preliminary numerical tests. Comput. Methods Appl. Mech. Eng. **94**(3), 339–351 (1992). https://doi.org/10.1016/0045-7825(92)90059-S
3. T.E. Tezduyar, M. Behr, S. Mittal, J. Liou, A new strategy for finite element computations involving moving boundaries and interfaces – the deforming-spatial-domain/space–time procedure: II. Computation of free-surface flows, two-liquid flows, and flows with drifting cylinders. Comput. Methods Appl. Mech. Eng. **94**(3), 353–371 (1992). https://doi.org/10.1016/0045-7825(92)90060-W
4. T.J.R. Hughes, G.M. Hulbert, Space–time finite element methods for elastodynamics: formulations and error estimates. Comput. Methods Appl. Mech. Eng. **66**, 339–363 (1988)
5. T.J.R. Hughes, A.N. Brooks, A multi-dimensional upwind scheme with no crosswind diffusion, in *Finite Element Methods for Convection Dominated Flows*, ed. by T.J.R. Hughes, AMD-vol.34 (ASME, New York, 1979), pp. 19–35
6. A.N. Brooks, T.J.R. Hughes, Streamline upwind/Petrov-Galerkin formulations for convection dominated flows with particular emphasis on the incompressible Navier-Stokes equations. Comput. Methods Appl. Mech. Eng. **32**, 199–259 (1982)
7. T.E. Tezduyar, S. Mittal, S.E. Ray, R. Shih, Incompressible flow computations with stabilized bilinear and linear equal-order-interpolation velocity-pressure elements. Comput. Methods Appl. Mech. Eng. **95**, 221–242 (1992). https://doi.org/10.1016/0045-7825(92)90141-6
8. T. Kuraishi, K. Takizawa, T.E. Tezduyar, Boundary layer mesh resolution in flow computation with the Space–Time Variational Multiscale method and isogeometric discretization. Math. Models Methods Appl. Sci. **32**(12), 2401–2443 (2022). https://doi.org/10.1142/S0218202522500567
9. T.E. Tezduyar, Computation of moving boundaries and interfaces and stabilization parameters. Int. J. Numer. Methods Fluids **43**, 555–575 (2003). https://doi.org/10.1002/fld.505
10. T.E. Tezduyar, M. Behr, S. Mittal, A.A. Johnson, Computation of unsteady incompressible flows with the finite element methods: space–time formulations, iterative strategies and massively parallel implementations, in *New Methods in Transient Analysis*, PVP-Vol.246/AMD-Vol.143 (ASME, New York, 1992), pp. 7–24
11. T. Tezduyar, S. Aliabadi, M. Behr, A. Johnson, S. Mittal, Parallel finite-element computation of 3D flows. Computer **26**(10), 27–36 (1993). https://doi.org/10.1109/2.237441

T. E. Tezduyar, K. Takizawa, *Space–Time Computational Flow Analysis*,
Modeling and Simulation in Science, Engineering and Technology,
https://doi.org/10.1007/978-3-031-88727-7

12. T.E. Tezduyar, S. Sathe, R. Keedy, K. Stein, Space–time finite element techniques for computation of fluid–structure interactions. Comput. Methods Appl. Mech. Eng. **195**, 2002–2027 (2006). https://doi.org/10.1016/j.cma.2004.09.014
13. T.E. Tezduyar, S. Sathe, Modeling of fluid–structure interactions with the space–time finite elements: solution techniques. Int. J. Numer. Methods Fluids **54**, 855–900 (2007). https://doi.org/10.1002/fld.1430
14. K. Takizawa, T.E. Tezduyar, Multiscale space–time fluid–structure interaction techniques. Comput. Mech. **48**, 247–267 (2011). https://doi.org/10.1007/s00466-011-0571-z
15. K. Takizawa, T.E. Tezduyar, Space–time fluid–structure interaction methods. Math. Models Methods Appl. Sci. **22**(Suppl. 02), 1230001 (2012). https://doi.org/10.1142/S0218202512300013
16. K. Takizawa, T.E. Tezduyar, T. Kuraishi, Multiscale ST methods for thermo-fluid analysis of a ground vehicle and its tires. Math. Models Methods Appl. Sci. **25**, 2227–2255 (2015). https://doi.org/10.1142/S0218202515400072
17. T.J.R. Hughes, Multiscale phenomena: Green's functions, the Dirichlet-to-Neumann formulation, subgrid scale models, bubbles, and the origins of stabilized methods. Comput. Methods Appl. Mech. Eng. **127**, 387–401 (1995)
18. T.J.R. Hughes, A.A. Oberai, L. Mazzei, Large eddy simulation of turbulent channel flows by the variational multiscale method. Phys. Fluids **13**, 1784–1799 (2001)
19. Y. Bazilevs, V.M. Calo, J.A. Cottrell, T.J.R. Hughes, A. Reali, G. Scovazzi, Variational multiscale residual-based turbulence modeling for large eddy simulation of incompressible flows. Comput. Methods Appl. Mech. Eng. **197**, 173–201 (2007)
20. Y. Bazilevs, I. Akkerman, Large eddy simulation of turbulent Taylor–Couette flow using isogeometric analysis and the residual–based variational multiscale method. J. Comput. Phys. **229**, 3402–3414 (2010)
21. T.J.R. Hughes, W.K. Liu, T.K. Zimmermann, Lagrangian–Eulerian finite element formulation for incompressible viscous flows. Comput. Methods Appl. Mech. Eng. **29**, 329–349 (1981)
22. V. Kalro, T.E. Tezduyar, A parallel 3D computational method for fluid–structure interactions in parachute systems. Comput. Methods Appl. Mech. Eng. **190**, 321–332 (2000). https://doi.org/10.1016/S0045-7825(00)00204-8
23. Y. Bazilevs, V.M. Calo, T.J.R. Hughes, Y. Zhang, Isogeometric fluid–structure interaction: theory, algorithms, and computations. Comput. Mech. **43**, 3–37 (2008)
24. Y. Bazilevs, K. Takizawa, T.E. Tezduyar, *Computational Fluid–Structure Interaction: Methods and Applications* (Wiley, Chichester, 2013). ISBN 978-0470978771
25. K. Takizawa, Y. Bazilevs, T.E. Tezduyar, Isogeometric discretization methods in computational fluid mechanics. Math. Models Methods Appl. Sci. **32**(12), 2359–2370 (2022). https://doi.org/10.1142/S0218202522020018
26. M.-C. Hsu, I. Akkerman, Y. Bazilevs, Finite element simulation of wind turbine aerodynamics: validation study using NREL Phase VI experiment. *Wind Energy* **17**, 461–481 (2014)
27. Y. Bazilevs, A. Korobenko, X. Deng, J. Yan, FSI modeling for fatigue-damage prediction in full-scale wind-turbine blades. J. Appl. Mech. **83**(6), 061010 (2016)
28. J. Yan, A. Korobenko, X. Deng, Y. Bazilevs, Computational free-surface fluid–structure interaction with application to floating offshore wind turbines. Comput. Fluids **141**, 155–174 (2016). https://doi.org/10.1016/j.compfluid.2016.03.008
29. M. Ravensbergen, A.M. Bayram, A. Korobenko, The actuator line method for wind turbine modelling applied in a variational multiscale framework. Comput. Fluids **201**, 104465 (2020). Available online, https://doi.org/10.1016/j.compfluid.2020.104465
30. N. Kozak, M.R. Rajanna, M.C.H. Wu, M. Murugan, L. Bravo, A. Ghoshal, M.-C. Hsu, Y. Bazilevs, Optimizing gas turbine performance using the surrogate management framework and high-fidelity flow modeling. Energies **13**, 4283 (2020)
31. Y. Bazilevs, K. Takizawa, M.C.H. Wu, T. Kuraishi, R. Avsar, Z. Xu, T.E. Tezduyar, Gas turbine computational flow and structure analysis with isogeometric discretization and a complex-geometry mesh generation method. Comput. Mech. **67**, 57–84 (2021). https://doi.org/10.1007/s00466-020-01919-w

32. Q. Zhu, J. Yan, A moving-domain CFD solver in FEniCS with applications to tidal turbine simulations in turbulent flows. Comput. Math. Appl. **81**, 532–546 (2021)
33. J. Yan, A. Korobenko, A.E. Tejada-Martinez, R. Golshan, Y. Bazilevs, A new variational multiscale formulation for stratified incompressible turbulent flows. Comput. Fluids **158**, 150–156 (2017). https://doi.org/10.1016/j.compfluid.2016.12.004
34. M. Ravensbergen, T.A. Helgedagsrud, Y. Bazilevs, A. Korobenko, A variational multiscale framework for atmospheric turbulent flows over complex environmental terrains. Comput. Methods Appl. Mech. Eng. **368**, 113182 (2020). https://doi.org/10.1016/j.cma.2020.113182
35. T.A. Helgedagsrud, Y. Bazilevs, K.M. Mathisen, O.A. Oiseth, ALE-VMS methods for wind-resistant design of long-span bridges. J. Wind Eng. Ind. Aerodyn. **191**, 143–153 (2019). https://doi.org/10.1016/j.jweia.2019.06.001
36. B. Augier, J. Yan, A. Korobenko, J. Czarnowski, G. Ketterman, Y. Bazilevs, Experimental and numerical FSI study of compliant hydrofoils. Comput. Mech. **55**, 1079–1090 (2015). https://doi.org/10.1007/s00466-014-1090-5
37. Q. Zhu, F. Xu, S. Xu, M.-C. Hsu, J. Yan, An immersogeometric formulation for free-surface flows with application to marine engineering problems. Comput. Methods Appl. Mech. Eng. **361**, 112748 (2020)
38. J. Yan, X. Deng, A. Korobenko, Y. Bazilevs, Free-surface flow modeling and simulation of horizontal-axis tidal-stream turbines. Comput. Fluids **158**, 157–166 (2017). https://doi.org/10.1016/j.compfluid.2016.06.016
39. Q. Zhu, J. Yan, A. Tejada-Martínez, Y. Bazilevs, Variational multiscale modeling of langmuir turbulent boundary layers in shallow water using isogeometric analysis. Mech. Res. Commun. **108**, 103570 (2020). https://doi.org/10.1016/j.mechrescom.2020.103570
40. J. Yan, S.S. Lin, Y. Bazilevs, G. Wagner, Isogeometric analysis of multi-phase flows with surface tension and with application to dynamics of rising bubbles. Comput. Fluids **179**, 777–789 (2019)
41. H. Cen, Q. Zhou, A. Korobenko, Variational multiscale framework for cavitating flows. Comput. Fluids **214**, 104765 (2021). https://doi.org/10.1016/j.compfluid.2020.104765
42. Z. Zhao, Q. Zhu, J. Yan, A thermal multi-phase flow model for directed energy deposition processes via a moving signed distance function. Comput. Methods Appl. Mech. Eng. **373**, 113518 (2021)
43. Q. Zhu, Z. Liu, J. Yan, Machine learning for metal additive manufacturing: predicting temperature and melt pool fluid dynamics using physics-informed neural networks. Comput. Mech. **67**, 619–635 (2021). https://doi.org/10.1007/s00466-020-01952-9
44. C. Wang, M.C.H. Wu, F. Xu, M.-C. Hsu, Y. Bazilevs, Modeling of a hydraulic arresting gear using fluid–structure interaction and isogeometric analysis. Comput. Fluids **142**, 3–14 (2017). https://doi.org/10.1016/j.compfluid.2015.12.004
45. D. Codoni, G. Moutsanidis, M.-C. Hsu, Y. Bazilevs, C. Johansen, A. Korobenko, Stabilized methods for high-speed compressible flows: toward hypersonic simulations. Comput. Mech. **67**, 785–809 (2021). https://doi.org/10.1007/s00466-020-01963-6.
46. D. Kamensky, M.-C. Hsu, D. Schillinger, J.A. Evans, A. Aggarwal, Y. Bazilevs, M.S. Sacks, T.J.R. Hughes, An immersogeometric variational framework for fluid-structure interaction: application to bioprosthetic heart valves. Comput. Methods Appl. Mech. Eng. **284**, 1005–1053 (2015)
47. F. Xu, E.L. Johnson, C. Wang, A. Jafari, C.-H. Yang, M.S. Sacks, A. Krishnamurthy, M.-C. Hsu, Computational investigation of left ventricular hemodynamics following bioprosthetic aortic and mitral valve replacement. Mech. Res. Commun. **112**, 103604 (2021). https://doi.org/10.1016/j.mechrescom.2020.103604
48. T.E. Tezduyar, K. Takizawa, C. Moorman, S. Wright, J. Christopher, Space–time finite element computation of complex fluid–structure interactions. Int. J. Numer. Methods Fluids **64**, 1201–1218 (2010). https://doi.org/10.1002/fld.2221
49. S. Xu, B. Gao, A. Lofquist, M. Fernando, M.-C. Hsu, H. Sundar, B. Ganapathysubramanian, An octree-based immersogeometric approach for modeling inertial migration of particles in channels. Comput. Fluids **214**, 104764 (2020)

50. K. Takizawa, B. Henicke, A. Puntel, T. Spielman, T.E. Tezduyar, Space–time computational techniques for the aerodynamics of flapping wings. J. Appl. Mech. **79**, 010903 (2012). https://doi.org/10.1115/1.4005073
51. K. Takizawa, T.E. Tezduyar, Y. Otoguro, T. Terahara, T. Kuraishi, H. Hattori, Turbocharger flow computations with the Space–Time Isogeometric Analysis (ST-IGA). Comput. Fluids **142**, 15–20 (2017). https://doi.org/10.1016/j.compfluid.2016.02.021
52. K. Takizawa, T.E. Tezduyar, A. Buscher, S. Asada, Space–time interface-tracking with topology change (ST-TC). Comput. Mech. **54**, 955–971 (2014). https://doi.org/10.1007/s00466-013-0935-7
53. K. Takizawa, T.E. Tezduyar, A. Buscher, S. Asada, Space–time fluid mechanics computation of heart valve models. Comput. Mech. **54**, 973–986 (2014). https://doi.org/10.1007/s00466-014-1046-9
54. K. Takizawa, T.E. Tezduyar, H. Mochizuki, H. Hattori, S. Mei, L. Pan, K. Montel, Space–time VMS method for flow computations with slip interfaces (ST-SI). Math. Models Methods Appl. Sci. **25**, 2377–2406 (2015). https://doi.org/10.1142/S0218202515400126
55. K. Takizawa, T.E. Tezduyar, T. Kuraishi, S. Tabata, H. Takagi, Computational thermo-fluid analysis of a disk brake. Comput. Mech. **57**, 965–977 (2016). https://doi.org/10.1007/s00466-016-1272-4
56. K. Takizawa, T.E. Tezduyar, T. Terahara, T. Sasaki, Heart valve flow computation with the integrated Space–Time VMS, Slip Interface, Topology Change and Isogeometric Discretization methods. Comput. Fluids **158**, 176–188 (2017). https://doi.org/10.1016/j.compfluid.2016.11.012
57. A.A. Johnson, T.E. Tezduyar, Simulation of multiple spheres falling in a liquid-filled tube. Comput. Methods Appl. Mech. Eng. **134**, 351–373 (1996). https://doi.org/10.1016/0045-7825(95)00988-4
58. T.E. Tezduyar, S. Sathe, J. Pausewang, M. Schwaab, J. Crabtree, J. Christopher, Air–fabric interaction modeling with the Stabilized Space–Time FSI technique, in *Proceedings of the Third Asian-Pacific Congress on Computational Mechanics (CD-ROM)*, Kyoto (2007)
59. T.E. Tezduyar, S. Sathe, J. Pausewang, M. Schwaab, J. Christopher, J. Crabtree, Interface projection techniques for fluid–structure interaction modeling with moving-mesh methods. Comput. Mech. **43**, 39–49 (2008). https://doi.org/10.1007/s00466-008-0261-7
60. T.E. Tezduyar, S. Sathe, M. Schwaab, J. Pausewang, J. Christopher, J. Crabtree, Fluid–structure interaction modeling of ringsail parachutes. Comput. Mech. **43**, 133–142 (2008). https://doi.org/10.1007/s00466-008-0260-8
61. T. Terahara, K. Takizawa, T.E. Tezduyar, A. Tsushima, K. Shiozaki, Ventricle-valve-aorta flow analysis with the Space–Time Isogeometric Discretization and Topology Change. Comput. Mech. **65**, 1343–1363 (2020). https://doi.org/10.1007/s00466-020-01822-4
62. T. Kuraishi, S. Yamasaki, K. Takizawa, T.E. Tezduyar, Z. Xu, R. Kaneko, Space–time isogeometric analysis of car and tire aerodynamics with road contact and tire deformation and rotation. Comput. Mech. **70**, 49–72 (2022). https://doi.org/10.1007/s00466-022-02155-0
63. T.E. Tezduyar, K. Takizawa, A chronological catalog of methods and solutions in the Space–Time Computational Flow Analysis: I. Finite element analysis. Comput. Mech. **75**, 793–831 (2025). https://doi.org/10.1007/s00466-024-02534-9
64. T.E. Tezduyar, K. Takizawa, A chronological catalog of methods and solutions in the Space–Time Computational Flow Analysis: II. Isogeometric analysis. Comput. Mech. **75**, 833–874 (2025). https://doi.org/10.1007/s00466-024-02535-8
65. T.E. Tezduyar, T.J.R. Hughes, Development of time-accurate finite element techniques for first-order hyperbolic systems with particular emphasis on the compressible Euler equations. NASA Technical Report NASA-CR-204772, NASA (1982). https://www.researchgate.net/publication/24313718/
66. T.E. Tezduyar, T.J.R. Hughes, Finite element formulations for convection dominated flows with particular emphasis on the compressible Euler equations, in *Proceedings of AIAA 21st Aerospace Sciences Meeting*, AIAA Paper 83-0125, Reno, Nevada (1983). https://doi.org/10.2514/6.1983-125

67. T.J.R. Hughes, T.E. Tezduyar, Finite element methods for first-order hyperbolic systems with particular emphasis on the compressible Euler equations. Comput. Methods Appl. Mech. Eng. **45**, 217–284 (1984). https://doi.org/10.1016/0045-7825(84)90157-9
68. T.E. Tezduyar, Finite elements in fluids: Stabilized formulations and moving boundaries and interfaces. Comput. Fluids **36**, 191–206 (2007). https://doi.org/10.1016/j.compfluid.2005.02.011
69. T.E. Tezduyar, M. Senga, D. Vicker, Computation of inviscid supersonic flows around cylinders and spheres with the SUPG formulation and YZβ shock-capturing. Comput. Mech. **38**, 469–481 (2006). https://doi.org/10.1007/s00466-005-0025-6
70. G.J. Le Beau, T.E. Tezduyar, Finite element computation of compressible flows with the SUPG formulation, in *Advances in Finite Element Analysis in Fluid Dynamics*, FED-Vol.123 (ASME, New York, 1991), pp. 21–27
71. G.J. Le Beau, S.E. Ray, S.K. Aliabadi, T.E. Tezduyar, SUPG finite element computation of compressible flows with the entropy and conservation variables formulations. Comput. Methods Appl. Mech. Eng. **104**, 397–422 (1993). https://doi.org/10.1016/0045-7825(93)90033-T
72. T.J.R. Hughes, L.P. Franca, M. Mallet, A new finite element formulation for computational fluid dynamics: VI. Convergence analysis of the generalized SUPG formulation for linear time-dependent multi-dimensional advective-diffusive systems. Comput. Methods Appl. Mech. Eng. **63**, 97–112 (1987)
73. T.E. Tezduyar, Y.J. Park, Discontinuity capturing finite element formulations for nonlinear convection-diffusion-reaction equations. Comput. Methods Appl. Mech. Eng. **59**, 307–325 (1986). https://doi.org/10.1016/0045-7825(86)90003-4
74. R.C. Almeida, A.C. Galeão, An adaptive Petrov–Galerkin formulation for the compressible Euler and Navier–Stokes equations. Comput. Methods Appl. Mech. Eng. **129**(1), 157–176 (1996)
75. T.J.R. Hughes, L.P. Franca, M. Balestra, A new finite element formulation for computational fluid dynamics: V. Circumventing the Babuška–Brezzi condition: a stable Petrov–Galerkin formulation of the Stokes problem accommodating equal-order interpolations. Comput. Methods Appl. Mech. Eng. **59**, 85–99 (1986)
76. S. Mittal, T.E. Tezduyar, A finite element study of incompressible flows past oscillating cylinders and aerofoils. Int. J. Numer. Methods Fluids **15**, 1073–1118 (1992). https://doi.org/10.1002/fld.1650150911
77. T. Kuraishi, K. Takizawa, T.E. Tezduyar, Tire aerodynamics with actual tire geometry, road contact and tire deformation. Comput. Mech. **63**, 1165–1185 (2019). https://doi.org/10.1007/s00466-018-1642-1
78. K. Takizawa, T.E. Tezduyar, Y. Otoguro, Stabilization and discontinuity-capturing parameters for space–time flow computations with finite element and isogeometric discretizations. Comput. Mech. **62**, 1169–1186 (2018). https://doi.org/10.1007/s00466-018-1557-x
79. S. Mittal, A. Ratner, D. Hastreiter, T.E. Tezduyar, Space–time finite element computation of incompressible flows with emphasis on flows involving oscillating cylinders. Int. Video J. Eng. Res. **1**, 83–96 (1991)
80. S. Mittal, T.E. Tezduyar, Massively parallel finite element computation of incompressible flows involving fluid-body interactions. Comput. Methods Appl. Mech. Eng. **112**, 253–282 (1994). https://doi.org/10.1016/0045-7825(94)90029-9
81. A.A. Johnson, T.E. Tezduyar, Mesh update strategies in parallel finite element computations of flow problems with moving boundaries and interfaces. Comput. Methods Appl. Mech. Eng. **119**, 73–94 (1994). https://doi.org/10.1016/0045-7825(94)00077-8
82. K. Takizawa, T.E. Tezduyar, R. Avsar, A low-distortion mesh moving method based on fiber-reinforced hyperelasticity and optimized zero-stress state. Comput. Mech. **65**, 1567–1591 (2020). https://doi.org/10.1007/s00466-020-01835-z
83. K. Stein, T. Tezduyar, R. Benney, Mesh moving techniques for fluid–structure interactions with large displacements. J. Appl. Mech. **70**, 58–63 (2003). https://doi.org/10.1115/1.1530635

84. A. Masud, T.J.R. Hughes, A space–time Galerkin/least-squares finite element formulation of the Navier-Stokes equations for moving domain problems. Comput. Methods Appl. Mech. Eng. **146**, 91–126 (1997)
85. S.K. Aliabadi, T.E. Tezduyar, Space–time finite element computation of compressible flows involving moving boundaries and interfaces. Comput. Methods Appl. Mech. Eng. **107**(1–2), 209–223 (1993). https://doi.org/10.1016/0045-7825(93)90176-X
86. M. Behr, T.E. Tezduyar, Finite element solution strategies for large-scale flow simulations. Comput. Methods Appl. Mech. Eng. **112**, 3–24 (1994). https://doi.org/10.1016/0045-7825(94)90016-7
87. T.E. Tezduyar, S.K. Aliabadi, M. Behr, S. Mittal, Massively parallel finite element simulation of compressible and incompressible flows. Comput. Methods Appl. Mech. Eng. **119**, 157–177 (1994). https://doi.org/10.1016/0045-7825(94)00082-4
88. G.P. Wren, S.E. Ray, S.K. Aliabadi, T.E. Tezduyar, Space–time finite element computation of compressible flows between moving components. Int. J. Numer. Methods Fluids. **21**, 981–991 (1995). https://doi.org/10.1002/fld.1650211015
89. S. Mittal, T.E. Tezduyar, Parallel finite element simulation of 3D incompressible flows: fluid-structure interactions. Int. J. Numer. Methods Fluids **21**, 933–953 (1995). https://doi.org/10.1002/fld.1650211011
90. J. Lighthill, *An Informal Introduction to Theoretical Fluid Mechanics* (Clarendon, Oxford, 1986)
91. V. Kalro, T.E. Tezduyar, Parallel finite element computation of 3D incompressible flows on MPPs, in *Solution Techniques for Large-Scale CFD Problems*, ed. by W.G. Habashi (John Wiley & Sons, 1995). Also in *Proceedings of the International Workshop on Solution Techniques for Large-Scale CFD Problems*, Montreal, Quebec (1994)
92. T. Tezduyar, S. Aliabadi, M. Behr, A. Johnson, V. Kalro, M. Litke, Flow simulation and high performance computing. Comput. Mech. **18**, 397–412 (1996). https://doi.org/10.1007/BF00350249
93. M. Behr, T. Tezduyar, The Shear-Slip mesh update method. Comput. Methods Appl. Mech. Eng. **174**, 261–274 (1999). https://doi.org/10.1016/S0045-7825(98)00299-0
94. M. Behr, T. Tezduyar, Shear-slip mesh update in 3D computation of complex flow problems with rotating mechanical components. Comput. Methods Appl. Mech. Eng. **190**, 3189–3200 (2001). https://doi.org/10.1016/S0045-7825(00)00388-1
95. T.E. Tezduyar, Finite element methods for flow problems with moving boundaries and interfaces. Arch. Comput. Methods Eng. **8**, 83–130 (2001). https://doi.org/10.1007/BF02897870
96. T.E. Tezduyar, Finite element methods for fluid dynamics with moving boundaries and interfaces, in *Encyclopedia of Computational Mechanics*, Volume 3: Fluids, ed. by E. Stein, R. de Borst, T.J.R. Hughes, Chapter 17 (Wiley, 2004). ISBN 978-0-470-84699-5. https://doi.org/10.1002/0470091355.ecm069
97. T.E. Tezduyar, Finite elements in fluids: special methods and enhanced solution techniques. Comput. Fluids **36**, 207–223 (2007). https://doi.org/10.1016/j.compfluid.2005.02.010
98. K. Kashiyama, K. Saitoh, M. Behr, T.E. Tezduyar, Parallel finite element methods for large-scale computation of storm surges and tidal flows. Int. J. Numer. Methods Fluids **24**, 1371–1389 (1997). https://doi.org/10.1002/(SICI)1097-0363(199706)24:12<1371::AID-FLD565>3.0.CO;2-7
99. S. Takase, K. Kashiyama, S. Tanaka, T.E. Tezduyar, Space–time SUPG finite element computation of shallow-water flows with moving shorelines. Comput. Mech. **48**, 293–306 (2011). https://doi.org/10.1007/s00466-011-0618-1
100. A.A. Johnson, T.E. Tezduyar, 3D simulation of fluid-particle interactions with the number of particles reaching 100. Comput. Methods Appl. Mech. Eng. **145**, 301–321 (1997). https://doi.org/10.1016/S0045-7825(96)01223-6
101. A.A. Johnson, T.E. Tezduyar, Parallel computing methods for large-scale 3D simulation of fluid-object interactions, in *Computational Mechanics – New Trends and Applications*, CD-ROM, ed. by S. Idelsohn, E. Onate, E. Dvorkin (1998)

102. A.A. Johnson, T.E. Tezduyar, Advanced mesh generation and update methods for 3D flow simulations. Comput. Mech. **23**, 130–143 (1999). https://doi.org/10.1007/s004660050393
103. V. Kalro, T. Tezduyar, A parallel finite element methodology for 3D computation of fluid–structure interactions in airdrop systems, in *Proceedings of the 4th Japan-US Symposium on Finite Element Methods in Large-Scale Computational Fluid Dynamics*, Tokyo (1998)
104. K. Stein, R. Benney, V. Kalro, T. Tezduyar, J. Leonard, M. Accorsi, Parachute fluid–structure interactions: 3-D computation, in *Proceedings of the 4th Japan-US Symposium on Finite Element Methods in Large-Scale Computational Fluid Dynamics*, Tokyo (1998)
105. K. Stein, R. Benney, V. Kalro, T.E. Tezduyar, J. Leonard, M. Accorsi, Parachute fluid–structure interactions: 3-D Computation. Comput. Methods Appl. Mech. Eng. **190**, 373–386 (2000). https://doi.org/10.1016/S0045-7825(00)00208-5
106. Y. Osawa, V. Kalro, T. Tezduyar, Multi-domain parallel computation of wake flows. Comput. Methods Appl. Mech. Eng. **174**, 371–391 (1999). https://doi.org/10.1016/S0045-7825(98)00305-3
107. Y. Osawa, V. Kalro, T. Tezduyar, A multi-domain computational method for wake flows, in *Modeling and Simulation Based Engineering, Proceedings of International Conference on Computational Engineering Science*, Atlanta, ed. by S. N. Atluri, P.E. O'Donoghue (1998)
108. T. Tezduyar, Y. Osawa, Methods for parallel computation of complex flow problems. Parallel Comput. **25**, 2039–2066 (1999). https://doi.org/10.1016/S0167-8191(99)00080-0
109. Y. Osawa, T. Tezduyar, 3D simulation and visualization of unsteady wake flow behind a cylinder. J. Vis. **2**, 127–134 (1999). https://doi.org/10.1007/BF03181515
110. T. Tezduyar, Y. Osawa, Fluid–structure interactions of a parachute crossing the far wake of an aircraft. Comput. Methods Appl. Mech. Eng. **191**, 717–726 (2001). https://doi.org/10.1016/S0045-7825(01)00311-5
111. A. Johnson, T. Tezduyar, Methods for 3D computation of fluid-object interactions in spatially-periodic flows. Comput. Methods Appl. Mech. Eng. **190**, 3201–3221 (2001). https://doi.org/10.1016/S0045-7825(00)00389-3
112. J.E. Akin, T.E. Tezduyar, M. Ungor, Computation of flow problems with the mixed interface-tracking/interface-capturing technique (MITICT). Comput. Fluids **36**, 2–11 (2007). https://doi.org/10.1016/j.compfluid.2005.07.008
113. I. Akkerman, Y. Bazilevs, D.J. Benson, M.W. Farthing, C.E. Kees, Free-surface flow and fluid–object interaction modeling with emphasis on ship hydrodynamics. J. Appl. Mech. **79**, 010905 (2012)
114. T. Tezduyar, Finite element interface-tracking and interface-capturing techniques for flows with moving boundaries and interfaces, in *Proceedings of the ASME Symposium on Fluid-Physics and Heat Transfer for Macro- and Micro-Scale Gas-Liquid and Phase-Change Flows (CD-ROM)*, ASME Paper IMECE2001/HTD-24206 (ASME, New York, 2001)
115. T.E. Tezduyar, S. Sathe, Enhanced-discretization space-time technique (EDSTT). Comput. Methods Appl. Mech. Eng. **193**, 1385–1401 (2004). https://doi.org/10.1016/j.cma.2003.12.029
116. K. Stein, T.E. Tezduyar, R. Benney, Automatic mesh update with the solid-extension mesh moving technique. Comput. Methods Appl. Mech. Eng. **193**, 2019–2032 (2004). https://doi.org/10.1016/j.cma.2003.12.046
117. R. Torii, M. Oshima, T. Kobayashi, K. Takagi, T.E. Tezduyar, Computer modeling of cardiovascular fluid–structure interactions with the Deforming-Spatial-Domain/Stabilized Space–Time formulation. Comput. Methods Appl. Mech. Eng. **195**, 1885–1895 (2006). https://doi.org/10.1016/j.cma.2005.05.050
118. S. Sathe, T.E. Tezduyar, Modeling of fluid–structure interactions with the space–time finite elements: contact problems. Comput. Mech. **43**, 51–60 (2008). https://doi.org/10.1007/s00466-008-0299-6
119. K. Takizawa, S. Wright, C. Moorman, T.E. Tezduyar, Fluid–structure interaction modeling of parachute clusters. Int. J. Numer. Methods Fluids **65**, 286–307 (2011). https://doi.org/10.1002/fld.2359

120. K. Takizawa, T. Spielman, T.E. Tezduyar, Space–time FSI modeling and dynamical analysis of spacecraft parachutes and parachute clusters. Comput. Mech. **48**, 345–364 (2011). https://doi.org/10.1007/s00466-011-0590-9
121. T.E. Tezduyar, S. Sathe, T. Cragin, B. Nanna, B.S. Conklin, J. Pausewang, M. Schwaab, Modeling of fluid–structure interactions with the space–time finite elements: arterial fluid mechanics. Int. J. Numer. Methods Fluids **54**, 901–922 (2007). https://doi.org/10.1002/fld.1443
122. R. Torii, M. Oshima, T. Kobayashi, K. Takagi, T.E. Tezduyar, Influence of wall elasticity in patient-specific hemodynamic simulations. Comput. Fluids **36**, 160–168 (2007). https://doi.org/10.1016/j.compfluid.2005.07.014
123. S.E. Wright, Advanced fluid–structure interaction techniques for modeling ringsail parachutes. Master's thesis, Rice University (2010)
124. K. Takizawa, C. Moorman, S. Wright, T. Spielman, T.E. Tezduyar, Fluid–structure interaction modeling and performance analysis of the Orion spacecraft parachutes. Int. J. Numer. Methods Fluids **65**, 271–285 (2011). https://doi.org/10.1002/fld.2348
125. T.E. Tezduyar, M. Schwaab, S. Sathe, Arterial fluid mechanics with the sequentially-coupled arterial FSI technique, in *Coupled Problems 2007*, CIMNE, Barcelona, ed. by E. Onate, M. Papadrakakis, B. Schrefler (2007)
126. T.E. Tezduyar, S. Sathe, M. Schwaab, B.S. Conklin, Arterial fluid mechanics modeling with the stabilized space–time fluid–structure interaction technique. Int. J. Numer. Methods Fluids **57**, 601–629 (2008). https://doi.org/10.1002/fld.1633
127. T.E. Tezduyar, M. Schwaab, S. Sathe, Sequentially-Coupled Arterial Fluid–Structure Interaction (SCAFSI) technique. Comput. Methods Appl. Mech. Eng. **198**, 3524–3533 (2009). https://doi.org/10.1016/j.cma.2008.05.024
128. T.E. Tezduyar, K. Takizawa, J. Christopher, Multiscale Sequentially-Coupled Arterial Fluid–Structure Interaction (SCAFSI) technique, in *International Workshop on Fluid–Structure Interaction — Theory, Numerics and Applications*, ed. by S. Hartmann, A. Meister, M. Schafer, S. Turek (Kassel University Press, 2009), pp. 231–252. ISBN 978-3-89958-666-4
129. T.E. Tezduyar, K. Takizawa, J. Christopher, C. Moorman, S. Wright, Interface projection techniques for complex FSI problems, in *Marine 2009, CIMNE*, Barcelona, ed. by T. Kvamsdal, B. Pettersen, P. Bergan, E. Onate, J. Garcia (2009)
130. T.E. Tezduyar, K. Takizawa, C. Moorman, S. Wright, J. Christopher, Multiscale sequentially-coupled arterial FSI technique. Comput. Mech. **46**, 17–29 (2010). https://doi.org/10.1007/s00466-009-0423-2
131. M.-C. Hsu, D. Kamensky, Y. Bazilevs, M.S. Sacks, T.J.R. Hughes, Fluid–structure interaction analysis of bioprosthetic heart valves: significance of arterial wall deformation. Comput. Mech. **54**, 1055–1071 (2014). https://doi.org/10.1007/s00466-014-1059-4
132. S.F. Hoerner, *Fluid Dynamic Drag*. Hoerner Fluid Dynamics (1993)
133. K. Takizawa, T. Spielman, C. Moorman, T.E. Tezduyar, Fluid–structure interaction modeling of spacecraft parachutes for simulation-based design. J. Appl. Mech. **79**, 010907 (2012). https://doi.org/10.1115/1.4005070
134. Y. Bazilevs, M.-C. Hsu, I. Akkerman, S. Wright, K. Takizawa, B. Henicke, T. Spielman, T.E. Tezduyar, 3D simulation of wind turbine rotors at full scale. Part I: Geometry modeling and aerodynamics. Int. J. Numer. Methods Fluids **65**, 207–235 (2011). https://doi.org/10.1002/fld.2400
135. K. Takizawa, B. Henicke, T.E. Tezduyar, M.-C. Hsu, Y. Bazilevs, Stabilized space–time computation of wind-turbine rotor aerodynamics. Comput. Mech. **48**, 333–344 (2011). https://doi.org/10.1007/s00466-011-0589-2
136. K. Takizawa, M. Fritze, D. Montes, T. Spielman, T.E. Tezduyar, Fluid–structure interaction modeling of ringsail parachutes with disreefing and modified geometric porosity. Comput. Mech. **50**, 835–854 (2012). https://doi.org/10.1007/s00466-012-0761-3
137. K. Takizawa, D. Montes, S. McIntyre, T.E. Tezduyar, Space–time VMS methods for modeling of incompressible flows at high Reynolds numbers. Math. Models Methods Appl. Sci. **23**, 223–248 (2013). https://doi.org/10.1142/s0218202513400022

138. K. Takizawa, T.E. Tezduyar, J. Boben, N. Kostov, C. Boswell, A. Buscher, Fluid–structure interaction modeling of clusters of spacecraft parachutes with modified geometric porosity. Comput. Mech. **52**, 1351–1364 (2013). https://doi.org/10.1007/s00466-013-0880-5
139. T.E. Tezduyar, K. Takizawa, Y. Bazilevs, Fluid–structure interaction and flows with moving boundaries and interfaces, in *Encyclopedia of Computational Mechanics Second Edition*, Part 2 Fluids, ed. by E. Stein, R. de Borst, T.J.R. Hughes (Wiley). Published online (December 2017). ISBN 9781119003793. https://doi.org/10.1002/9781119176817.ecm2069
140. K. Takizawa, H. Takagi, T.E. Tezduyar, R. Torii, Estimation of element-based zero-stress state for arterial FSI computations. Comput. Mech. **54**, 895–910 (2014). https://doi.org/10.1007/s00466-013-0919-7
141. T. Sasaki, K. Takizawa, T.E. Tezduyar, Medical-image-based aorta modeling with zero-stress-state estimation. Comput. Mech. **64**, 249–271 (2019). https://doi.org/10.1007/s00466-019-01669-4
142. K. Takizawa, T.E. Tezduyar, C. Boswell, R. Kolesar, K. Montel, FSI modeling of the reefed stages and disreefing of the Orion spacecraft parachutes. Comput. Mech. **54**, 1203–1220 (2014). https://doi.org/10.1007/s00466-014-1052-y
143. K. Takizawa, T.E. Tezduyar, R. Kolesar, FSI modeling of the Orion spacecraft drogue parachutes. Comput. Mech. **55**, 1167–1179 (2015). https://doi.org/10.1007/s00466-014-1108-z
144. T.J.R. Hughes, J.A. Cottrell, Y. Bazilevs, Isogeometric analysis: CAD, finite elements, NURBS, exact geometry, and mesh refinement. Comput. Methods Appl. Mech. Eng. **194**, 4135–4195 (2005)
145. Y. Bazilevs, V.M. Calo, Y. Zhang, T.J.R. Hughes, Isogeometric fluid–structure interaction analysis with applications to arterial blood flow. Comput. Mech. **38**, 310–322 (2006)
146. Y. Bazilevs, T.J.R. Hughes, NURBS-based isogeometric analysis for the computation of flows about rotating components. Comput. Mech. **43**, 143–150 (2008)
147. J.A. Cottrell, T.J.R. Hughes, Y. Bazilevs, *Isogeometric Analysis. Toward Integration of CAD and FEA* (Wiley, Hoboken, 2009)
148. K. Takizawa, T.E. Tezduyar, Space–time computation techniques with continuous representation in time (ST-C). Comput. Mech. **53**, 91–99 (2014). https://doi.org/10.1007/s00466-013-0895-y
149. T. Kuraishi, Z. Xu, K. Takizawa, T.E. Tezduyar, S. Yamasaki, High-resolution multi-domain space–time isogeometric analysis of car and tire aerodynamics with road contact and tire deformation and rotation. Comput. Mech. **70**, 1257–1279 (2022). https://doi.org/10.1007/s00466-022-02228-0
150. S.T. Zalesak, Fully multideimensional flux-corrected transport algorithms for fluids. J. Comput. Phys. **31**, 335–362 (1979)
151. K. Takizawa, T.E. Tezduyar, S. McIntyre, N. Kostov, R. Kolesar, C. Habluetzel, Space–time VMS computation of wind-turbine rotor and tower aerodynamics. Comput. Mech. **53**, 1–15 (2014). https://doi.org/10.1007/s00466-013-0888-x
152. K. Takizawa, K. Schjodt, A. Puntel, N. Kostov, T.E. Tezduyar, Patient-specific computer modeling of blood flow in cerebral arteries with aneurysm and stent. Comput. Mech. **50**, 675–686 (2012). https://doi.org/10.1007/s00466-012-0760-4
153. T. Kanai, K. Takizawa, T.E. Tezduyar, K. Komiya, M. Kaneko, K. Hirota, M. Nohmi, T. Tsuneda, M. Kawai, M. Isono, Methods for computation of flow-driven string dynamics in a pump and residence time. Math. Models Methods Appl. Sci. **29**, 839–870 (2019). https://doi.org/10.1142/S021820251941001X
154. T. Kuraishi, F. Zhang, K. Takizawa, T.E. Tezduyar, Wind turbine wake computation with the ST-VMS method, isogeometric discretization and multidomain method: I. Computational framework. Comput. Mech. **68**, 113–130 (2021). https://doi.org/10.1007/s00466-021-02022-4

155. T. Kuraishi, F. Zhang, K. Takizawa, T.E. Tezduyar, Wind turbine wake computation with the ST-VMS method, isogeometric discretization and multidomain method: II. Spatial and temporal resolution. Comput. Mech. **68**, 175–184 (2021). https://doi.org/10.1007/s00466-021-02025-1
156. F. Zhang, T. Kuraishi, K. Takizawa, T.E. Tezduyar, Wind turbine wake computation with the ST-VMS method and isogeometric discretization: directional preference in spatial refinement. Comput. Mech. **69**, 1031–1040 (2022). https://doi.org/10.1007/s00466-021-02129-8
157. Y. Bazilevs, T.J.R. Hughes, Weak imposition of Dirichlet boundary conditions in fluid mechanics. Comput. Fluids **36**, 12–26 (2007)
158. Life tower: wind power tower with tsunami evacuation shelter. http://cosmosunfarm.co.jp/lifetower.html
159. K. Takizawa, T.E. Tezduyar, H. Hattori, Computational analysis of flow-driven string dynamics in turbomachinery. Comput. Fluids **142**, 109–117 (2017). https://doi.org/10.1016/j.compfluid.2016.02.019
160. S.B. Raknes, X. Deng, Y. Bazilevs, D.J. Benson, K.M. Mathisen, T. Kvamsdal, Isogeometric rotation-free bending-stabilized cables: statics, dynamics, bending strips and coupling with shells. Comput. Methods Appl. Mech. Eng. **263**, 127–143 (2013)
161. K. Takizawa, T.E. Tezduyar, T. Terahara, Ram-air parachute structural and fluid mechanics computations with the space–time isogeometric analysis (ST-IGA). Comput. Fluids **141**, 191–200 (2016). https://doi.org/10.1016/j.compfluid.2016.05.027
162. K. Takizawa, T.E. Tezduyar, S. Asada, T. Kuraishi, Space–time method for flow computations with slip interfaces and topology changes (ST-SI-TC). Comput. & Fluids **141**, 124–134 (2016). https://doi.org/10.1016/j.compfluid.2016.05.006
163. T. Kuraishi, K. Takizawa, T.E. Tezduyar, Space–Time Isogeometric flow analysis with built-in Reynolds-equation limit. Math. Models Methods Appl. Sci. **29**, 871–904 (2019). https://doi.org/10.1142/S0218202519410021
164. K. Takizawa, T.E. Tezduyar, T. Kanai, Porosity models and computational methods for compressible-flow aerodynamics of parachutes with geometric porosity. Math. Models Methods Appl. Sci. **27**, 771–806 (2017). https://doi.org/10.1142/S0218202517500166
165. Y. Otoguro, K. Takizawa, T.E. Tezduyar, Space–time VMS computational flow analysis with isogeometric discretization and a general-purpose NURBS mesh generation method. Comput. Fluids **158**, 189–200 (2017). https://doi.org/10.1016/j.compfluid.2017.04.017
166. K. Takizawa, Y. Bazilevs, T.E. Tezduyar, M.-C. Hsu, T. Terahara, Computational cardiovascular medicine with isogeometric analysis. J. Adv. Eng. Comput. **6**, 167–199 (2022). https://doi.org/10.55579/jaec.202263.381
167. Y. Otoguro, K. Takizawa, T.E. Tezduyar, A general-purpose NURBS mesh generation method for complex geometries, in *Frontiers in Computational Fluid–Structure Interaction and Flow Simulation: Research from Lead Investigators under Forty – 2018*, ed. by T.E. Tezduyar. Modeling and Simulation in Science, Engineering and Technology (Springer, 2018), pp. 399–434. ISBN 978-3-319-96468-3. https://doi.org/10.1007/978-3-319-96469-0_10
168. K. Takizawa, Y. Otoguro, T.E. Tezduyar, Variational multiscale method stabilization parameter calculated from the strain-rate tensor. Math. Models Methods Appl. Sci. **33**(8), 1661–1691 (2023). https://doi.org/10.1142/S0218202523500380
169. T.E. Tezduyar, Adaptive determination of the finite element stabilization parameters, in *Proceedings of the ECCOMAS Computational Fluid Dynamics Conference 2001 (CD-ROM)*, Swansea, Wales (2001)
170. Y. Otoguro, K. Takizawa, T.E. Tezduyar, Element length calculation in B-spline meshes for complex geometries. Comput. Mech. **65**, 1085–1103 (2020). https://doi.org/10.1007/s00466-019-01809-w
171. K. Takizawa, Y. Ueda, T.E. Tezduyar, A node-numbering-invariant directional length scale for simplex elements. Math. Models Methods Appl. Sci. **29**, 2719–2753 (2019). https://doi.org/10.1142/S0218202519500581

172. T. Kanai, K. Takizawa, T.E. Tezduyar, T. Tanaka, A. Hartmann, Compressible-flow geometric-porosity modeling and spacecraft parachute computation with isogeometric discretization. Comput. Mech. **63**, 301–321 (2019). https://doi.org/10.1007/s00466-018-1595-4
173. T. Kuraishi, K. Takizawa, T.E. Tezduyar, Space–time computational analysis of tire aerodynamics with actual geometry, road contact, tire deformation, road roughness and fluid film. Comput. Mech. **64**, 1699–1718 (2019). https://doi.org/10.1007/s00466-019-01746-8
174. T. Terahara, K. Takizawa, T.E. Tezduyar, Y. Bazilevs, M.-C. Hsu, Heart valve isogeometric sequentially-coupled FSI analysis with the space–time topology change method. Comput. Mech. **65**, 1167–1187 (2020). https://doi.org/10.1007/s00466-019-01813-0
175. M.-C. Hsu, D. Kamensky, F. Xu, J. Kiendl, C. Wang, M.C.H. Wu, J. Mineroff, A. Reali, Y. Bazilevs, M.S. Sacks, Dynamic and fluid–structure interaction simulations of bioprosthetic heart valves using parametric design with T-splines and Fung-type material models. Comput. Mech. **55**, 1211–1225 (2015). https://doi.org/10.1007/s00466-015-1166-x
176. T. Terahara, K. Takizawa, T.E. Tezduyar, Heart valve computational flow analysis with boundary layer and leaflet contact representation, in *Frontiers in Computational Fluid–Structure Interaction and Flow Simulation: Research from Lead Investigators under Forty – 2023*, ed. by T.E. Tezduyar. Modeling and Simulation in Science, Engineering and Technology (Springer, 2023), pp. 437–475. ISBN 978-3-031-36941-4. https://doi.org/10.1007/978-3-031-36942-1_13
177. M.E. Moghadam, Y. Bazilevs, T.-Y. Hsia, I.E. Vignon-Clementel, A.L. Marsden, M. of Congenital Hearts Alliance (MOCHA), A comparison of outlet boundary treatments for prevention of backflow divergence with relevance to blood flow simulations. Comput. Mech. **48**, 277–291 (2011). https://doi.org/10.1007/s00466-011-0599-0
178. K. Takizawa, N. Kostov, A. Puntel, B. Henicke, T.E. Tezduyar, Space–time computational analysis of bio-inspired flapping-wing aerodynamics of a micro aerial vehicle. Comput. Mech. **50**, 761–778 (2012). https://doi.org/10.1007/s00466-012-0758-y
179. K. Takizawa, T.E. Tezduyar, H. Uchikawa, T. Terahara, T. Sasaki, K. Shiozaki, A. Yoshida, K. Komiya, G. Inoue, Aorta flow analysis and heart valve flow and structure analysis, in *Frontiers in Computational Fluid–Structure Interaction and Flow Simulation: Research from Lead Investigators under Forty – 2018*, ed. by T.E. Tezduyar. Modeling and Simulation in Science, Engineering and Technology (Springer, 2018), pp. 29–89. ISBN 978-3-319-96468-3. https://doi.org/10.1007/978-3-319-96469-0_2
180. J. Kiendl, M.-C. Hsu, M.C.H. Wu, A. Reali, Isogeometric Kirchhoff–Love shell formulations for general hyperelastic materials. Comput. Methods Appl. Mech. Eng. **291**, 280–303 (2015)
181. P. Tonon, R.A.K. Sanches, K. Takizawa, T.E. Tezduyar, A linear-elasticity-based mesh moving method with no cycle-to-cycle accumulated distortion. Comput. Mech. **67**, 413–434 (2021). https://doi.org/10.1007/s00466-020-01941-y
182. T. Kuraishi, K. Takizawa, T.E. Tezduyar, A general-purpose IGA mesh generation method: NURBS Surface-to-Volume Guided Mesh Generation. Comput. Mech. **75**, 159–169 (2025). https://doi.org/10.1007/s00466-024-02496-y
183. Y. Otoguro, H. Mochizuki, K. Takizawa, T.E. Tezduyar, Space–time variational multiscale isogeometric analysis of a tsunami-shelter vertical-axis wind turbine. Comput. Mech. **66**, 1443–1460 (2020). https://doi.org/10.1007/s00466-020-01910-5
184. L. Aydinbakar, K. Takizawa, T.E. Tezduyar, T. Kuraishi, Space–time VMS isogeometric analysis of the Taylor–Couette flow. Comput. Mech. **67**, 1515–1541 (2021). https://doi.org/10.1007/s00466-021-02004-6
185. A. Korobenko, J. Yan, S.M.I. Gohari, S. Sarkar, Y. Bazilevs, FSI simulation of two back-to-back wind turbines in atmospheric boundary layer flow. Comput. Fluids **158**, 167–175 (2017). https://doi.org/10.1016/j.compfluid.2017.05.010
186. Y. Liu, K. Takizawa, Y. Otoguro, T. Kuraishi, T.E. Tezduyar, Flow computation with the space–time isogeometric analysis and higher-order basis functions in time. Math. Models Methods Appl. Sci. **32**(12), 2445–2475 (2022). https://doi.org/10.1142/S0218202522500579

187. Y. Liu, K. Takizawa, T.E. Tezduyar, T. Kuraishi, Y. Zhang, Carrier-domain method for high-resolution computation of time-periodic long-wake flows. Comput. Mech. **71**, 169–190 (2023). https://doi.org/10.1007/s00466-022-02230-6
188. Y. Liu, K. Takizawa, T.E. Tezduyar, High-resolution 3D computation of time-periodic long-wake flows with the Carrier-Domain Method and Space–Time Variational Multiscale method with isogeometric discretization. Comput. Mech. **74**, 1–22 (2024). https://doi.org/10.1007/s00466-023-02419-3
189. T. Kuraishi, Z. Xu, K. Takizawa, T.E. Tezduyar, T. Kakegami, Space–time isogeometric analysis of tire aerodynamics with complex tread pattern, road contact, and tire deformation. Comput. Mech. **75**, 575–591 (2025). https://doi.org/10.1007/s00466-024-02520-1
190. Y. Liu, Y. Otoguro, K. Takizawa, T.E. Tezduyar, Space–Time Isogeometric Analysis of NREL 5MW wind turbine rotor and tower aerodynamics. Comput. Mech. **75**, 1483–1499 (2025). https://doi.org/10.1007/s00466-024-02574-1

Index

T. E. Tezduyar, K. Takizawa, *Space–Time Computational Flow Analysis*,
Modeling and Simulation in Science, Engineering and Technology,
https://doi.org/10.1007/978-3-031-88727-7

The manufacturer's authorised representative in the EU is Springer Nature Customer Service Centre GmbH, Europaplatz 3, 69115 Heidelberg, Germany. If you have any concerns regarding our products, please contact ProductSafety@springernature.com

Printed and bound by CPI Group (UK) Ltd, Croydon, CR0 4YY

05/10/2026

02228127-0007